Introduction to ENVIRONMENTAL SCIENCE

Kendall Hunt publishing company

Dennis Wilson

Cover image © Shutterstock.com

Kendall Hunt
publishing company

www.kendallhunt.com
Send all inquiries to:
4050 Westmark Drive
Dubuque, IA 52004-1840

ISBN: 978-1-7924-1952-2

Published in the United States of America

CONTENTS

Chapter 1

INTRODUCTION

Chapter Topics

1.1 What Is the Environment?

Discussion prompt: We often hear the term "environment" in the media, especially during debates over public policy. What does the term "environment" mean to you?

The **environment** is defined as all of the living and nonliving things around us. It is an all-encompassing term that includes the air we breathe, the water we drink, and the food we eat. It is also the build environment of our homes, schools, and places of work. It includes the clouds, the continents, the oceans, the plants, and animals around us. The environment includes the landscapes of the world, from forests, grasslands, deserts, and tundra. In short, it is our entire world.

The planet Earth is unique. It is the only place in the Solar System that is capable of supporting life as we know it. As far as we know, this blue-green marble is the only place in the entire universe capable of supporting human life. While it may seem huge from our perspective, the Earth has only a finite amount of space and resources. And human beings are not the only species that depends on the Earth for our survival. We share this world with millions of other species and these species exist in an interconnected web of relationships in which they depend on each other for food, water, and a breathable atmosphere.

Earth sustains life, but things could have turned out differently. The Earth could have ended up like one of its neighbors. Today, the planet Mars is a frozen desert with an atmosphere too thin to support life as we know it, but may not have always been so. Over the past few decades, we have landed several robotic probes on Mars and those rovers have found evidence that the Red Planet was once warmer and wetter than it is today. Conditions that could have sustained life. Unfortunately, Mars' small size and low gravity was unable to hold onto most of its atmosphere, which slowly bled out into space.

The Earth is unique in the Solar System in its ability to sustain life as we know it.

Ilya Nikolaevic/Shutterstock.com

From a distance, Venus and Earth could be twins. The two planets are close to the same size and are made of the same rocky material. Up close, however, the worlds could not be more different. While the Earth has massive oceans filled with abundant water, Venus is a hellish world. With an atmospheric pressure 90 times that of sea level on the Earth and a runaway greenhouse effect that produces temperatures hot enough to melt lead, Venus is probably the closest to our conception of Hell than any other world. Oh, and it rains sulfuric acid. A human astronaut who stepped foot on Venus would be killed instantly.

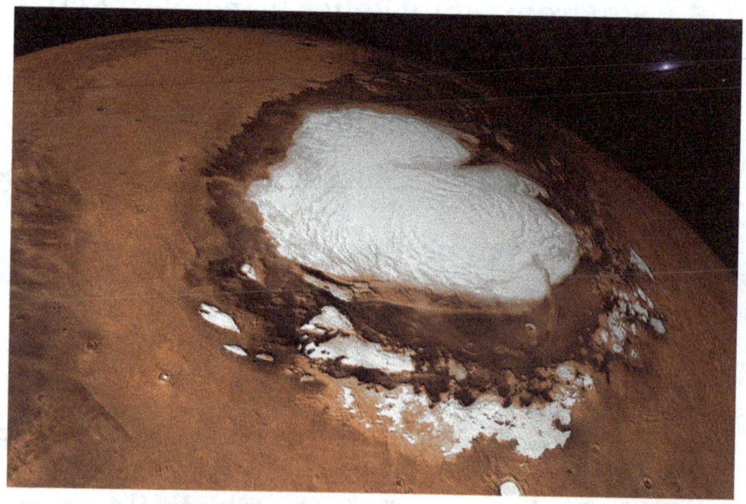

Mars is a small, frozen desert compared to Earth.

Pike-28/Shutterstock.com

Earth and Venus formed at about the same time and both planets once had very similar atmospheres. Early in the history of the Solar System, volcanoes on Earth and Venus released gases such as carbon dioxide, nitrogen, and methane. So how did things turn out so differently for each world?

The surface of Venus is a hellish landscape inhospitable to life.

Jurik Peter/Shutterstock.com

The answer may be from what covers 70% of the Earth's surface: Water. We know the oceans absorb a lot of carbon dioxide, which eventually mixes with sediments to form carbonate minerals. These minerals became incorporated into our crust. Meanwhile, at most, Venus had only very shallow seas, which could not absorb as much carbon dioxide. So, as the volcanoes continued to pump out carbon dioxide, the gas built up in the atmosphere of Venus. And we know carbon dioxide can trap heat energy near the surface of the planet. So, the more carbon dioxide filled the atmosphere, the higher the temperatures on the surface. This caused the shallow seas to evaporate, which in turn reduced the planet's ability to clear carbon dioxide from the atmosphere. It became a wild feedback loop, the hotter the planet became, the less it could remove the gas that was trapping the heat!

We call Earth the Goldilocks planet. Mars is too cold, Venus is too hot, but Earth is just right for life as we know it. However, all three planets may have been more hospitable to life in the past. It did not take much differences to change things, however.

This is why studying the environment is important. We need to know what factors affect habitability, both on the global and the local level. A sudden rise in water temperature can trigger fish kill, killing thousands of fish. Too much nitrogen and phosphorus released into a bay will deplete the oxygen content in the water, creating dead zones where life cannot exist. A temperature inversion in the atmosphere can trap pollutants over a city, making the air dangerous to breathe. Improper use of pesticides can endanger species we wish to protect.

ESB Professional/Shutterstock.com

Biologists collecting water samples from a river.

Environmental science is the study of the environment and our interactions with it. The aim is to understand how human activity can affect the environment and how changes in the environment can affect us. We study the environment to learn about how it sustains life and how our actions can affect it. But we also study the environment because we depend on it for our survival as a species and for the resources it provides to keep our modern civilization going.

1.2 Natural Resources

The Earth is big, but finite. And like it or not, we are still a one-planet species. In order to survive, humans need three main things: food, clothing, and shelter. In addition, our technologically advanced society requires the production of energy, *a lot* of energy, in order to manufacture goods, provide transportation, and exchange information. None of these things are free and in order to obtain or produce these things, we need to make use of whatever resources we can find on this planet.

Natural resources are materials and sources of energy from which we can get the things we need to survive. We can categorize natural as being either renewable or nonrenewable. **Renewable resources** are those that can be replenished within short periods. Some renewable resources are inexhaustible. They will not be depleted no matter how much we make use for them. Sunlight, for example, is an inexhaustible resource. The Sun will continue to shine for about another 4 billion years, at which point it will swell up into a red giant and probably incinerate the Earth. But in the meantime, no matter much of the Sun's light we harness, it will still keep coming. We cannot do anything to prevent the Sun producing more light. Wind and wave energy, both of which are indirect forms of solar energy, are also **inexhaustible resources**.

Exhaustible resources, on the other hand, take longer to replenish themselves. Trees, fertile soil, drinkable water, and species diversity can take several years to renew themselves. As a result, we can consume them faster than they can be replenished. You're probably familiar with Dr. Seuss' story The Lorax, which is a parable about poor management of an exhaustible resource. The Once-ler needs the Truffula trees in order to make his thneeds (I do love the names Dr. Seuss came up with). At first, things go just fine from the Once-ler's perspective. He finds a ready market for this thneeds and his business grows. However, he ignores warnings from the Lorax and continues to cut down the trees at a faster and faster rate. When the final Truffula tree is cut down, not only is the once pristine environment ruined, but his business has to close. Without the resource he needs to make his thneeds, his company folds. Indiscriminate exploitation of an exhaustible resource is not only bad for the environment, it's also a poor business model!

Alberto Masnovo/Shutterstock.com

Renewable resources.

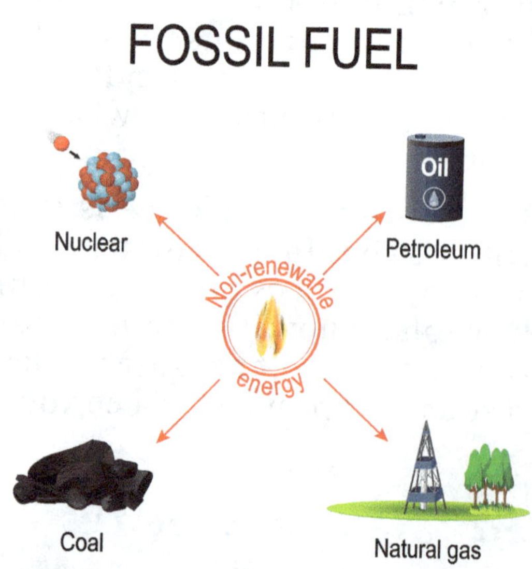

Designua/Shutterstock.com

Fossil fuels are nonrenewable resources.

While *The Lorax* is fiction, we can find real-world examples of poor resource management. On Easter Sunday in 1722, Dutch explorers landed on an island in the Pacific Ocean. The natives called it Rapa Nui, but given the date of their arrival, the Dutch gave the island its more famous name: Easter Island. The island is best known for the gigantic stone heads that line its beaches. The Easter Islanders apparently had a mania for building these statues. Just moving these massive stone blocks from the quarry to beach takes a lot of manpower and organization, especially for a society that had not yet invented the wheel. But when the Dutch arrived, instead of a large, thriving society, they found a tiny society of less than 3,000 inhabitants who were still recovering from a famine that devastated their numbers. So, what happened?

There were a number of factors involved in the fall of the Easter Island civilization, but one of the biggest was deforestation. In order to transport those huge stone heads to the beaches, they used log rollers. The more statues they built, the more trees they cut down. Once the trees were gone, they didn't have the materials they needed to build canoes, which in turn imperiled their ability to fish. Without this major source of food, the people of Rapa Nui starved.

Keith Michael Taylor/Shutterstock.com

Easter Island or Rapa Nui was once home to a thriving civilization that collapsed in part due to loss of resources.

The history of Rapa Nui is a lesson in what happens to a society that becomes dependent on a resource that can be depleted. Currently, our modern global society is dependent on fossil fuels, which is an example of nonrenewable resources. **Nonrenewable resources** are those resources that cannot be replenished in any appreciable timescale. Fossil fuels, such as coal, petroleum, and natural gas take millions of years to form. At the rate we are extracting them, most geologists estimate that they will be depleted in a matter of decades (for petroleum) or centuries (for coal). Once the current reserves of these fuels are gone, there won't be new reserves forming for millions of years.

Other examples of nonrenewable resources include mineral resources such as metals.

1.3 Ecosystem Services

In addition to natural resources, the environment also produces **ecosystem services**, which are benefits for us that arise from the normal functioning of ecosystems. Examples of ecosystem services include

- Pollination: Many of our agricultural crops are dependent on the actions of insects, birds, and even some bats that transfer the pollen from flower to flower, enabling these plants to reproduce.
- Pest and disease management: Ecosystems provide habitats for the predators and parasites that help control pests and the diseases that many of them carry.
- Water purification: Part of the hydrologic cycle (the movement of water through the environment) cleanses the water from pollutants and disease-causing microorganisms.
- Food and fiber products: These are grown from various plants and animals, some of which are domesticated, while others are wild.
- Genetic resources: Many of our domesticate plants and animals could lack resistance to diseases and parasites. Their wild relatives could have traits that hold the key to preventing crop failures. These traits could be introduced into our domesticated species through crossbreeding or genetic engineering.
- Recreation and tourism: Think about it. Nobody wants to go hiking, camping, fishing, or hunting in a barren wasteland. It is the natural beauty and wild species that brings people to our national parks like Yellowstone and Yosemite.
- Erosion prevention: Forests and out natural vegetation help hold the soil in place, reducing erosion.
- Protection from natural disasters: Coral reefs, mangrove forests, kelp forests, and coastal wetlands serve to reduce flooding and heavy wave action from hurricanes and other storm systems.
- Carbon sequestration and storage: Rain forests and the oceans absorb excess carbon dioxide from the atmosphere and help stabilize the climate.
- Air pollution control: Ecosystems help break disperse and break down pollutants in the air.
- Nutrient cycling and soil fertility: Through the natural cycle of death and decay, decomposing microorganisms free up nutrients such as nitrogen and phosphorus and return them to the soil, so that plants can take them up again.

bsd/Shutterstock.com

1.4 Human Impact on the Environment

For most of our existence as a species, humans lived in small bands of hunter gatherers. Living in limited, scattered populations, our impact on the environment was fairly limited. But, about 10,000 years ago, something changed. Instead of living in nomadic clans, some people began forming permanent settlements where they planted crops for the first time. This transition from hunting and gathering to the first farming communities is called the **agricultural revolution**. It caused the first surge in our population. Raising crops and livestock produces a more dependable source of food and support a larger population. As agriculture spread across the world, the human population grew to a few hundred million people.

The agricultural revolution was the beginning of what we call human civilization.

For thousands of years, the global population remained relatively stable, growing slowly over the millennia. It took us until the 1800 CE to reach 1 billion people. After that, it only took us 130 years to double it to 2 billion in 1930 and then, only 30 years to reach 3 billion by 1960. Today, the global population is about 7.5 billon. This new surge in population was triggered by the next big revolution in human existence: The **Industrial Revolution**, where we transitioned from a rural, agriculture-based society to an urban, manufacturing society. The Industrial Revolution brought not only the ability to mass produce consumer goods, but also mechanization to farming as well as improvements in medicine and sanitation. These advances increased the average human life expectancy and enabled us to again support an even larger population.

The Industrial Revolution started us into the age of mass production.

The Industrial Revolution, however, was powered by fossil fuels, first coal and then petroleum and natural gas. Our modern society comes with an insatiable demand for energy. With that, our **ecological footprint**, the amount of biological productive land and water needed to produce the resources needed and dispose of the wastes generated, has increased exponentially. The per person ecological footprint varies from country to country and as a general rule, the wealthier and more industrialized a society is, bigger its per capita ecological footprint.

Globally, our ecological footprint currently exceeds the capacity of the planet by about 68%. Unfortunately, we don't have an extra two-thirds of an Earth to provide use with additional natural resources and ecological services. So, our goal for sustainability must be to meet the needs of our growing population and provide a decent standard of living for all individuals while consuming less resources.

creativemarc/Shutterstock.com

Our ecological footprint represents our impact on the environment.

Think of the world's resources as a bank account. You start with a certain amount of money in your account. Over time, the account earns interest, which will increase your balance over time. However, if you withdraw money from your account faster than it accumulates, eventually, your balance will hit zero. Likewise, if we consume resources faster than they can be replenished, eventually, the resources will run.

We can't cut down all the trees today if we wanted to have trees tomorrow.

1.5 Environmental Policy Principles, Activism, and Ethics

If you've had to deal with large organization, whether it's a government agency, an online retailer, or a college, you will eventually run into policy. A policy is a written set of principles adopted by an organization in order to ensure that problems are dealt with and issues are resolved in a consistent manner. After all, you don't want to call your credit company with a question about your account and get a random answer invented on the spot by the customer service rep on the phone. You want a consistent and reliable answer that will be the same each you have to deal with the issue.

Environmental policy is policy adopted by the government or other institution relating to environmental issues such as resource management or protecting people from the effects of pollution. Often, environmental policy decisions are made by elected or appointed government officials. These policy makers can be influenced by science, lobbyists, or environmental activists.

We defined environmental science as the study of the natural world and our interactions with it. Like other scientific disciplines, environmental science should have the goal of discovering factual information and increasing our understanding of the world regardless of personal biases or political views. A good scientist will follow the facts wherever they lead, even if they do not support the conclusions they would prefer. It is therefore important to keep science and activism as separate practices. An environmental activist is someone who tries to influence environmental policy or personal behaviors. That is not to say that an environmental scientist cannot also be an activist. Many in fact do so based on some of the things they uncover in their research. For example, Diane Fosey's studies of gorillas led to her becoming an activist to protect these endangered animals. However, a scientist should not let her desire for policy change influence her observations and conclusions in her research.

iQoncept/Shutterstock.com

We use ethics to guide us in discerning right from wrong, what is fair and unfair.

Ethics is the branch of philosophy that deals with questions of right and wrong. We all have ethical standards by which we make decisions about whether an action is good or bad. For example, the categorical imperative, also knowns as the Golden Rule, is an ethical standard that says we should treat others as we would want to be treated. The Principle of Utility is another ethical standard, which states that something is right if it produces the greatest good for largest number of people. Or, as Mr. Spock said in the classic "Star Trek: The Wrath of Khan," "the needs of many outweigh the needs of the few or the one."

Two extreme examples of ethical standards are universalism and relativism. Universalists say that standards for right and wrong are always the same, regardless of culture or social context. On the other hand, relativists state that ideas of right and wrong can vary depending on circumstance. A universalist would say that it is always wrong to tell a lie, while a relativist would say that a little white lie might be okay if it avoided a greater problem. For example, imagine it's Thanksgiving and your mother has spent hours preparing a huge meal for the entire family. But when you take a bite of the turkey, it tastes like a 5,000-year-old mummy. It is bone dry. Your mother asks you whether you like it. A universalist would say you must tell him or her the truth, even if it hurts his or her feelings. However, it might be better in this case to take the relativist position and tell him or her it's fine and ask him or her to pass the gravy. You don't want to be the one who ruins Thanksgiving dinner. Leave that to your drunk uncle who watches nothing but cable news.

When it comes to the environment, there are a number of ethical questions we can raise, such as

- Should we conserve resources for the benefit of future generations?
- Do other species have as much right to exist as we do?
- What criteria should we use to determine what is an acceptable risk to public health when dealing with pollutants?
- Should we sacrifice short-term economic gains in order to avoid long-term damage?

In order to address questions like these we can adopt a number of different views in terms of our relationship to the environment and what should be given priority.

afazuddin/Shutterstock.com

Anthropocentrism is the view that only humans and human interests should be given priority. Decisions about right and wrong should only consider the benefit to human beings. A more expansive view is **biocentrism**, which gives value to both human and nonhuman life. Finally, **ecocentrism** takes an even more holistic approach, giving value to entire ecosystems and says consideration should be given to both the living and nonliving factors in the environment.

Harmony Gerber/Shutterstock.com

John Muir memorial in Yosemite National Park.

As the Industrial Revolution made our impact on the environment more apparent, many people began adopting more biocentric and ecocentric views. For example, John Muir, the founder of the Sierra Club, argued that nature should be kept in a pristine, unaltered state whenever possible. Muir's view became known as the **preservation ethic**, which states that we should adopt a hands-off approach to nature and leave in its current state, unaltered by human activity. Muri's promotion of land preservation influenced Teddy Roosevelt's decision to create many of America's first national parks and national forests. An avid big game hunter, Roosevelt was forward thinking enough to realize that if we lost all the forests and other habitats for his favorite game animals, they would die out and there would be none left for him to shoot. This made him one of our most proactive presidents in setting aside protected lands.

Gifford Pinchot served as the first chief of the US Forest Service and later served as governor of Pennsylvania. A contemporary of John Muir, Pinchot shared many of the same goals, but the two often disagreed about the best way to protect the environment. Pinchot favored the **conservation ethic**, or the idea that natural resources should be used by people but they should not be exploited faster than they can replenish themselves. He reasoned that our society needed timber to make wood and paper products, but we should maintain the forests so that we have trees today and in the future. Today, there is a state park in York County, Pennsylvania, named after him.

Everett Historical/Shutterstock.com

Gifford Pinchot (right) with Teddy Roosevelt.

Aldo Leopold also had an influence on environmental ethics. Originally, he adopted an anthropocentric view and argued that the government should encourage the shooting of predators such as bears and mountain lions in order to increase the population of game animals such as deer and elk. However, he soon realized that eliminating predators caused prey populations to explode and then overgraze their environment. So, he switched to more ecocentric view to promote a balance between predator and prey populations that became known as the **land ethic**.

Diego G Diaz/Shutterstock.com

Protesters against the Dakota Access Pipeline at the Standing Rock Reservation is an example of demands for environmental justice.

More recently, environmental activist began promoting the idea do **environmental justice**, the idea that the impact of pollution and other environmental hazards should be assumed equally and be inflicted disproportionately on one population to the benefit of another. This emerged during the 1970s when studies showed that hazardous waste sites and incinerators were being sited more often in poor and minority communities. Wealthier communities had the political clout to keep these hazards out of their community, pushing them into low-income areas. Environmental justice seeks to redress this inequality.

1.6 The Scientific Method

Whenever you're faced with a decision, you always want to have accurate information and that's true for environmental policy decisions. The most reliable way to gain being understanding of the world around is the **scientific method**, which is a formalized technique for testing ideas through observation, testing, and analysis.

The scientific method begins with making initial observations, which will lead the scientist to develop an initial explanation for the phenomena called a **hypothesis**. A good hypothesis will make predictions that can be tested. The results of the testing can then be compared to the predictions. Though it may seem counterintuitive, when scientists run experiments, they are actually trying to disprove their own hypothesis. So, the results they get will either indicate that the hypothesis is false and therefore the hypothesis will be rejected. Or, the results will align with the predictions made by the hypothesis in which case, it fails to reject the hypothesis. No hypothesis is ever considered "proven" with 100% certainty because any hypothesis could possibly be disproven by a later experiment. Instead, scientists speak in terms of their confidence level that the data support the hypothesis.

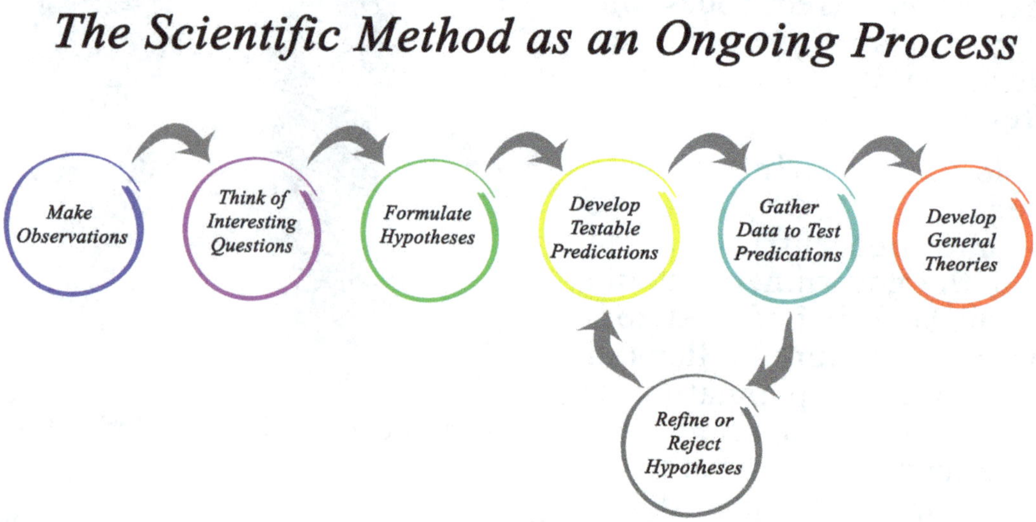

arka38/Shutterstock.com

Any experiment involves the use of two kinds of variables: independent and dependent variables. The **independent variable** is the condition that is changed in order to test the hypothesis, while the **dependent variable** is the one that is measured as the result of the changes in the dependent variable. For example, suppose I wanted to measure the effect a certain fertilizer had on the growth of algae. My independent variable would be the amount of fertilizer applied, while the dependent variable is the amount of algae growth.

Ihor Bondarenko/Shutterstock.com

Ideally, in a controlled experiment, I would want to have several tanks of water. In each tank, I would apply a different amount of fertilizer and then measure how much algae grew in each tank. This would enable me to plot the **data**, or the information measured, into a graph to illustrate how the algae growth varied with the amount of fertilizer. I would also want to have at least one tank that received no fertilizer so that I could measure how the algae grow without the presence of the fertilizer. The tanks that receive the fertilizer would make up my **treatment group**, while the tank that received no fertilizer is my **control group**. A control group is important in order to compare your experimental results with the normal growth without the fertilizer. It is the only way to demonstrate that the fertilizer and not some other factor that influences the chance in growth.

Allans Photo/Shutterstock.com

In studying the effects fertilizer had on algae growth, one would need at least two groups. A treatment group that received the fertilizer and the control that didn't receive any. Only by comparing the growth rates in both groups can we make any conclusions on how the fertilizer affects growth.

Gorodenkoff/Shutterstock.com

Scientific experiments have greatly increased our understanding of life and the world around us.

The above is an example of a **manipulative experiment**, in which a single variable is to be changed under controlled conditions. Many people think that manipulative experiments are the only way in which science is done, but there are many cases in which the phenomena being studied are too big or too distant in space or time to be studied by a controlled experiment in the lab. In these cases, scientists learn by doing **observational science**. For example, we can't build a star in a lab and study it up close. Fortunately, our galaxy is filled with billions of stars of different sizes, ages, and temperatures. By collecting data from many stars in various stages of stellar evolution, astronomers can build a model for how stars form and change over time.

A scientific experiment must be repeatable, that is, you or someone else conducting the same study should get similar results. One case might be a fluke, but if several repeats of the experiment yield similar results, then the hypothesis is better supported.

Once the scientist has completed his or her study, the usual practice is for him or her to write a paper and submit it to a peer-reviewed journal. **Peer review** is the process by which scientists in the same field evaluate each other's work. When a journal receives a proposed paper, it sends it out to several other scientists (the "peers") who review it and recommend whether or not to publish it. If the peers conclude that the research followed good scientific methodology and the conclusions are supported by the data, they recommend publication. Otherwise, the paper is rejected and the scientist has to do more work before resubmitting.

Unfortunately, there are many places where "scientific" papers can be published without peer review. Many of these are pay-for-publishing journals that will publish almost any paper if the researcher pays a fee. Also, some organizations with a political agenda may self-publish what looks like legitimate research that is actually slanted to reach a predetermined conclusion.

Today, few people dispute the fact that there is a strong link between smoking tobacco and certain forms of cancer. However, for decades, doubt was sown by an organization called the Tobacco Institute, which published many convincing-looking papers. All of these papers came to the same conclusion: "Gosh, we just can't find any link between tobacco and cancer." This was the complete opposite of what nearly every other researcher in the field concluded. How were they able to come to different results? Well, the Tobacco Institute was funded by the tobacco industry and we now know that they were under orders by their sponsors to come to predetermined conclusions no matter what the data said.

More recently, a research paper in an in pay-to-publish journal created a sensation in the news media by announcing that you can lose weight by eating chocolate. Wouldn't that be great? Sounds too good to be true, right? Well, it was. The researcher behind it came forward to admit that the entire paper was a hoax. He did it to highlight how easy it is to get bogus studies published and how readily the media can hype what seem like sensational results. Sadly, most journalists are not trained to discern the difference between sound science and what we can label "junk science."

Science plays a big role in our understanding of the world around us. It is therefore imperative that we have a scientifically literate society that is capable of discerning good science from junk science, especially when making decisions relating to their health and well-being.

One final word on science and relates to how people use the word theory. Many nonscientists use the word to mean a guess or a hunch based on incomplete information. However, this is now how scientists use the term. In science, a **theory** is a broad explanation for a phenomenon that has been well-tested, shown to be supported repeatedly by experiment, and has gained wide acceptance. It is not a single hunch or guess. They way nonscientists use the term theory is more akin to how scientists use the word theory. Another misconception is that if a hypothesis is validated by experiment, it may be "promoted" to being a theory. That is not accurate either. A theory is a broader explanation, while a hypothesis is generally narrower in scope. Indeed, a single theory may encompass several hypotheses into it.

Keep that in mind when someone dismisses a scientific principle as "just a theory." Nearly everything you are taught in a science class is based on theory. Of course, just because a theory has gained broad acceptance doesn't always mean it is true, but it takes a lot of evidence to overturn a theory that has been repeatedly validated by observations. Probably the best-known example of such a major shift in our understanding was the Copernican Revolution, when the geocentric (Earth-centered) model for our solar system eventually gave way to the heliocentric (Sun-centered) model.

1.7 Sustainability and Sustainable Development

If there was one key term that could sum up a sound environmental policy, it would be **sustainability**. A system is considered **sustainable** if it can continue on indefinitely in its current form. The principles of sustainability include

- Living within the means of our planet's resources
- Conserving resources so that future generations will have them
- Keeping ecosystems functioning and providing natural resources and ecosystem services

iQoncept/Shutterstock.com

It's pretty obvious our dependence on a nonrenewable resource like fossil fuels is not sustainable. Because they are finite, they will run out and we will no longer have use of them. Getting onto a path toward sustainability will include both an increase in energy efficiency and switching to renewable sources of energy.

This is one aspect of **sustainable development**, the goal of which is to provide a growing standard of living while still living within our planet's means and not at

garagestock/Shutterstock.com

the expense of future generations. Other principles of sustainable development include soil and water conservation, protecting endangered species and habitats, reducing air and water pollution, improving our waste management, and combating global warming.

1.8 Discussion Questions

1. What things come to mind when you think of the *environment*?
2. Define *ecosystem services*. How do these services affect our quality of life?
3. Describe how better resource management could have changed things for the people of Easter Island/Rapa Nui?
4. What is the difference between an inexhaustible resource and an exhaustible one?
5. How does our dependence on nonrenewable resources like fossil fuels affect our expectations of the future?
6. How did the agricultural revolution and Industrial Revolution affect the human population and our relationship to the environment?
7. How did John Muir's commitment to the preservation ethic influence Teddy Roosevelt's decisions with respect to establishing some of the first national parks and national forests?
8. Describe the steps in the scientific method and how it might apply to studying the effects pollutants might have on wildlife.
9. What is a theory? How does it differ from a hypothesis?
10. What are the goals of sustainable development?
11. How do the conditions on Venus and Mars differ with the Earth in terms of being able to support life?
12. How do journal publishers use peer review to ensure that the scientific papers that they publish describe research that followed good scientific methods?

1.9 Key Terms

Agricultural revolution
Anthropocentrism
Biocentrism
Conservation ethic
Control group
Data
Dependent variable
Ecocentrism
Ecological footprint
Ecosystem services
Environmental justice
Environmental science
Environment
Ethics
Exhaustible resources
Hypothesis
Independent variable
Industrial Revolution
Inexhaustible resources
Land ethic
Manipulative experiment
Natural resources
Nonrenewable resources
Observational science
Peer review
Preservation ethic
Renewable resources
Scientific method
Sustainability
Sustainable
Sustainable Development
Theory
Treatment group

Chapter 2

ENVIRONMENTAL SYSTEMS

Chapter Topics

2.1 Environmental Systems

In this chapter, we will discuss the basics of systems. A **system** is defined as a series of components that interact with each other through the exchange of matter, energy, or information. Every system also has an input and an output. For example, in biology, the digestive system consists of your teeth, esophagus, stomach, intestines, and other organs that aid in the digestion of food. The input is food and the output is waste.

Likewise, we can describe the natural world as a series of systems. An **ecosystem** is a group of living organisms in a given area and the nonliving components, such as soil, water, air, sunlight, that they interact with. For example, the Chesapeake Bay, the largest bay in the United States, is an ecosystem. It's a huge estuary where freshwater from the land mixes with saltwater from the ocean. It receives inputs from rivers like the Susquehanna and the Potomac and its output includes the impacts those inputs have on aquatic life in the bay.

Unfortunately, the health of the Chesapeake Bay has declined greatly in recent decades. Nitrogen and phosphorus carried by the rivers have resulted in a condition called **hypoxia** or a lack of dissolved oxygen in the water. This lack of oxygen results in areas of the bay being called **dead zones**, where aquatic organisms cannot survive. Oysters used to be abundant in the Chesapeake, but as a result of hypoxia combined with overharvesting, populations of oysters have plummeted.

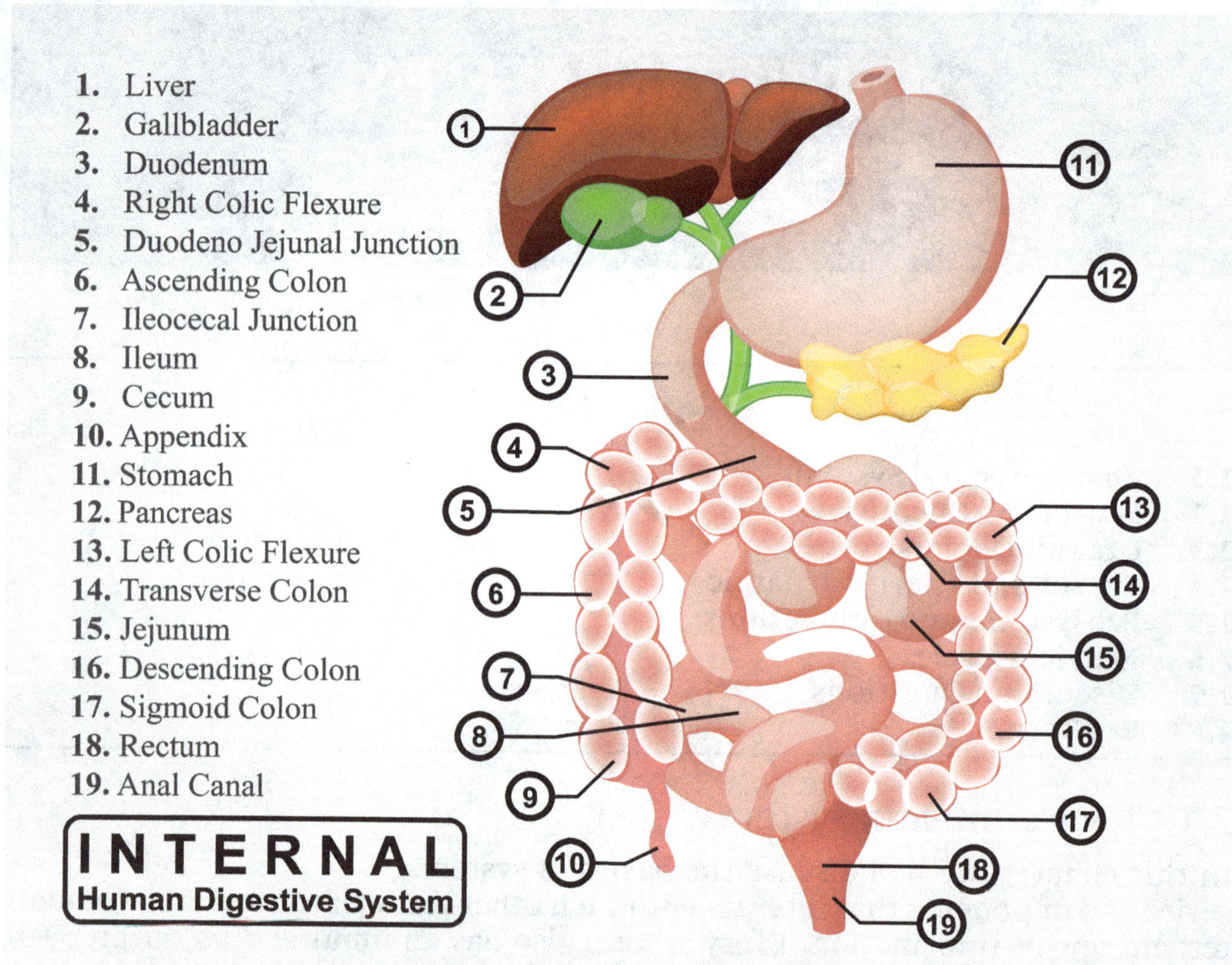

rendix_alextian/Shutterstock.com

The digestive system is one example of a system with an input and an output.

Many systems experience a feedback loop, where the output is fed back to the system as the new input. Hopefully, this doesn't happen with your digestive system, but if you've ever held a microphone to close to the speakers, that obnoxious screeching noise you hear is the result of a **feedback loop**, where the sound is fed into the microphone, gets kicked up by the amplifier, and comes out of the speaker only reenter the microphone, repeating the process over and over again.

Feedback loops can be either positive or negative. In a **negative feedback loop**, the result of the feedback is to drive the system in the opposite direction. Organisms make use of negative feedback in order to maintain a **homeostasis**, or stable internal conditions. For example, your body regulates your blood sugar by releasing insulin or glucagon. If your blood sugar level is too high, your body produces insulin to lower your blood sugar. If it's too low, your body produces glucagon to raise it. Because your blood sugar level is in a constant state of flux, it is said to be in a state of **dynamic equilibrium**, where the system is stable even though levels fluctuate around an average number. A person with diabetes has difficulty maintaining this equilibrium.

On the other hand, a **positive feedback loop** pushes the system in the same direction as the stimulant. As a result, conditions for the system tends to move away from a condition of dynamic equilibrium and instead toward a more extreme state. For example, ice is more reflective of sunlight than liquid water or solid ground, so a surface covered in ice tends to remain frozen. However, if some of the ice starts to melt and exposes these less reflective materials, the surface absorbs more energy from the sun and the temperature rises. As the temperature rises, more ice melts, exposing more heat-absorbing liquid water and solid ground surfaces. This in turn causes more temperature rise and more ice melting. Because organisms tend to thrive when they maintain homeostasis, a positive feedback loop can be disruptive in nature.

INSULIN AND GLUCAGON
regulate blood glucose levels

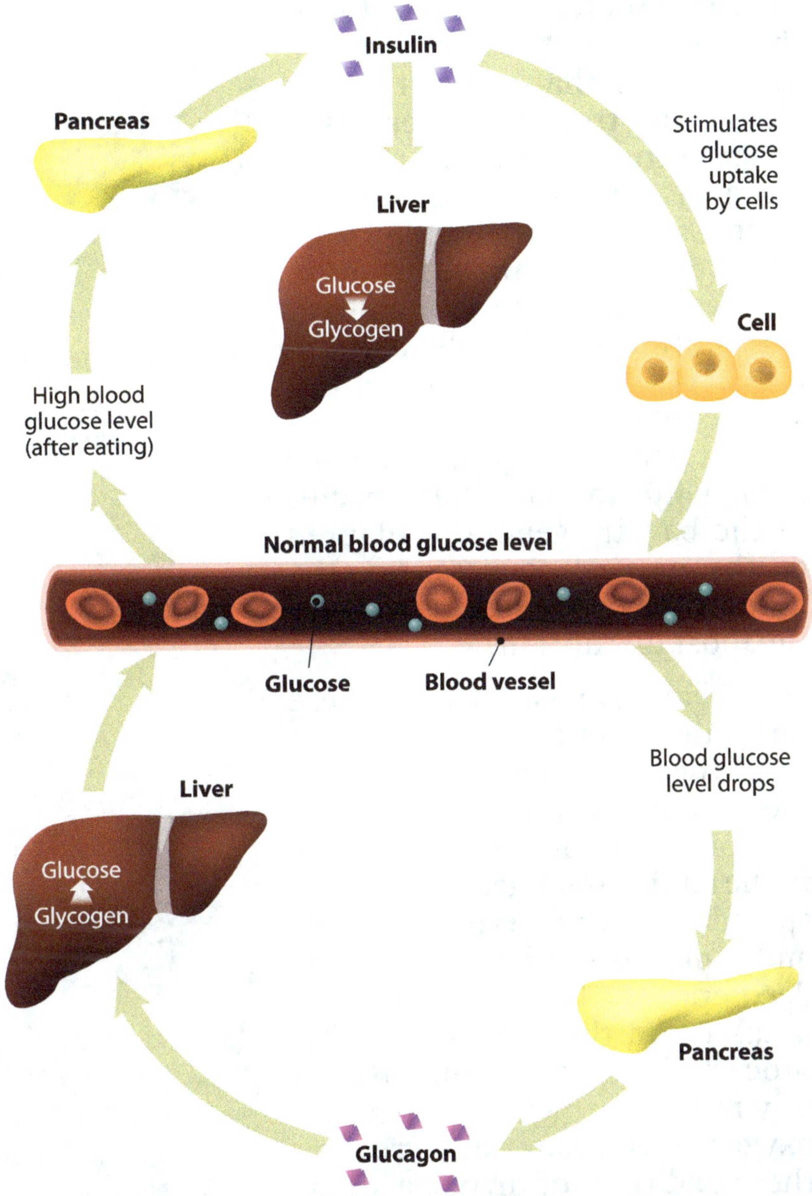

Designua/Shutterstock.com

The body's ability to regulate blood sugar results from a negative feedback loop.

Ecosystems do not exist in isolation. They all exchange matter, energy, and information with their surroundings. For example, the Chesapeake Bay gets inputs in the form of **runoff**, rainwater that moves across the surface, eventually collecting in rivers, streams, and other bodies of water. The area in which the surface water all tends to flow into a single body of water is called a watershed. The Chesapeake Bay watershed encompasses a large area that includes portions of six states: Pennsylvania, New York, Maryland, Delaware, Virginia, and West Virginia.

As the runoff flows over the land in those states, it can pick up all kinds of pollutants. Of particular concern is nitrogen and phosphorus. Studies have found that the majority of the nitrogen and phosphorus that enter the bay have come from fertilizers used in agriculture and animal manure on farms. Other sources include wastewater treatment plants, urban and suburban fertilizer use on lawns, and (in the case of nitrogen) atmospheric deposition. Of course, there are also natural sources of nitrogen and phosphorus entering the bay, such wastes from wild animals, but species in the bay are generally adapted to the natural sources. So, the human-generated sources cause the most disruption.

So, what happens when these nutrients enter the bay? Well, just as nitrogen and phosphorus promote plant grown on the farm, they also promote the growth of algae in the bay. As the algae blooms and spreads, some of it eventually dies and sinks to the bottom. There, decomposing bacteria begins to break down the dead algae. This process of bacterial decomposition is what consumes the dissolved oxygen in the water and creating the condition of hypoxia. This process is referred to as **eutrophication**, which means "overfeeding," as in the algae and bacteria are being overfed as a result of the nitrogen and phosphorus.

Hypoxia is a global problem with over 500 cases of dead zones forming along populated coastal regions.

Laura Pl/Shutterstock.com

Glaciers melting and exposing land produces a positive feedback loop in which the surface absorbs more heat as the ice continues to melt.

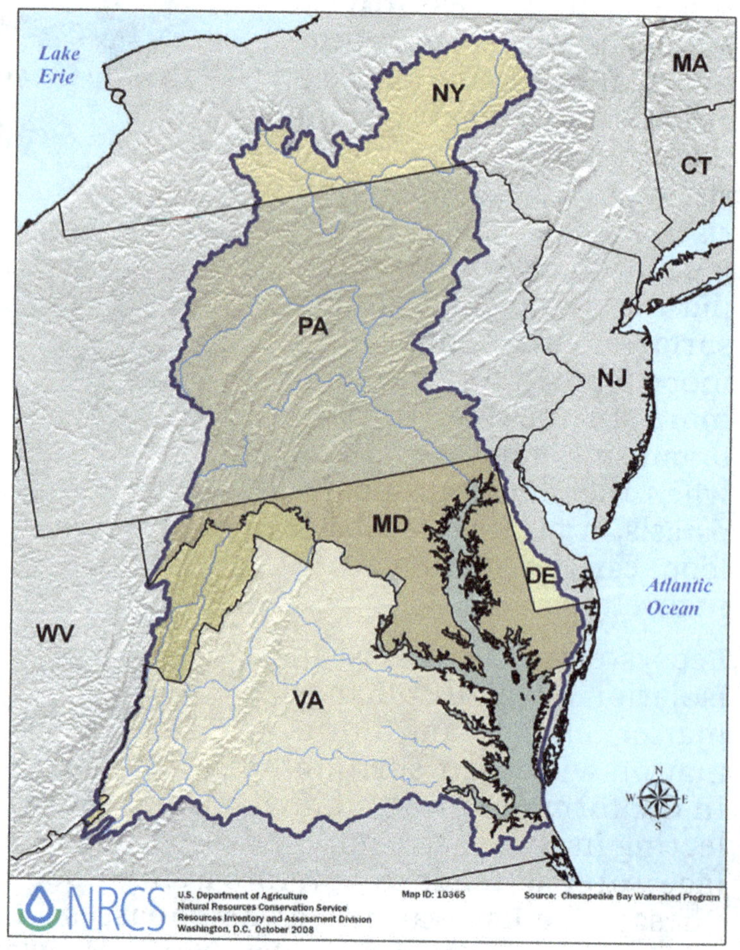

Source: https://www.nrcs.usda.gov/Internet/FSE_MEDIA/stelprdb1046486.png

The Chesapeake Bay Watershed.

In order to address the problem in the Chesapeake Bay, the EPA has adopted a program to reduce the amount of human-made sources of nitrogen and phosphorus. These include upgrading wastewater treatment plants to remove the nutrients from the water they release to the rives, upgrading stormwater systems, and encouraging farming practices that reduce fertilizer runoff. These include conservation tillage, reconstructing wetlands, and planting trees and other buffer plants along the stream banks to capture the nutrients before they enter the streams.

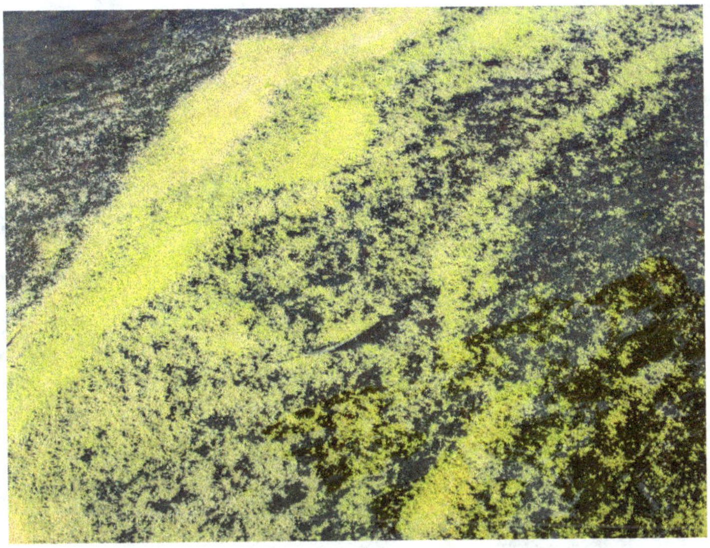

blojfo/Shutterstock.com

Algal bloom in seawater. Too much algae growth can lead to eutrophication.

2.2 Matter and Chemistry

Nitrogen and phosphorus promote the growth of algae. CFCs break down ozone in the stratosphere. The release of sulfur dioxide into the air contributes to acid rain. Drinking water with a low pH can erode lead-based solder in old plumbing. These are all examples of environmental problems that relate to chemistry.

Chemistry is the field of science that studies the interaction of matter and **matter** is defined as any material substance that has mass and occupies space. Matter can exist in three main states: solid, liquid, or gas. One of the most fundamental laws in chemistry is the **law of conservation of matter** that states that matter cannot be created or destroyed. It can only be transformed from one form to another. For example, if you take a pile of wood and set it on fire, it is converted into gases, primarily carbon dioxide and water vapor, and ash. None of the matter has been destroyed. It's only been transformed. The law of conservation of matter means that harmful materials like nuclear waste or toxic heavy metals cannot be wished away. There is no magic hole we can throw these things down so that they will disappear forever. Our only options are to try to transform them into something less harmful or, if that isn't possible, isolate them from the environment so that they cannot do harm.

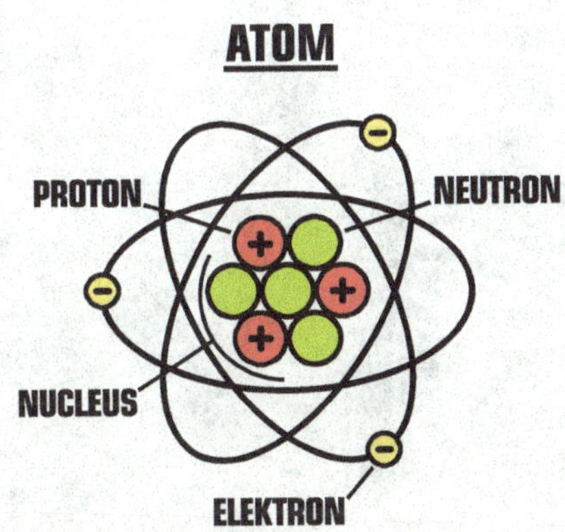

ShadeDesign/Shutterstock.com

By way of a review of basic chemistry, matter is made up of elements. Elements are chemical substances that cannot be broken down into other substances. There are 92 naturally occurring elements and over twenty that have been created artificially. Elements are organized according to their properties and atomic structures in the **periodic table**.

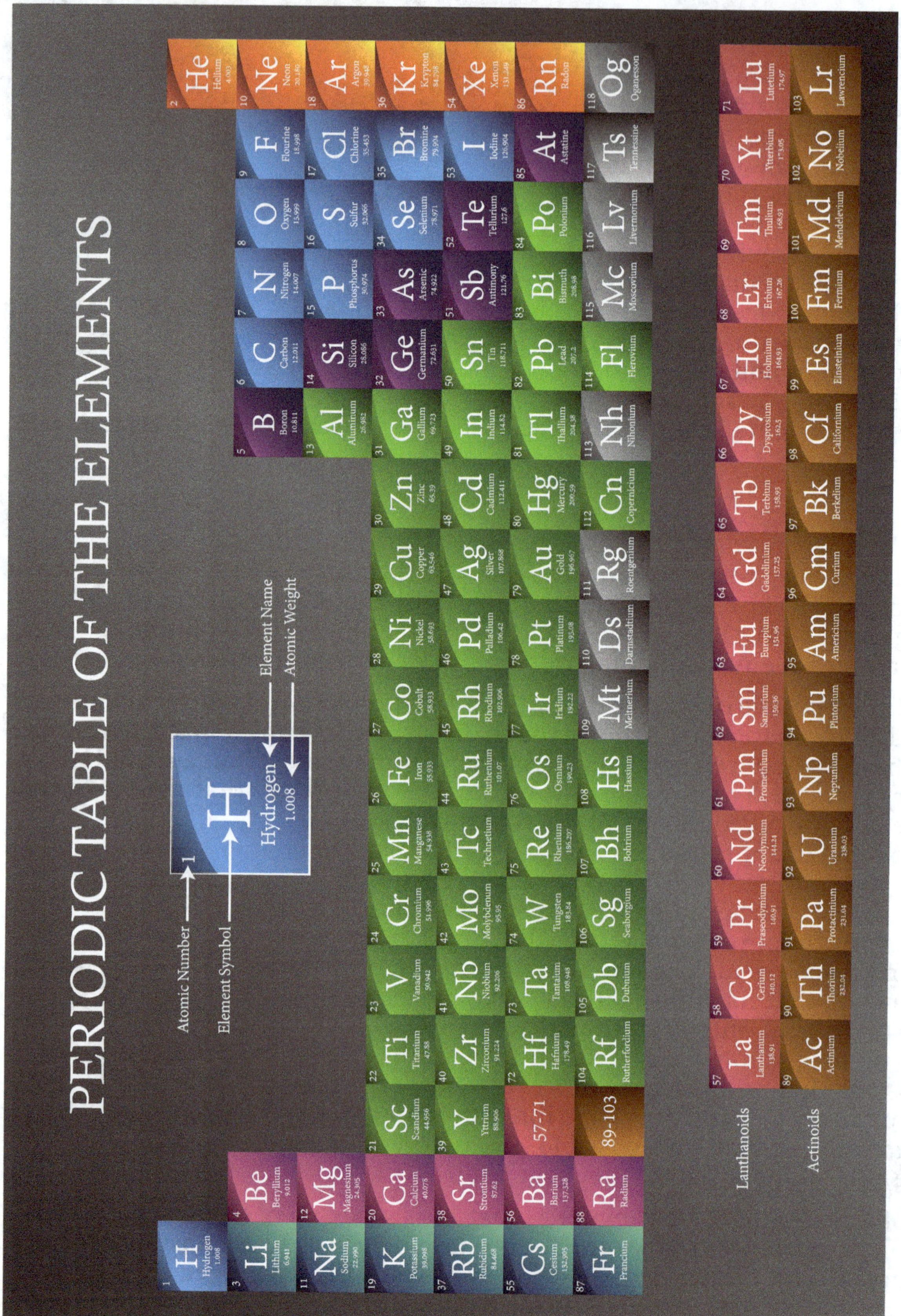

Hemin Xylan/Shutterstock.com

Periodic table of the elements.

The smallest unit of matter that still has an element's chemical properties is called the **atom**. Atoms consist of a central mass called a **nucleus**. Two kinds of particles make up the nucleus: **protons**, which have positive electric charge and **neutrons**, which have no charge. A third particle, the negatively charged **electron**, orbits around the nucleus in various energy states called shells.

The number of protons in the nucleus of the atom, called the **atomic number**, defines what element that atom is. For example, hydrogen has an atomic number of one, so it has only one proton in the nucleus. Helium has an atomic number of two and has two protons in its nucleus. Carbon has an atomic number six, so it has six protons in the nucleus. If the atom has the same number of electrons as protons, the charges cancel each out and the atom has no net charge.

The number of protons plus the number of neutrons in the atom is called the **atomic mass** (the electrons, being very light, do not contribute much to the mass of the atom). Atoms of the same element can have different atomic masses, this is, a different number of neutrons. These atoms are called **isotopes**. For example, the most common form of hydrogen has an atomic mass of one, which means it has just one proton by itself. However, there is also an isotope of hydrogen that has one proton and one neutron in the nucleus, giving it an atomic mass of two. Isotopes are referred to by their atomic mass, often with a superscript in their symbol. The symbol for hydrogen is the letter H, so hydrogen-2 (sometimes called deuterium or heavy hydrogen) is given the symbol H^2.

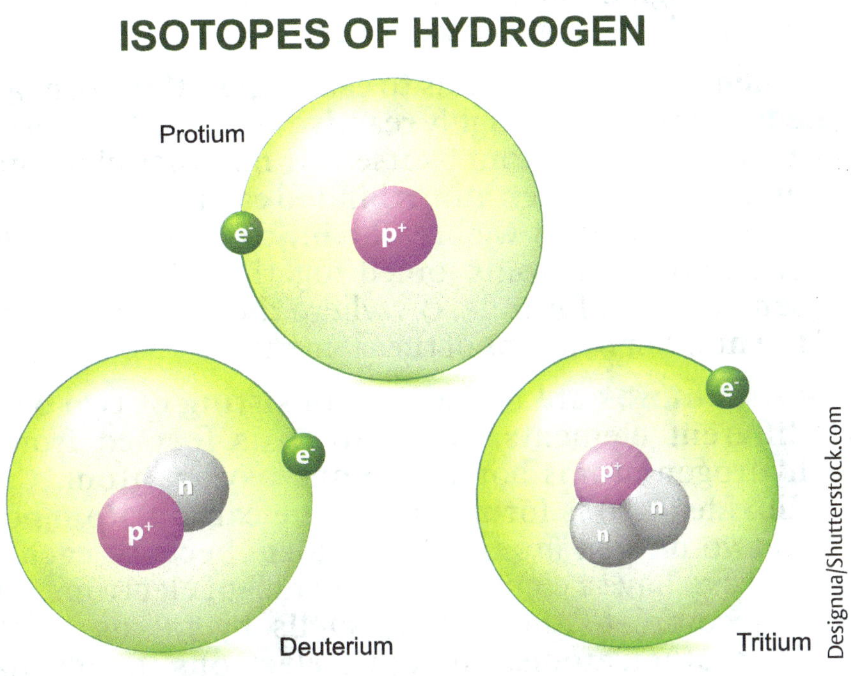

Isotopes of hydrogen.

Some isotopes are unstable. They have too much energy in the nucleus. To make themselves more stable, they emit radiation in the form of particles and electromagnetic energy over time, transforming them into more stable atoms, a process called **radioactive decay**. These isotopes are referred to as **radioisotopes**. Uranium, the heaviest of the natural elements, decays in a series of stages, eventually transforming in a stable isotope of lead. The **half-life** of the isotope, the time it takes for half of the radioisotope present to decay, governs the rate at which a radioisotope decays, emits radiation, and transforms into a stable atom. So, if you start with a certain amount of a radioisotope, after one half-life, 50% of the material will have decayed away and 50% would remain. After two half-lives, only 25% of the original material would remain and after three half-lives, we are down to

12.5%. And so on down to infinity. Each isotope has its own half-life, which can range from a fraction of a second to millions of years. Uranium-235, the fuel used in nuclear power plants, has a half-life of 700 million years.

Because the radiation emitted by these radioisotopes can cause damage to living tissue, ranging from radiation sickness, to inducing certain forms of cancer, and death, they must be handled carefully. Disposal of radioactive waste is one of the major concerns related to nuclear power.

Atoms can also gain or lose elections, giving them a net charge. Taking away electrons gives the atom a net positive charge while adding electrons gives them a net negative charge. For example, if you take away the one electron in a hydrogen atom, you're left with a single, positively charged proton in the nucleus. Atoms (or molecules) that have a net charge are called **ions** and are given a plus or minus sign indicating their net charge. So, the hydrogen ion would be represented by H^+.

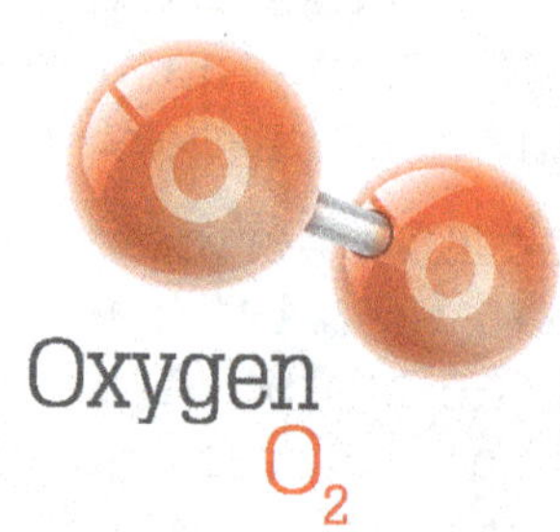

LoopAll/Shutterstock.com

Oxygen molecule.

When atoms form bonds to each other, they form a **molecule**. Oxygen is a highly reactive element. You never find a single oxygen atom by itself. If no other elements are present, the oxygen atoms bind to each other. Therefore, oxygen gas, what we breathe in, consists of a molecule of two oxygen atoms joined together. Oxygen is represented by the formula, O_2, where the subscript indicates the number of atoms of the element.

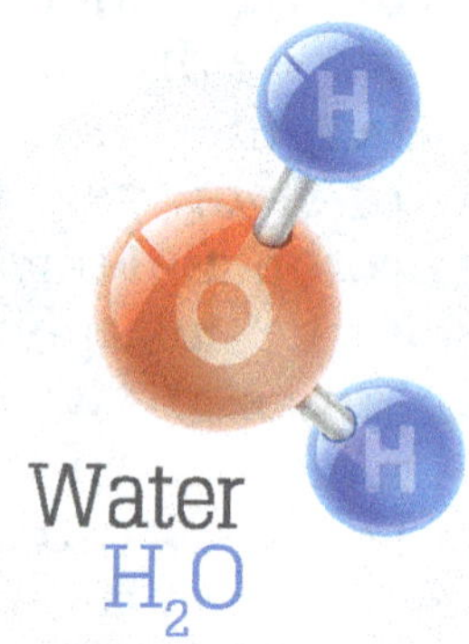

LoopAll/Shutterstock.com

Water molecule.

Compounds are molecules consisting of two or more different elements. Water (H_2O) is formed from two hydrogen atoms bonded to one oxygen atom. Carbon dioxide (C_2O) is formed from one carbon atom bonded to two oxygen atoms. Bonds form because each atom seeks to have a complete outer shell of electrons. Atoms can complete their outer shells by taking electrons from other atoms, giving up electrons, or by sharing electrons. Compounds can be formed from either ionic bonds or covalent bonds.

In **ionic bonds**, electrons are transferred from one atom to another. For example, sodium (Na) can have a complete outer shell by giving up an electron while chlorine can have a complete outer shell by adding an electron. This transfer for an electron makes the sodium atom a positive ion (Na^+) while the chlorine atom is now a negative ion (Cl^-). Because the two ions have opposite charges, they are attracted to each other and form an ionic bond. This new compound is called sodium chloride or ordinary table salt.

LoopAll/Shutterstock.com

Carbon dioxide molecule.

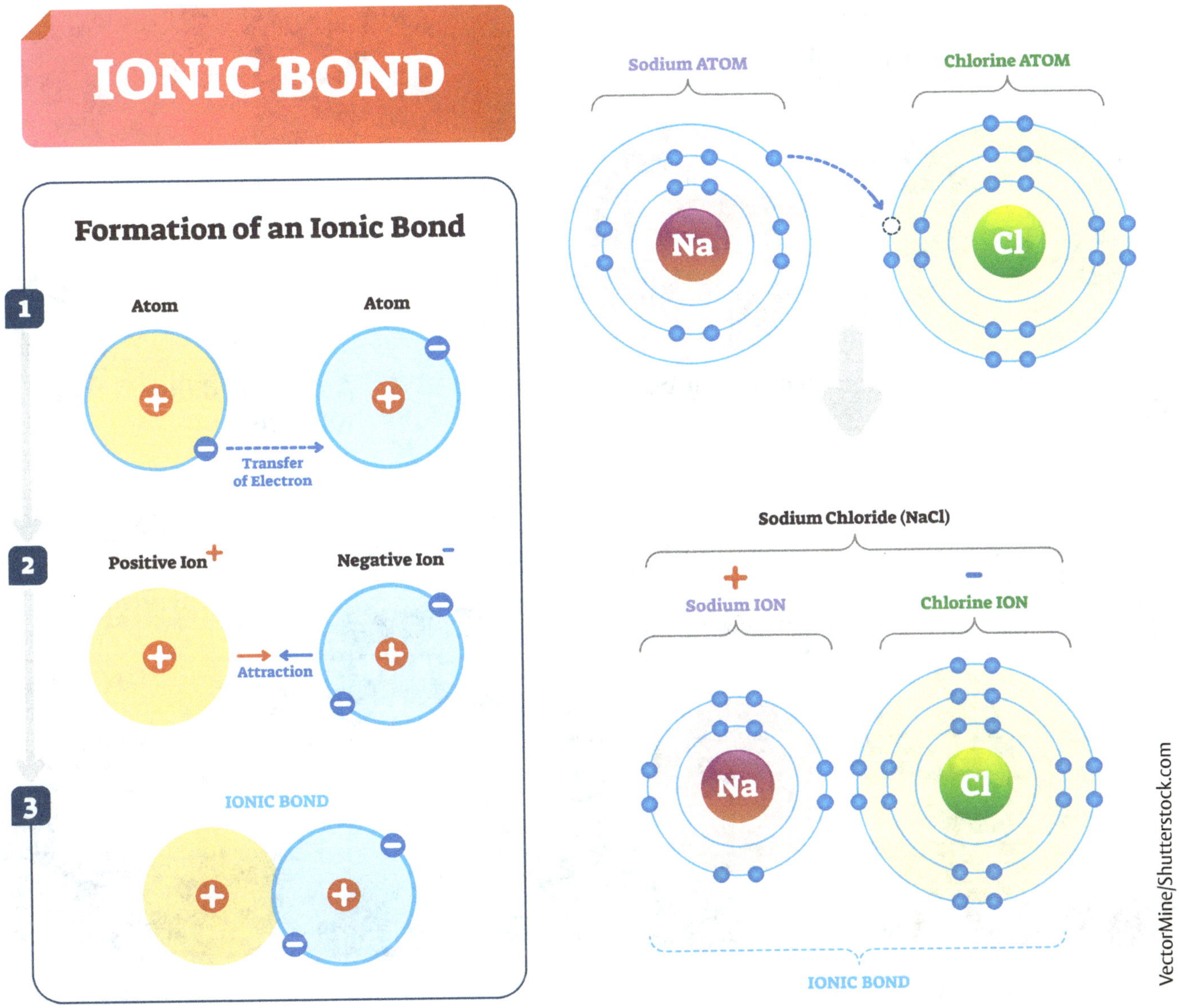

Ionic bond forming salt from sodium and chlorine atoms.

Atoms form **covalent bonds** by sharing electrons. For example, a hydrogen atom needs one additional electron to complete its outer shell while an oxygen atom needs two. Two hydrogen atoms can therefore complete their outer shell by each sharing a single electron with an oxygen atom.

Materials can be combined without chemical bonds in a **solution**. A solution can be either a liquid, gas, or solid. Seawater is a solution in which salt and other minerals are dissolved in water. The atmosphere is a gaseous solution of nitrogen and oxygen. Metal alloys, such as steel or bronze, are solid solutions.

In an aqueous (water-based) solution, there is a tendency for some of the water molecules to separate into hydrogen (H^+) and hydroxide (OH^-) ions. The ratio of the hydrogen to hydroxide ions in a solution is given by the **pH scale**, which is an inverse logarithmic scale. I'll spare you any rehash of painful memories you may

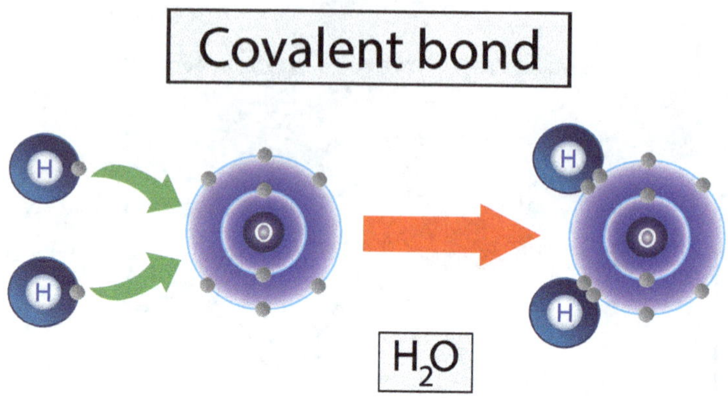

Nasky/Shutterstock.com

Water molecule formed by covalent bonds.

Inna Bigun/Shutterstock.com

A solution differs from a mixture.

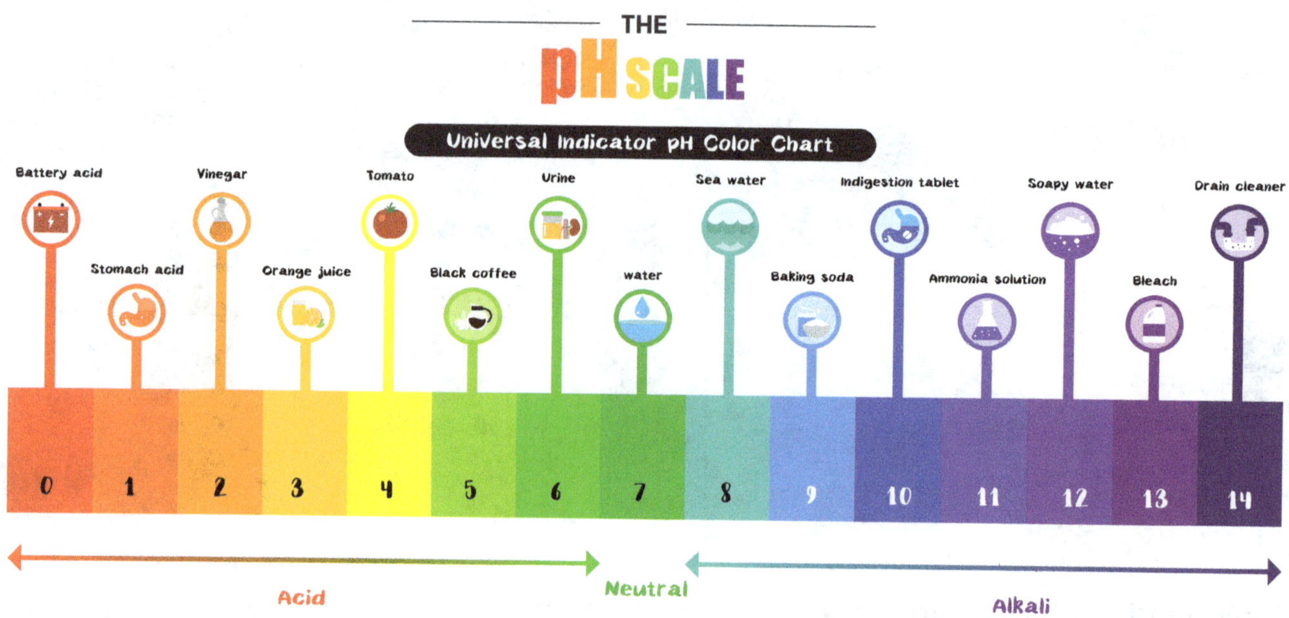

trgrowth/Shutterstock.com

The pH scale.

have had about logarithms in high school and just say that each step on the pH scale charges the ratio by a factor of 10. The lower the pH, the higher the concentration of hydrogen ions. Therefore, a pH of 4 has 10 times the concentration of hydrogen ions as a pH of 5.

Pure distilled water has an equal ratio of hydrogen and hydroxide ions and has a pH of 7. Any solution with a pH less than 7 is an **acid** while any solution with a pH greater than 7 is a **base**. A solution with a pH of exactly 7 is considered **neutral**, neither acidic nor basic. The pH scale ranges from 0 (the strongest acid) to 14 (the strongest base). Both strong acids and strong bases are caustic and can cause chemical burns, so great care should go into handling them.

2.3 Organic Chemistry

Carbon has a number of interesting properties. It can form bonds with up to four different atoms and it can also form covalent bonds with other carbon atoms. This enables carbon to form long complex molecules consisting of long chains, rings of atoms, and other shapes. Carbon has an entire field of chemistry dedicated to studying its chemistry. Molecules formed from carbon bonds, or **organic compounds**, can very complex and are essential to life as we know it.

Hydrocarbons, molecules formed by bonding carbon and hydrogen atoms, have a lot of industrial uses. The simplest hydrocarbon, methane or natural gas, consists of one carbon atom bonded to two hydrogen atoms. Other hydrocarbons include ethane, naphthalene, benzene, and gasoline. All fossil fuels and petroleum products such as plastics are made up mostly of hydrocarbons.

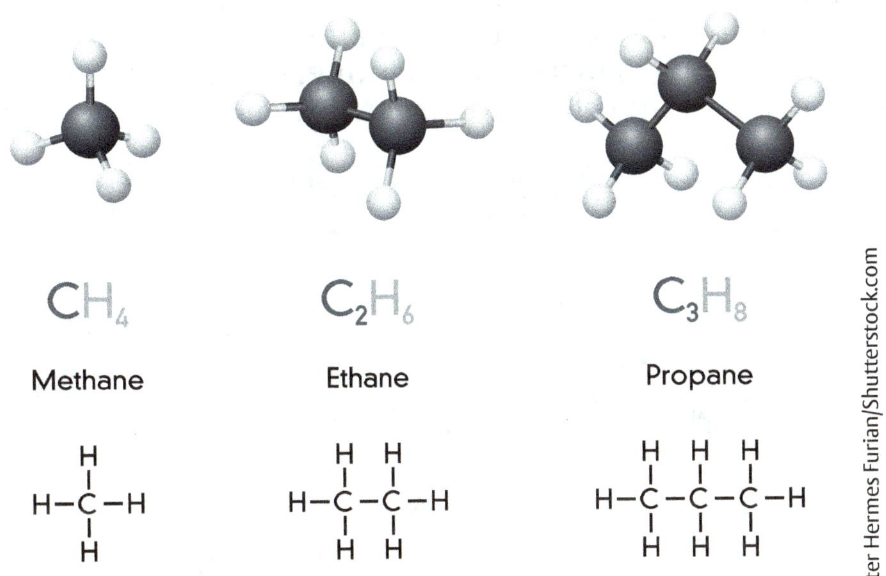

Some simply hydrocarbons.

Many of the organic compounds found in living organisms consist of **polymers**, long chains made up of similar, smaller molecules linked together. Often, these large polymers are given the name **macromolecules** given their large size.

Proteins consist of chains of smaller molecules called **amino acids**. There are 20 different amino acids that are essential to life and a single protein can contain hundreds of them. Proteins are the workhorses of life. They make up the structural material in our skin and muscles. They also make up components to our immune system and can act as enzymes that help trigger certain chemical reactions.

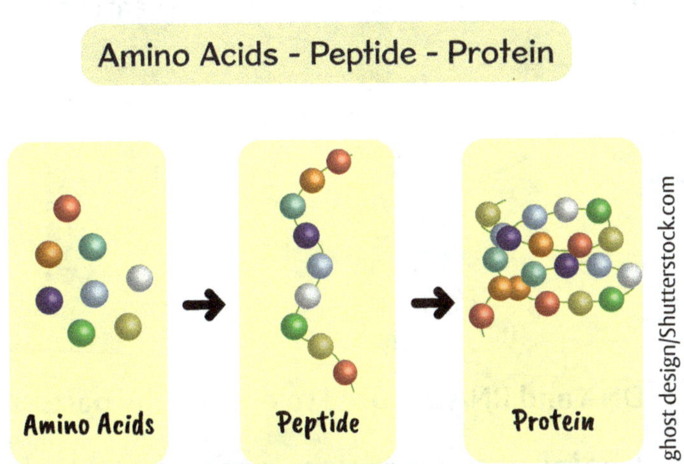

Proteins are polymers made up of chains of amino acids.

Polymers made up of chains of nucleotides, called **nucleic acids** get their name from their original discovery in the nucleus of living cells.

Each individual nucleotide has three components: a phosphate group, a sugar, and a nitrogenous base. The two types of nucleic acids, deoxyribonucleic acid (DNA) and ribonucleic acid (RNA) contain the hereditary information we inherent from our parents and the blueprints for building proteins. DNA uses for nitrogenous bases in its nucleotides: adenine, guanine, cytosine, and thymine, which are represented by the letters A, G, C, and T. The nucleotides in DNA bind together to form a long strand. A single DNA molecule consists of two strands winding around each other in the double helix formation. As the strands wind, the nitrogenous bases pair up. A always pairs with T and G always with C. A **gene** is a section of a DNA molecule that encodes the instructions for making a specific protein.

When the time comes to build a protein molecule, the gene unzips itself from the double helix and builds a single strand of RNA. RNA differs DNA in that it always found in a single strand, has a different sugar molecule, and uses the nitrogenous base uracil (U) instead of thymine. Once completed, the RNA molecule exits the nucleus of the cell and instructs the cell to build the protein.

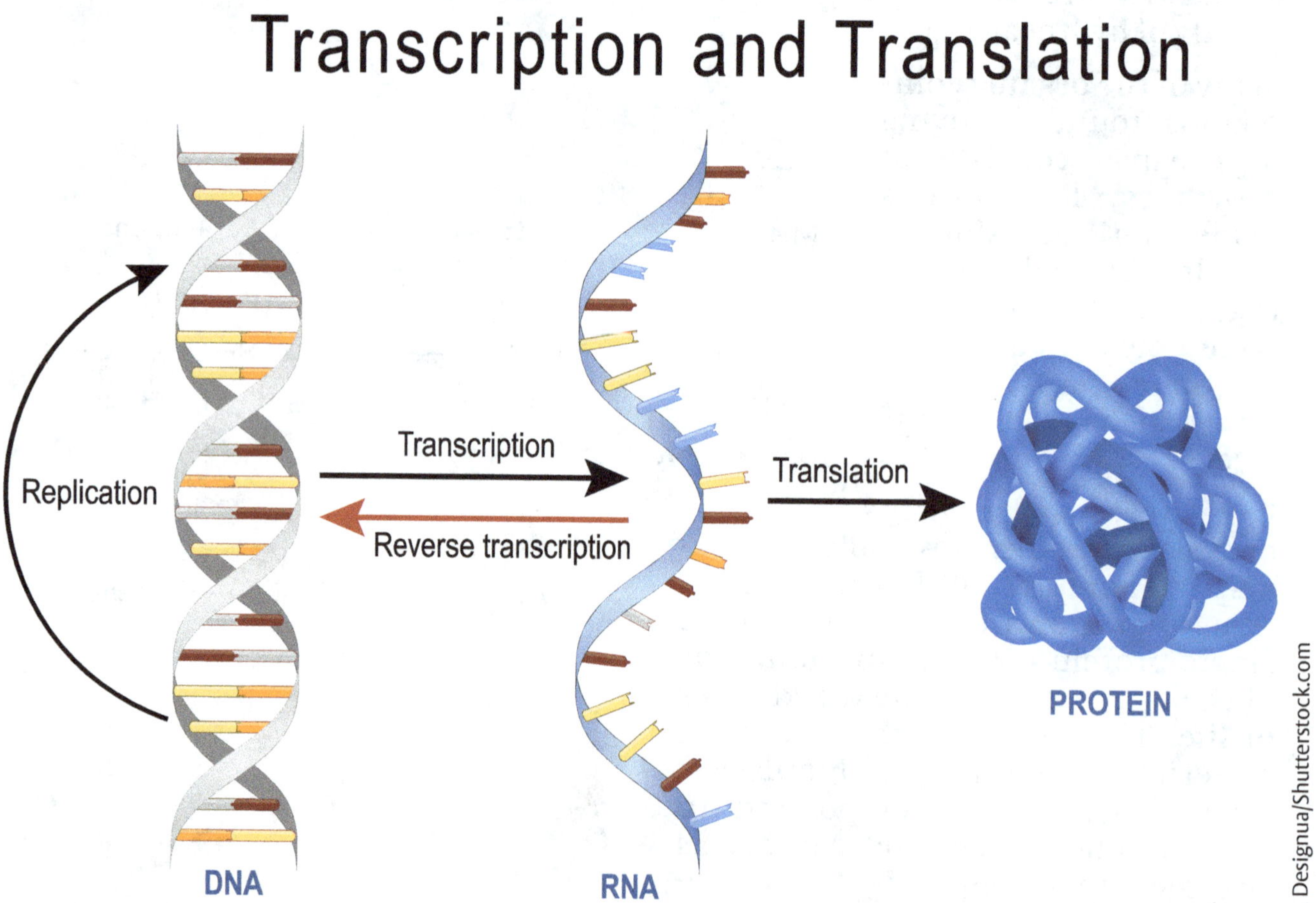

DNA and RNA are used to encode the patterns for protein synthesis.

Chains of sugar molecules form molecules called **carbohydrates**. Carbohydrates, such as glucose (blood sugar) can be used to store and transport energy. Another carbohydrate, cellulose, forms the cell walls around plant cells and gives plants their toughness. Chitin is a carbohydrate found in the exoskeletons of insects and other arthropods.

Carbohydrates

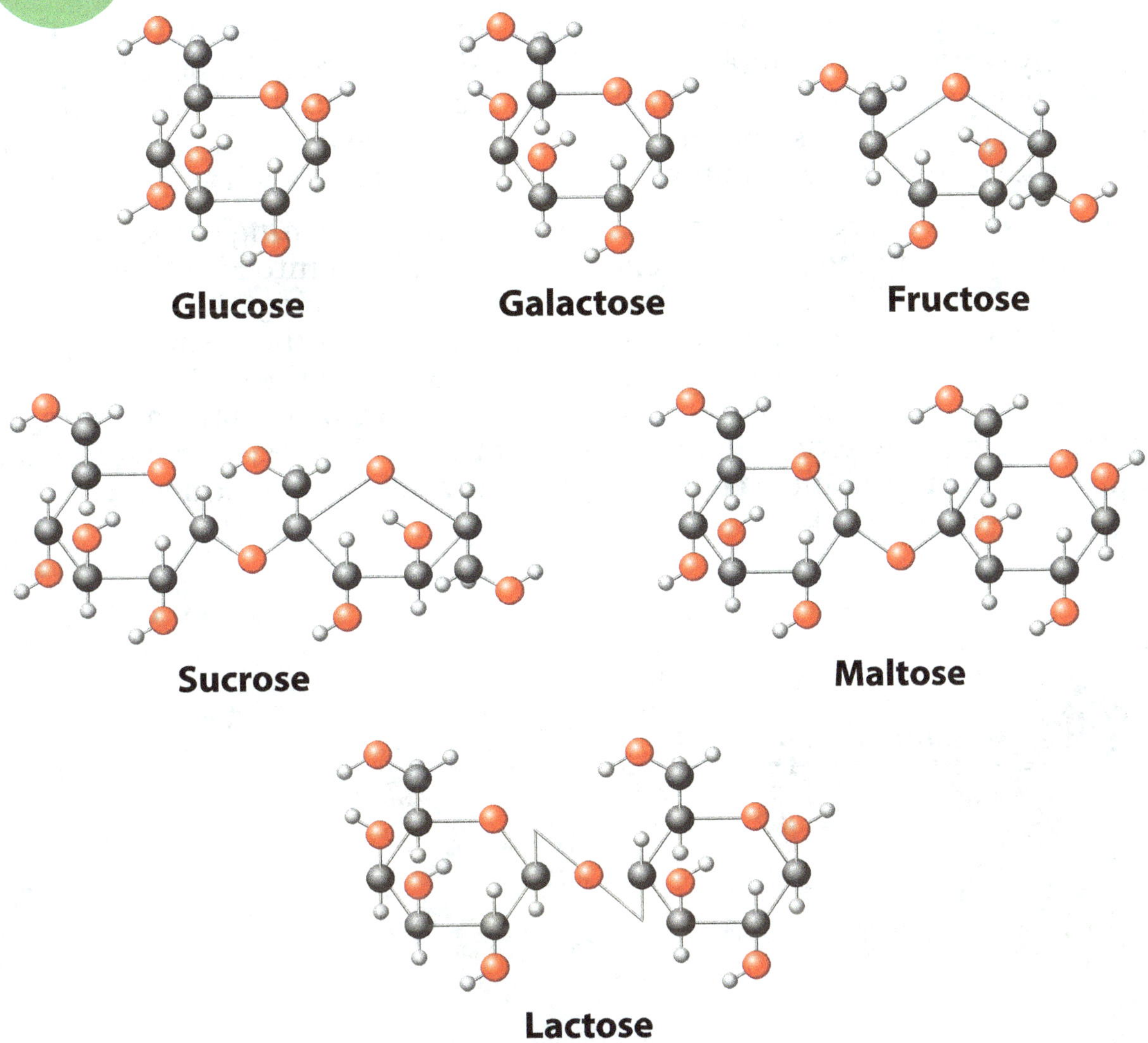

chromatos/Shutterstock.com

Carbohydrates are made up of chains of sugar molecules.

While not polymers, **lipids** are another class of organic macromolecules important to life. All lipids do not dissolve in water. You can see this property in a salad dressing made of vinegar and vegetable oil. Vinegar is mostly water, so if you let the dressing sit, the lipids in the vegetable oil will separate themselves from the vinegar. That's why you have to shake them up before putting them on your salad. Lipids include molecules that store energy, such as fats and oils. They also include waxes used for structural support and certain hormones.

malialeon/Shutterstock.com

Lipids, such as fats, oils, and waxes, are molecules that do not dissolve in water.

2.4 Energy and Thermodynamics

To affect any change in the position, temperature, or composition of matter (what physicists call "work") you need the intangible phenomenon called **energy**. There are two main kinds of energy:

Potential energy or energy of position is stored energy that is not currently performing any work. For example, if you take a weight and hold it above the ground, you give it potential energy due to the force of gravity pulling on it. An archer pulling on the string produces potential energy in the tension in the bow. Potential energy can also be stored as chemical energy, such in a fuel such as gasoline.

Kinetic energy is energy motion or energy that is performing work. Letting the weight drop converts the potential energy of its position into kinetic energy as it falls to the ground. Likewise, as the archer releases the string, the bow returns to its normal shape, transferring its energy to the arrow as it flies. Burning gas in an engine releases the energy stored in the chemical bonds in the molecules and converts it into heat energy that in turn can be converted into mechanical energy that drives the engine. Any action that produces motion, a change in form, or a change in temperature is the result of converting potential energy into kinetic energy.

NoPainNoGain/Shutterstock.com

An archer uses tension in a bow to transform potential energy into kinetic energy.

Just as a car engine converts the chemical energy in fuel into the kinetic energy of motion, your cells convert the chemical energy in glucose (sugar) molecules into the energy your body to power your muscles and perform all the metabolic activities that keep you alive.

Energy conversions are governed by the laws of thermodynamics. **The First Law of Thermodynamics** states that energy cannot be created or destroyed, only converted from one form to another. This means we can't have a free lunch or a perpetual motion machine. If we want to perform work, we have to put energy into our system. For this reason, the First Law is often called the **Law of Conservation of Energy**.

The Second Law of Thermodynamics, also known as the Law of Entropy, states that it is the nature of energy to go from a state of more-ordered to a state of less-ordered. Look at this way: If you drop coffee mug, if breaks. The broken pieces are less ordered that the mug as a whole. Likewise, burning a campfire converts the wood into a more disordered state of ash and gases. We call the measure of disorder in our system, **entropy**. One implication of the Second Law is that the total amount of entropy in the universe is increasing. Therefore, if we want to move a system from a less-order (high entropy) to a state of higher order (low entropy), we will increase entropy in another way. This increase in entropy is in the form of waste energy. If you drive your car for a while, you will no doubt notice that the engine gets hot. That heat radiating off the energy is wasted energy that did not contribute to the motion of your car. For any energy conversation system, some of the energy will be lost to waste. There is no such thing as an energy conversation system with 100% efficiency. You always get less energy (and work) out of the system than you put into it. Or, as my old physics professor would say: The universe is a rigged casino. You can't win and you can't even break even.

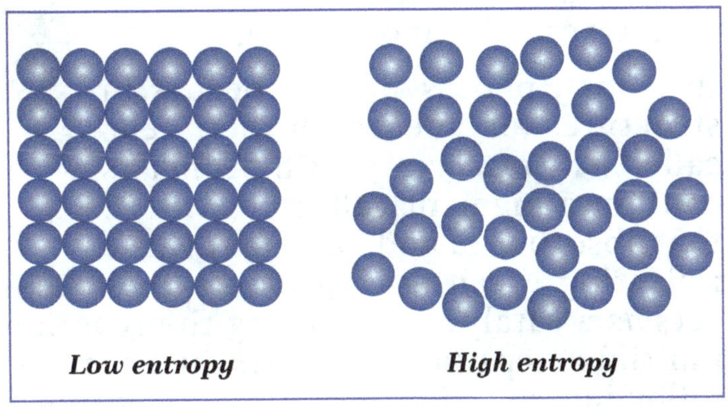

Fouad A. Saad/Shutterstock.com

Low vs. high entropy.

Another implication of the laws of thermodynamics is that if you take a closed system, where no energy from the outside can enter, it will eventually break down and stop functioning. If would be great if our cars could run forever on a single tank of gas, wouldn't it? Unfortunately, you have to keep buying gas. This also means that if the Earth was a closed system, it couldn't support life. Eventually, all the available energy on the planet would dissipate into waste and it would be dead.

Fortunately, the Earth is not a closed system. It does receive a constant input of energy from an outside source and that source is the Sun. The Sun emits a steady output of electromagnetic radiation, ranging from radio waves to the highly energetic gamma waves. Most of what gets through the Earth's atmosphere is from a narrow band in the middle that we call visible light (It's no coincidence that our eyes are adapted see these wavelengths). It is this energy that powers most of the life on Earth.

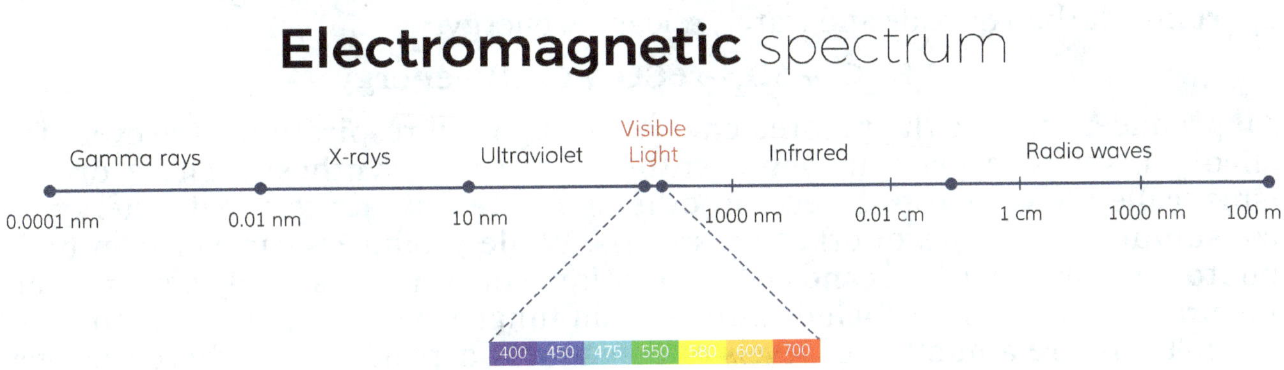

AlexVector/Shutterstock.com

The electromagnetic spectrum.

2.5 Ecosystems and Interactions

So, how do ecosystems use sunlight to power life? Through specialized organisms called autotrophs (self-feeders) or **producers**. **Autotrophs** are organisms that use sunlight to build sugar molecules in a process called **photosynthesis**. They include green plants and certain species of bacteria called cyanobacteria.

Plants are autotrophs or producers.

In plants, photosynthesis occurs in structures (organelles) in their cells called **chloroplasts**. Chloroplasts contain a pigment call **chlorophyll** that absorbs red and blue light while reflecting green, giving plants their characteristic green color. Chlorophyll acts as a catalyst, promoting the reaction of photosynthesis. While there are multiple steps to photosynthesis, the basic process can be summed up in the following equation in which water and carbon dioxide are combined using energy from the sun to produce glucose (sugar) and releasing oxygen as a waste product.

$$6CO_2 + 6H_2O + \text{sun's energy} \rightarrow C_6H_{12}O_6 + 6O_2$$

The numbers before each formula indicates the number of molecules involved in a reaction to produce a single glucose molecule. Plants absorb water through their roots and take in carbon dioxide from the air and make the sugar in their chloroplasts. They then store that sugar in their tissues until it is needed.

Now, not all producers use sunlight. For example, sunlight can't penetrate deep in the ocean, so bacteria in those depths use heat energy from hydrothermal vents and hydrogen sulfide (H_2S) to build sugar in a reaction called **chemosynthesis**.

$$6CO_2 + 6H_2O + 3H_2S + \text{thermal energy} \rightarrow C_6H_{12}O_6 + 3H_2SO_4$$

Once the sugar molecules are built, the energy stored in them can be released through the reaction of **cell respiration**. The reaction of cell respiration looks like photosynthesis run in reverse, where sugar is reacted with oxygen, breaking it apart into carbon dioxide and water, releasing energy.

$$C_6H_{12}O_6 + 6O_2 \rightarrow 6CO_2 + 6H_2O + \text{energy}$$

All producers release their stored energy through cell respiration. However, not all organisms can perform photosynthesis or chemosynthesis. These organisms, called **heterotrophs** (feed on others) or **consumers**, get their energy by consuming the tissues of other organisms. While producers can perform both photo- or chemosynthesis and cell respiration, consumers can only perform cell respiration. Consumers include animals and fungi as well as many microorganisms. There are a number of levels of consumers depending on where they get

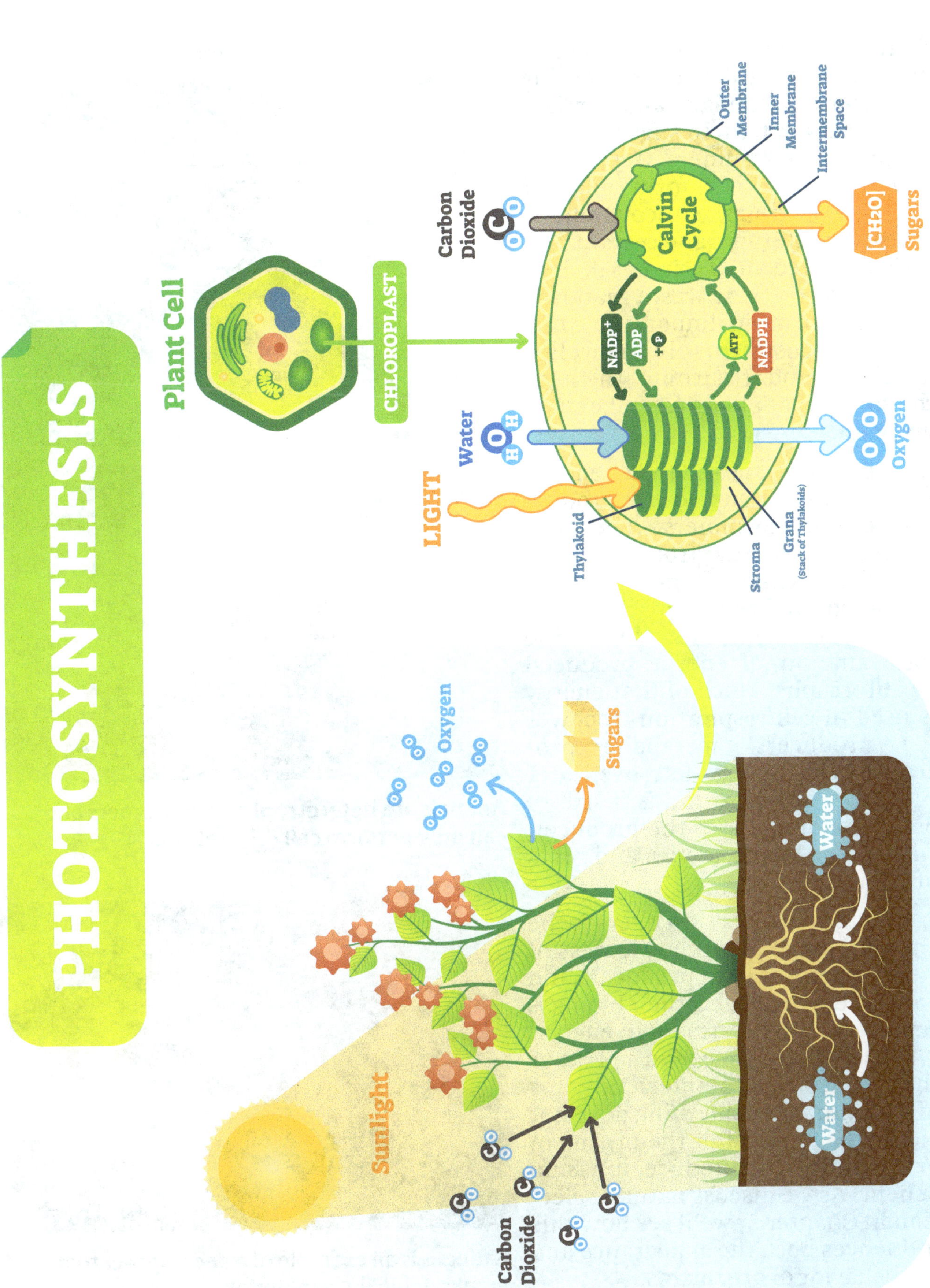

VectorMine/Shutterstock.com

their sugar. Primary consumers or herbivores, get their sugar by eating producers. Secondary and tertiary consumers get their sugar by eating primary consumers. **Detrivores** and **decomposers** form another important group of consumers. These include fungi and certain bacteria that feed on the tissues of dead organisms or their waste material. They perform an important function in ecosystems by recycling nutrients and returning them to the soil so that plants can take them up again.

In most ecosystems, then, energy flows from the sun to the producers to the consumers. The conversion of energy from the sun into the chemical bonds of sugar is called **primary production. Gross primary production** is the total amount of energy produced by autotrophs. Much of that energy is used in cell respiration to power the producers' metabolic processes. The energy left over, **net primary production**, is used to build the biomass of the producer, such as the leaves, stems, and roots of a plant.

Secondary production refers to the energy used to build the biomass of consumers. For both producers and consumers then, all of their biomass represents an investment in energy. Whether it's the flower on a plant or the muscles, tails, or internal organs of an animal, if something does not aid in the survival of the organism or its ability to reproduce, it has no benefit. As we discuss natural selection in Chapter 3, we'll see how this influences both the appearance and the behavior of organisms.

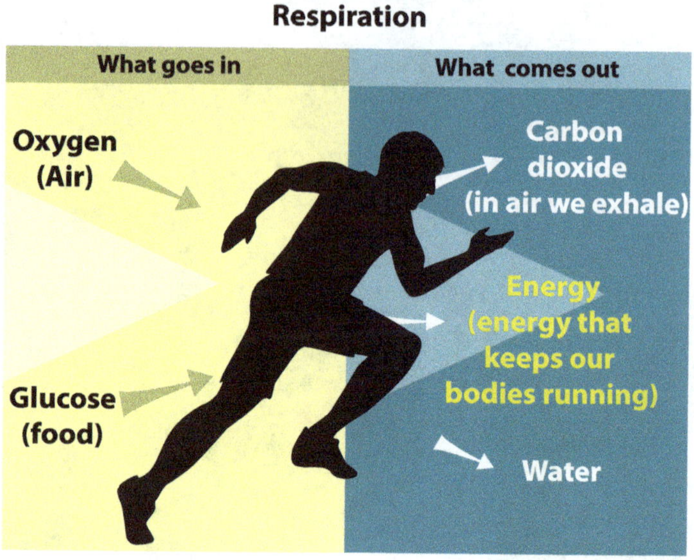

Sakurra/Shutterstock.com

Susan Schmitz/Shutterstock.com

Animals are heterotrophs or consumers and can only perform cell respiration.

Bukhta Yurii/Shutterstock.com

A fungus is an example of a decomposer that breaks down dead material.

Productivity is the rate at which energy is used to produce biomass. Some types of ecosystems have higher rates of productivity than others. For example, tropical forests, coral reefs, and freshwater marshes while deserts, tundra, and the open ocean are areas of low productivity. **Ecotones**, or transitional areas between ecosystems are also areas of high productivity. For example, the shallow marshy area between a dryland ecosystem and pond ecosystem would support more plants such as lily pads and cat o' nine tails than the open water or the land. These plants would in turn attract insects and fish to breed in these areas, which in turn will attract predators like frogs and birds to feed on them.

Ecosystems interact with each other and form larger complexes called **landscapes**. A single example could include an evergreen forest, a deciduous forest, a river, a grassland, and a marshland. At the borders of these ecosystems, there are patches of ecotones, such as meadows in the forest or copes of trees in the grassland. Conservation biologists study the loss of biodiversity in landscapes and how can prevent it or restore what was lost. They collect data in the field and from satellite imaging and compile this data in software like **geographic information systems (GIS)**. The data collected in GIS may include information like the geology, hydrology, plant and animal species, and the impact of human development. Conservation biologists also use the data collected in computer models, simplified mathematical representations of complex natural phenomena. Models simulate the natural ecosystem and can be used to make predictions about future impacts. Ecosystems, however, have too much complexity to be fully represented in a computer model, so modeling is an imperfect science. However, as biologists gather more data about the ecosystem, they can feed into their models and improve them over time. The predictions are tested against the real world, which can then be used to update and refine future models.

Valerii_M/Shutterstock.com

A mountain landscape.

2.6 Nutrient Cycles

In Section 2.1, we saw how the movement of nutrients through an ecosystem can influence life. Water, carbon, nitrogen, and phosphorus are all critical for life and since they all exist in finite amounts on Earth, they are constantly recycled through the environment. What one organism releases as a waste product becomes a vital resource form another. These **nutrient cycles** or **biogeochemical cycles** track the movement of these nutrients through hydrosphere, atmosphere, lithosphere, and biosphere.

As a rule, nutrients move from one **reservoir** or pool to another. Living organisms, the oceans, the soil, and the air are all examples of reservoirs. Once the nutrient enters the reservoir, it remains there for a certain amount of time called the **residence time**, which varies based on the nature of both the nutrient and the reservoir. Reservoirs that have a long residence time and tend to receive more than they release back into the environment are called **sinks**. Those with shorter residence times and tend to release more than they receive are called **sources**. The rate at which nutrients flow from one reservoir to another is called the **flux**.

The **hydrologic cycle** encompasses the movement of water as it moves throughout the world. Life as we know it cannot exist without water. It is the medium for all biochemical reactions. It's also the transport medium that carries many other nutrients as well as sediments and pollution. Water is so vital for life that astronomers looking for signs of life on other worlds focus on places where liquid water can exist.

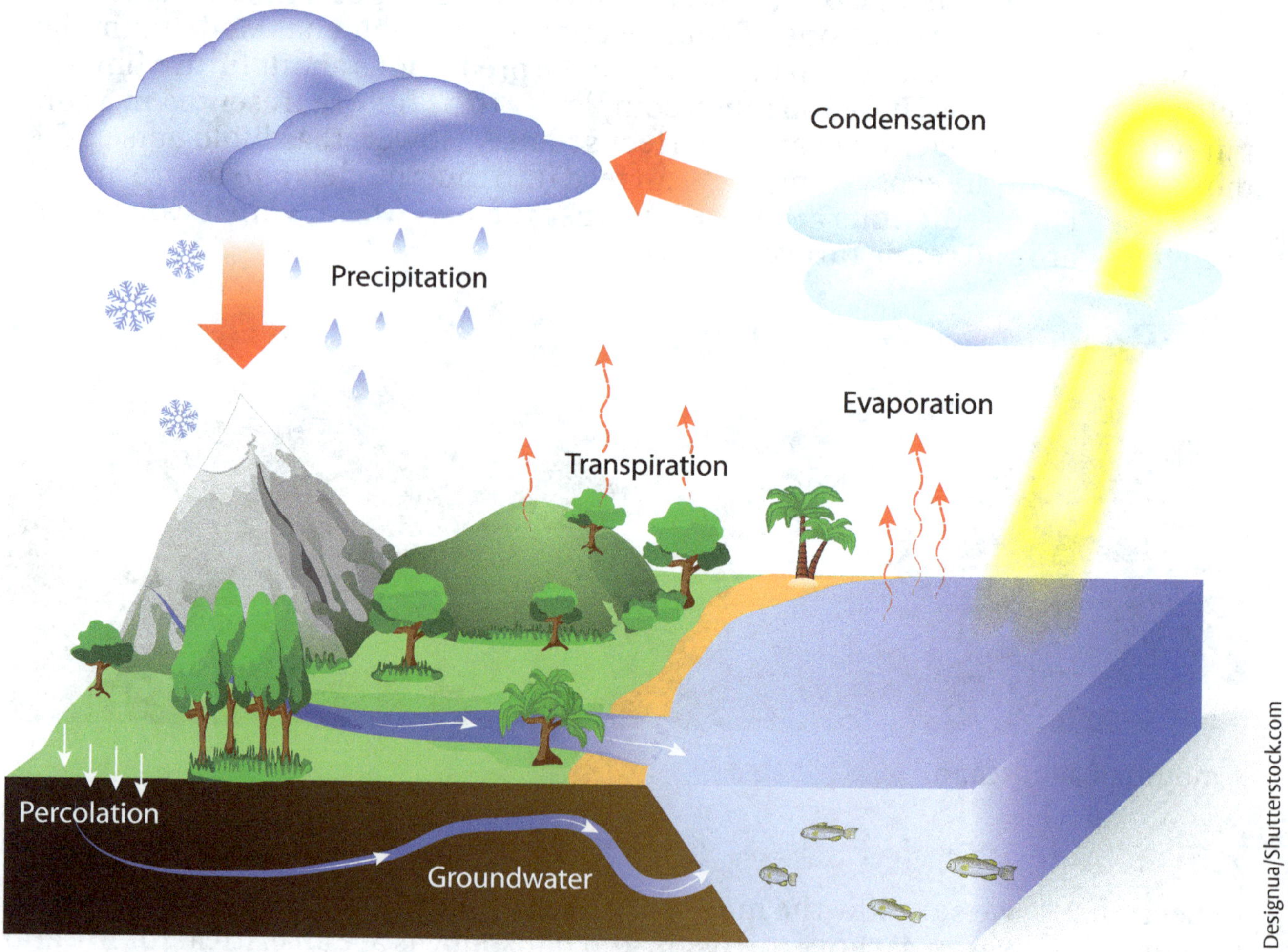

The hydrologic cycle.

Obviously, the oceans are the largest reservoir of water, containing more than 97% of the world's water. We have an entire chapter on water, but the basics of the hydrologic cycle are evaporation, condensation, and precipitation. Water moves from the liquid to the vapor phase through evaporation, then it condenses in the atmosphere to form clouds, and then returns to the surface as precipitation. Water also moves across the surface of the land, forming rivers that flow into lakes and the oceans. It also seeps into the ground to form groundwater.

Transpiration is evaporation of water through tiny pores in the leaves of plants called stomata. Ever wonder how plants move nutrients from the soil to their leaves without having a heart to pump fluids around? They use transpiration and some special properties of water.

Because of the shape of the water molecule and the fact that the oxygen atom has more protons than the hydrogen atoms do, the electrons experience a stronger pull toward the oxygen molecule. Thus, they spend more time around the oxygen atom than either of the hydrogen atom. This gives one end of the molecule a slight negative charge and the other end a slight positive charge, even though the molecule as a whole is neutral. Molecules with this uneven distribution of charge are called **polar molecules**. Because opposite charges attract, the hydrogen atoms in a water molecule are attracted to the oxygen atoms in other water molecules, forming a bond called a **hydrogen bond**. While individual hydrogen bonds are weak, the cumulative effect of multiple hydrogen bonds causes the water molecules to stick together through **adhesion**. This stickiness explains waters high boiling temperature, its effectiveness as a solvent, and few of its other unique properties.

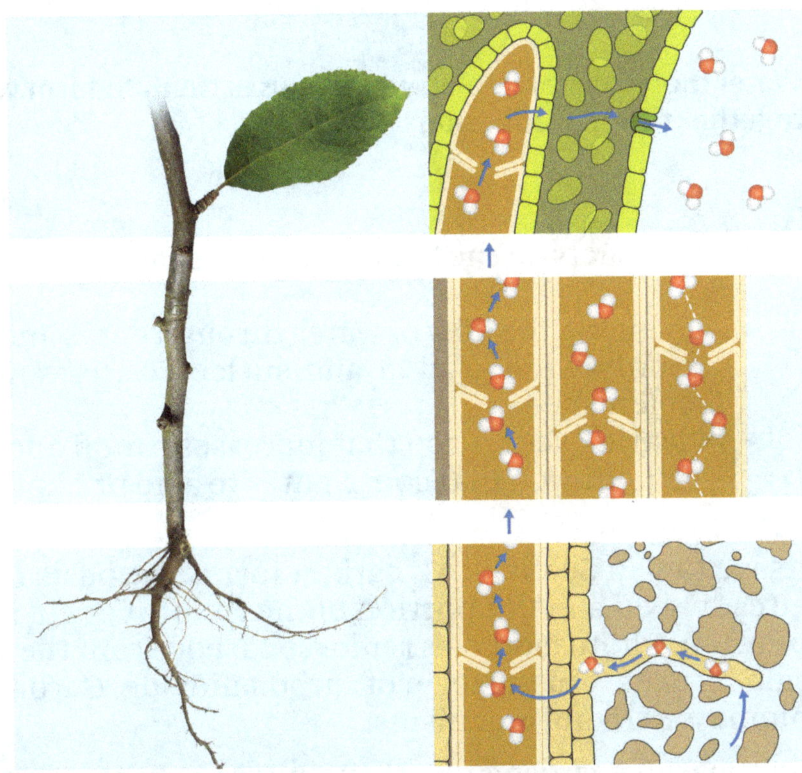

Aldona Griskeviciene/Shutterstock.com

Transpiration allows for water and nutrient transport up through a plant.

As water escapes from the leaves of plants through transpiration, each water molecule pulls on the one behind it as if they are all connected by an invisible string that runs all the way down to the roots. As the water molecules pull each up with through the plant, they drag the nutrients absorbed from the soil with them, thus providing plants with the nutrients they need.

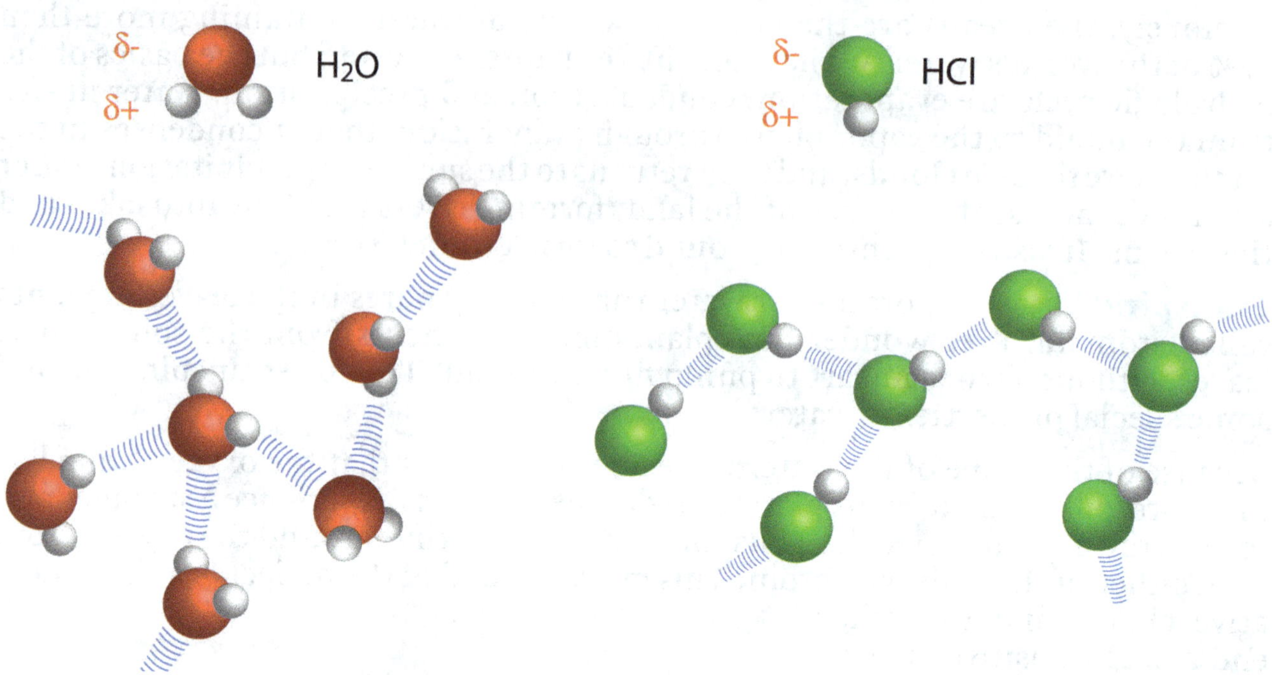

Water molecules are polar, which causes them to form weak hydrogen bonds and stick together through adhesion.

Human impacts on the hydrologic cycle include:

- Changing the flow of water through dams and levees.
- Overdrawing ground and surface water sources faster than they can be replenished.
- Clearing vegetation that increases runoff and erosion.
- Air pollution that contributes to acid precipitation.

As we saw in Section 2.3, carbon forms the basis of all organic molecules. We've already seen a major portion of the **carbon cycle** in photosynthesis and cell respiration. Photosynthesis removes carbon from the atmosphere while cell respiration returns it in the form of carbon dioxide. Carbon is also incorporated into the biomass of living organisms.

As aquatic organisms die, they sink to the bottom and get buried under sediments. Over time they could end up being subjected to enough heat and pressure to convert them into fossil fuels. Other sediments containing carbon can be converted into carbonate minerals. These make sedimentary rocks the largest single reservoir of carbon.

The second-largest reservoir is the oceans, which absorb carbon dioxide from the atmosphere and other sources of carbon. This carbon could be incorporated into the shells of marine organisms or become buried in sediments and converted into carbonate minerals.

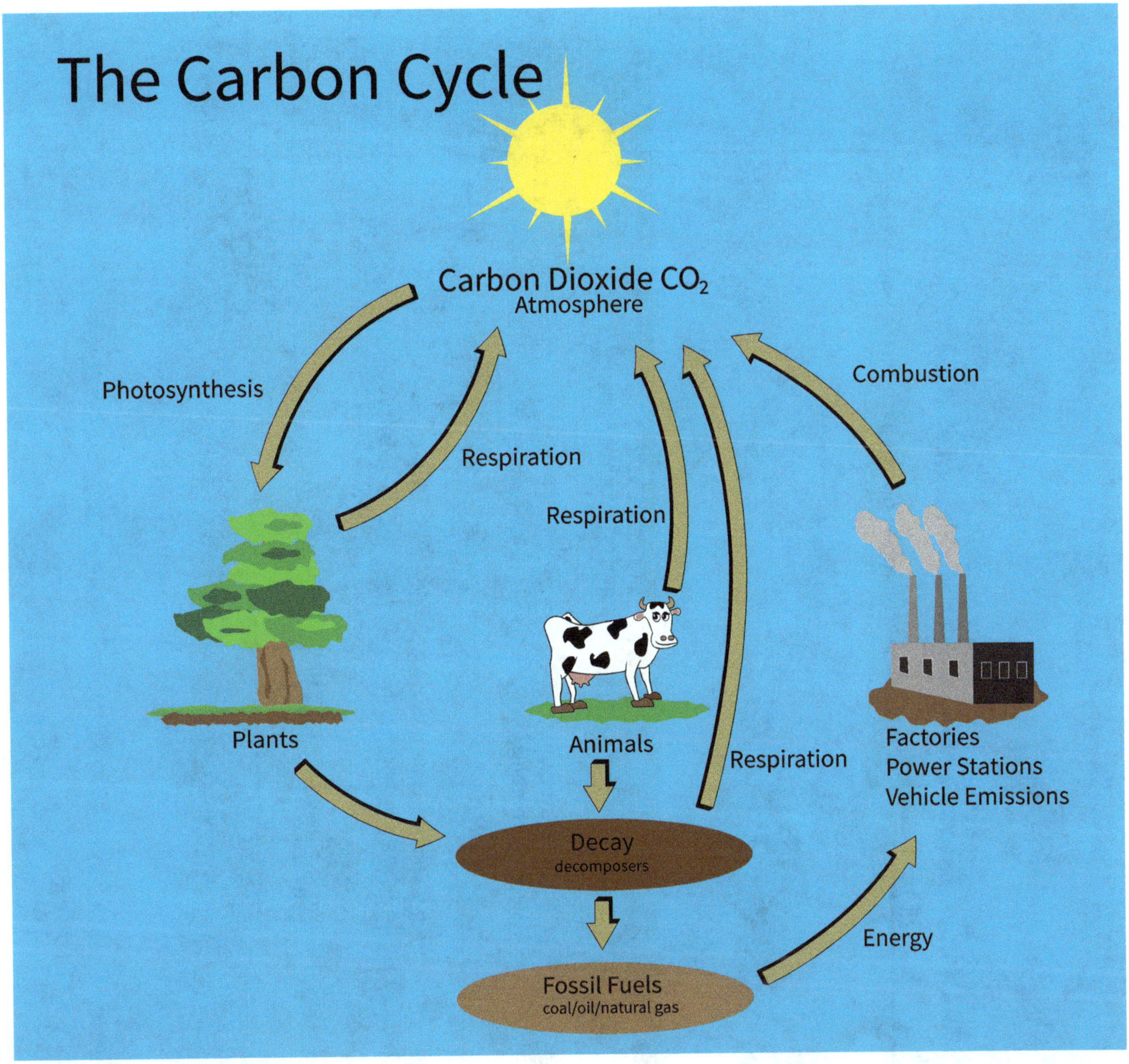

Steve Cymro/Shutterstock.com

The carbon cycle.

The largest single human impact on the carbon cycle comes from the burning of fossil fuels. In effect, we are taking carbon that has been sequestered from the environment for millions of years and returning to the atmosphere as carbon dioxide. We will see in a later chapter how the increase in atmospheric carbon dioxide is influencing climate change.

Nitrogen is an essential component in nucleic acids and proteins. While nitrogen is abundant, making up 78% of the atmosphere, there is a problem. Nitrogen gas is chemically inert and doesn't react with much other chemicals. Producers can't take in nitrogen gas so it first must be converted into a chemical form that they can absorb. The conversion of nitrogen gas into biologically available nitrogen is the first step in the **nitrogen cycle**.

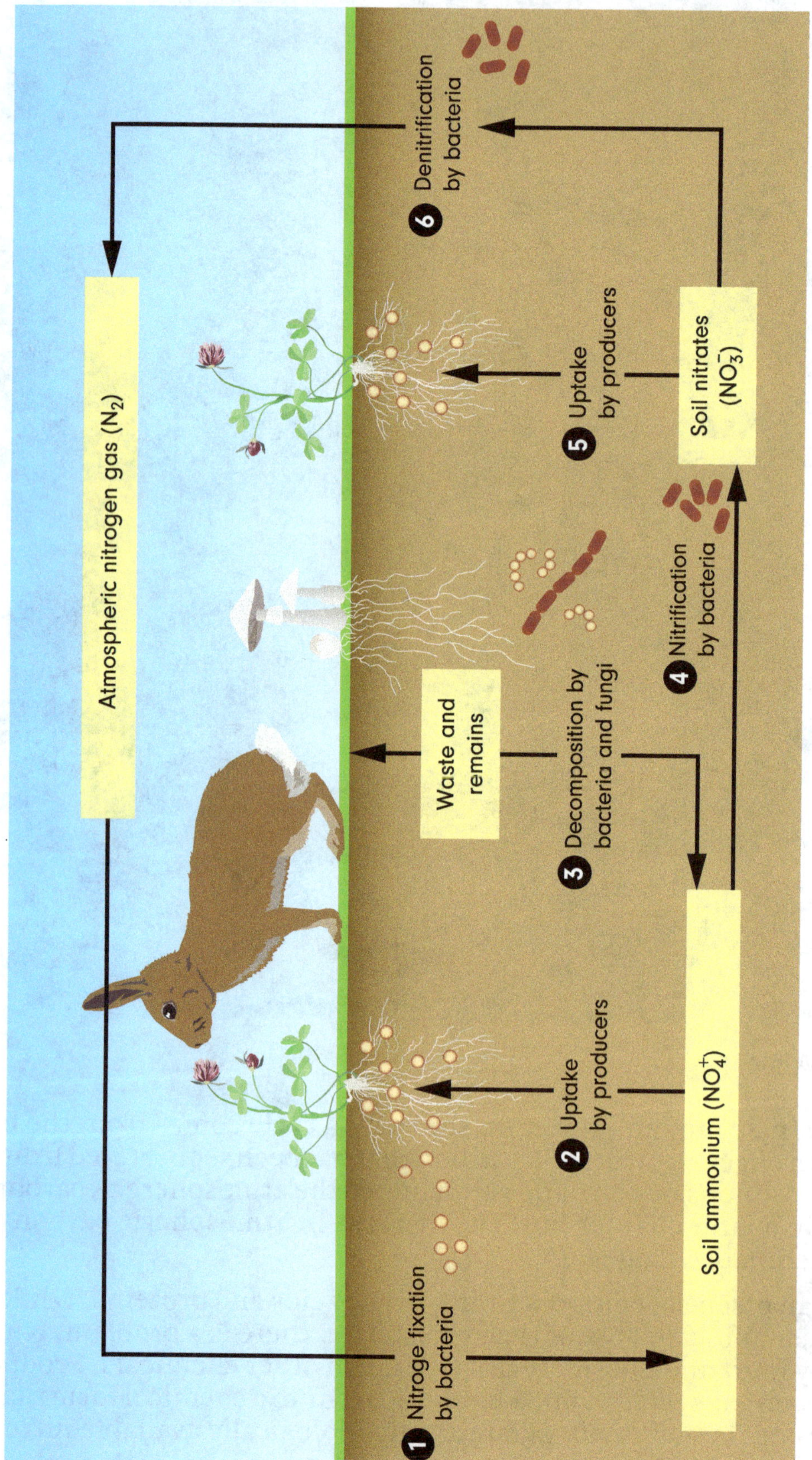

The nitrogen cycle.

Nitrogen fixation, or the conversation of nitrogen gas into biologically available nitrogen, can be accomplished in two ways. The first way is through lightning strikes, which trigger a reaction that converts nitrogen gas into ammonia (NH_3). Nitrogen fixing bacteria can also convert nitrogen gas into ammonia. These bacteria live in the roots of legumes. Any kind of bean plant, including soybeans, peas, string beans, and peanuts, is a legume. Farmers often plant legumes in alternating seasons in order to increase the amount of biologically available nitrogen in the soil. Once formed, the ammonia dissolves into water, forming the ammonium ion (NH_4^+), which plants can take up and use. Animals then get their nitrogen by eating plants or other animals. As plants and animals die, their nitrogen is decomposed and returned to the soil.

Another type of soil bacterium, **nitrifying bacteria**, converts the ammonium ion first into nitrite ions (NO_2^-) and then to nitrate ions (NO_3^-). Plants can also absorb nitrite and nitrate ions, and make use of them. Finally, **denitrifying bacteria** (yes, the third type of bacteria!) convert the nitrate ions into nitrogen gas, releasing it back into the atmosphere, completing the nitrogen cycle.

Because nitrogen must first be "fixed" to become biologically available, the amount of nitrogen is often the biggest limiting factor in the growing of crops. Through industrial fixation, we can produce synthetic fertilizers that provide crops with the nitrogen they need. However, as we saw in Section 2.1, much of this fertilizer is carried off the farm in the form of runoff, which then enters aquatic ecosystems, trigging eutrophication, and formation of hypoxic dead zones.

Like nitrogen, phosphorus is needed to form nucleic acids, proteins, and other important organic molecules. Compared to the nitrogen cycle, however, the **phosphorus cycle** is much simpler as there is no significant atmospheric reservoir for phosphorus. The largest phosphorus sources and sinks are rocks, soil, sediments, and the oceans. Phosphorus begins with rocks. Weathering by air and water releases phosphate ions (PO_4^-) that are dissolved into water. Some of these phosphates become sediments, settle into the bottom of the oceans and become incorporated into the crust again. Producers absorb phosphates from the soil or water while consumers get their phosphate by eating producers or other consumers. As organisms die, decomposers break them down and return the phosphorus to the soil.

As with nitrogen, the biggest human impact on the phosphorus cycle comes from synthetic fertilizers, which get picked up by runoff, carried to the oceans where they contribute to eutrophication and the formation of hypoxic dead zones. Also, some laundry and dishwashing detergents contain phosphates, although they are being phased out in locations where hypoxia is a big concern.

Emre Terim/Shutterstock.com

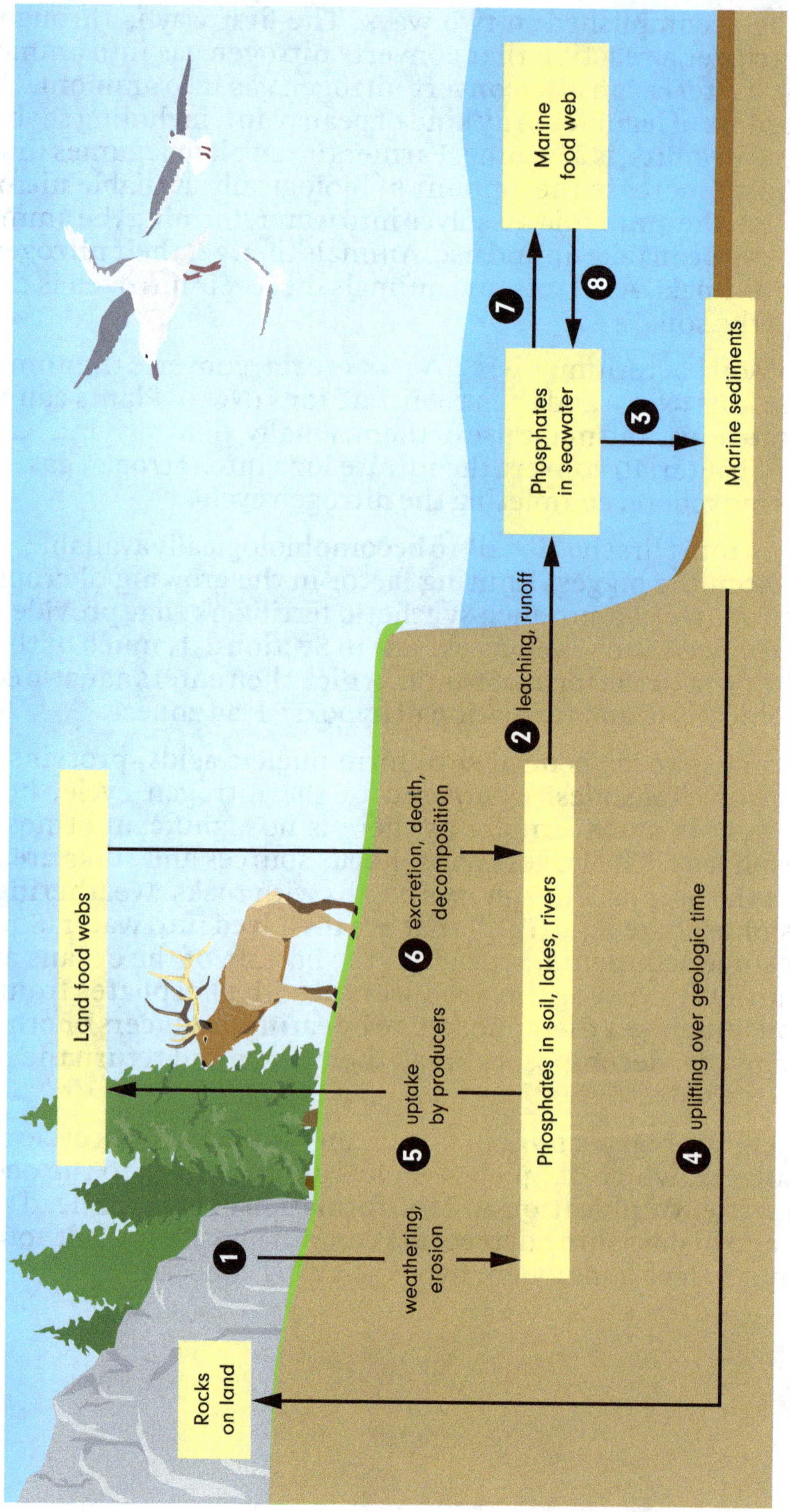

The phosphorus cycle.

2.7 Discussion Questions

1. What is the definition of a system? Give at least three examples of environmental systems.
2. What is the difference between a negative feedback loop and a positive feedback loop? Which one is more common in nature?
3. Describe the causes of hypoxia in bodies of water like the Chesapeake Bay.
4. Describe the difference between a covalent bond and an ionic bond.
5. What is a polymer?
6. What roles do proteins play in living organisms?
7. What is the difference between kinetic and potential energy? Give examples of each and describe how one can be converted into another.
8. An inventor tells you he has created a perpetual motion machine that runs forever without any energy input. Use the laws of thermodynamics to explain why this would be impossible.
9. Describe the interrelationship between producers and consumers.
10. What role do detrivores and decomposers play in an ecosystem?
11. What factors might influence the difference in productivity between a tropical rainforest and a desert or tundra?
12. How does transpiration aid plants in taking up nutrients?
13. What concerns are raised by human influence in the carbon, nitrogen, and phosphorus cycles, respectively?

2.8 Key Terms

Acid
Amino acid
Atom
Atomic mass
Atomic number
Autotroph
Base
Biogeochemical cycle
Carbohydrates
Carbon cycle
Cell respiration
Chemistry
Chemosynthesis
Chlorophyll
Chloroplast
Consumer
Covalent bond
Dead zone
Decomposer
Denitrifying bacteria
Detrivore
Dynamic equilibrium
Ecosystem
Ecotone
Electron
Energy
Entropy
Eutrophication
Feedback loop
Flux
Gene
Geographic information system
Half-life
Heterotroph
Homeostasis
Hydrocarbon
Hydrogen bond
Hydrologic cycle
Hypoxia
Ion
Ionic bond
Isotope
Kinetic energy
Landscape
Law of Conservation of Matter
Laws of Thermodynamics
Lipid
Macromolecule
Matter
Molecule
Negative feedback loop
Net primary production
Neutron
Nitrifying bacteria
Nitrogen cycle
Nitrogen fixation
Nucleic acid

Nucleus
Nutrient cycle
Organic compound
Periodic table
pH scale
Phosphorus cycle
Photosynthesis
Polar molecule
Polymer
Positive feedback loop
Potential energy
Primary production
Producer
Productivity
Protein
Proton
Reservoir
Residence time
Radioactive decay
Radioisotopes
Runoff
Secondary production
Sink
Solution
Source
System
Transpiration

Chapter 3

EVOLUTION AND BIODIVERSITY

Chapter Topics

3.1 Principles of Natural Selection

Life on Earth exists in staggering amounts of diversity. There are multitudes of species occupying almost every conceivable niche. From the tropical rain forests of Costa Rica to deep underwater hydrothermal vents to lakes that have been covered in ice in Antarctica for thousands of years, we find life in almost any place we look on Earth. Even in environments once thought too extreme for life, from waters saturated with salt, to temperatures below freezing or above the boiling point, to a pH near that of battery acid, we have found organisms able to survive and thrive. As Ian Malcolm in Jurassic Park would say, “Nature finds a way.”

Wherever we look, we find that species are adapted to the conditions in which they live. Take the polar bear for example. It has a number of traits that are good for living in the Arctic Circle. The white color blends in with the ice and is good for sneaking up on prey. Its stocky frame retains heat in the bitter cold. If we were to take a polar bear, however, and drop him in the Amazon rainforest, he wouldn’t last long. The white color is no longer good camouflage and he’d overheat quickly in the tropical climate. Conversely, a jaguar has good traits for the tropical region. Her color pattern blends in with the forest canopy and her sleek form with little body fat helps her maintain a comfortable temperature. Take our jaguar to the Arctic and, again, the traits that work so well at the equator are a hindrance. She would freeze to death pretty quickly. We don’t find polar bears in the tropics or jaguars in the Arctic for the simple reason that they wouldn't last long there if they somehow made their way to these different climates.

Any trait that enhances survival and reproduction is called an **adaptation**.

But do species seek out the environment that suits them according to their traits or does the environment help shape them to promote the development of adaptations? That was the question that vexed a young Charles Darwin as he pondered the specimens that he collected during his voyage around the world on the HMS Beagle. It led him to develop his theory of natural selection. But before we discuss natural selection, we have to answer a more fundamental question: What is a species?

Philosophers have been classifying organisms since the time of Aristotle. The usual method has been to group them by similar characteristics. The father of modern **taxonomy**, the classification of organisms, was a Swedish naturalist named Carolus Linnaeus. The system he developed, with some modification, is still used today. In the **Linnaean Classification System**, organisms are grouped in a nested hierarchy by common characteristics. The largest group Linnaeus used was the **Kingdom**, however, in recent times, a larger grouping, the **Domain** was added, so we'll start there.

There are two main domains: eukarya and prokarya. **Prokaryotes** are all single cell organisms that lack a membrane-enclosed nucleus containing their DNA. They also lack many of the specialized organelles found in eukaryote cells. There are two kingdoms in the domain prokarya: bacteria and archaea, both of which are small, single-celled organisms.

Vaclav Sebek/Shutterstock.com

Polar bears have adaptations than enable them to survive in the arctic region.

Enrique Aguirre/Shutterstock.com

Jaguars have adaptations better suited for the tropical rainforest.

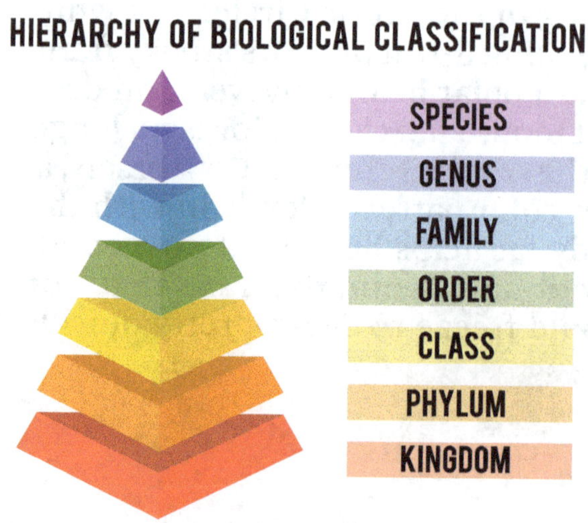

alinabel/Shutterstock.com

Eukaryotes, on the other hand, have larger cells with their DNA separated from the rest of the cell in a membrane-enclosed nucleus. They also have a number of specialized structures called organelles, which are analogous to the organs in your body. Organelles include mitochondria, the endoplasmic reticulum, the Golgi apparatus, and (in plants) chloroplasts. There are at least four kingdoms in the domain eukarya. Three for them, the animal, plant, and fungus kingdoms, are well-defined. However, there is still debate as to where the fourth group, protists, should be in a single kingdom or divided into two or more kingdoms. Plants, animals, and fungi are all multicellular organisms. On other hand, protists include single cell organisms like the ameba and multicellular organisms like kelp.

Difference between prokaryotic and eukaryotic cell

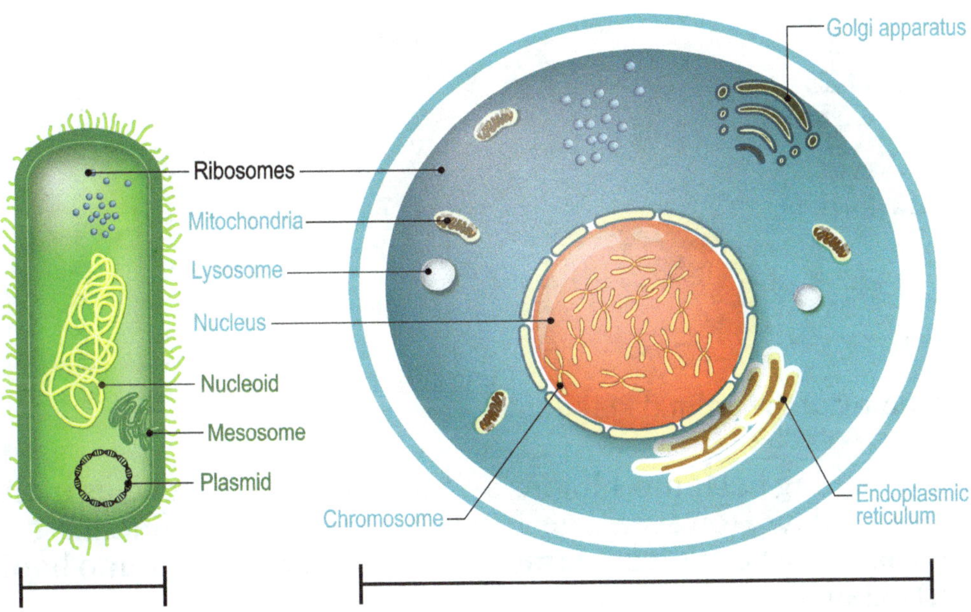

Designua/Shutterstock.com

The next step down from Kingdom is the **Phylum**. There are several phyla in the animal kingdom. The phylum chordata (chordates) include all animals that have a spinal cord that runs down their back. Fish, amphibians, reptiles, birds, and mammals are all chordates.

After Phylum comes the **Class**. Humans belong to the class of mammals. Mammals are all warm-blooded animals that give birth to live young and nurse them with milk. After Class comes **Order** and humans belong to the order of primates, which includes monkeys, apes, and lemurs. Next comes the **Family** and humans are in the Hominidae, a group that also includes all of the apes. After Family is **Genus** and then Species. The standard convention in Linnaean Taxonomy is to give the binomial (two-part) name that includes both the genus and species. Our genus is Homo and our species is sapiens, so we are classified as Homo sapiens, which literally means "wise man" because we made up this system and have a high opinion of ourselves.

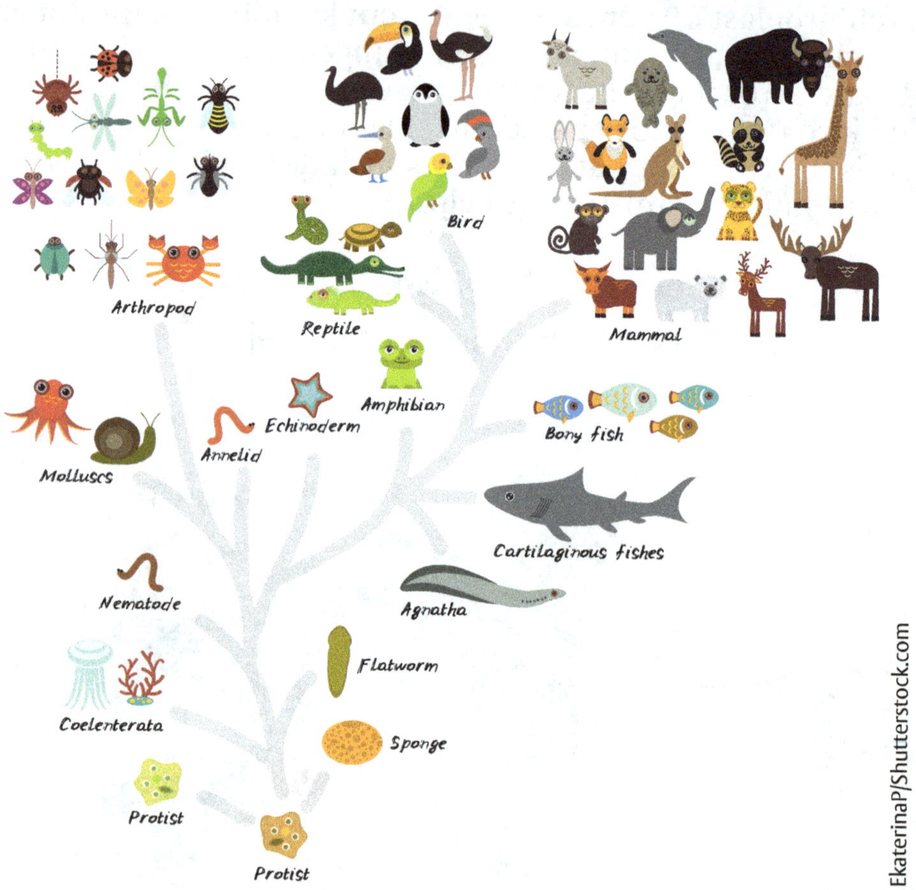

Taxonomy of the animal kingdom.

All living humans belong to Homo sapiens and we are the only species in the genus Homo that exists today. However, the fossil record shows evidence of extinct human species that have been given names like Homo erectus, Homo habilis, and Homo neanderthalensis.

This is, of course, a simplified description of the classification system. Over the years, many intermediate groupings have been added with names like Tribe, Suborder, and Suprafamily. There are even populations within the same species with slight variations from one another that are called subspecies.

Getting back to our question, what is a species? Because the Linnaean Classification System groups organisms by common characteristics, that is one requirement. The other requirement is that should be able to mate and produce fertile offspring. So, we will define a **species** as a population or group of populations of organisms that share common characteristics and can mate and produce fertile offspring. This isn't a perfect definition. It works best with animals, less so with plants, and, of course, doesn't work at all for single cell organisms that reproduce asexually. But it works well enough for our purposes.

Korawat photo shoot/Shutterstock.com

AkulininaOlga/Shutterstock.com

Mules and ligers are sterile hybrids of related species.

We can see different species of animals that can interbreed however. For example, if you mate a male donkey and a female horse, the offspring is called a mule. If you reverse the genders, it's called a hinny. However, there is a problem with mules and hinnies in that they are almost always sterile. You can't take two mules and mate them to produce more mules. You have to go back to a horse and a donkey. Likewise, a pairing between a male tiger and a female lion will produce a tigon, while a female tiger and a male lion will produce a liger. And tigons and ligers are also almost always sterile. However, the fact that these similar species can produce even infertile offspring is a strong indication that they are closely related. It points to them having a recent common ancestor that diverged from another far enough to be considered separate species but are still similar enough to be at least partially interfertile.

A **population** is a group of the same species that live in the same area and are capable of mating with one another. Populations, however, can become separated from one another. Over time, mutations can cause them to diverge from one another and if they experience different environmental conditions, different traits may be favored for survival. If they diverge far enough from each other, they may no longer be able to interbreed and produce fertile offspring, thus becoming two separate species. This is the basic principle of **natural selection** that Charles Darwin and Alfred Russell Wallace both independently developed in the nineteenth century.

Everett Historical/Shutterstock.com

Charles Darwin as an old man.

Evolution means change over time and **biological evolution** specifically refers to changes in the appearance and behavior of populations over the course of many generations. There is a common misconception that Darwin "discovered" evolution. In fact, the idea that species can change over time was already being discussed decades before Darwin began his voyage on the HMS Beagle. Geologists had already recognized that certain fossils only appeared in specific rock layers and not others. The presence of certain fossils in various strata was so consistent that they became known as "index fossils." Geologists began using these index fossils to determine the relative dates of different layers of sedimentary rock. What they lacked was a mechanism

Morphart Creation/Shutterstock.com

Jean-Baptiste Lamarck, a French naturalist who supported the idea that species evolved through inherited acquired characteristics.

neftali/Shutterstock.com

Charles Darwin at about the age he sailed on the HMS Beagle.

to explain why these fossils differed from each other and from life today.

Before Darwin, the French naturalist Jean-Baptiste Lamarck proposed a mechanism in which organisms would acquire traits in response to the environment. Under Larmackism, the ancestors of modern giraffes would stretch their necks to reach higher leaves in the trees. Their offspring would then inherit these longer necks. Each generation would then stretch their necks further and pass ever lengthening necks on to their descendants. The central flaw to Lamarckism is pretty evident. If you lose your hand in accident, your children won't be born missing a hand. Generally, such gross-acquired characteristics cannot be inherited. Modern genetics and a better understanding of the rules of inheritance that was unavailable in the early nineteenth century have ruled out Lamarckism as a viable mechanism for evolution.

Even at an early age, Charles Darwin had a strong curiosity about the world around him. In fact, it ran in his family. His grandfather, Erasmus Darwin, was already a well-known naturalist (terms like "biologist" would come into common usage for many years later). However, Charles' father wanted him to become a physician, so he set him to the University of Edinburgh, at the time, the most prestigious medical school in Britain. However, Charles hated medicine and would rather spend his time hiking in the words collecting beetles and rocks. So, he soon dropped out and then his father decided he should join the clergy and sent him to the seminary. After, graduating, Charles decided he'd rather become a naturalist like his grandfather. However, at this point, he needed to convince his father that he was serious about this occupation, so in 1831 at age 25, he took a job on board the HMS Beagle as the ship's naturalist.

The Beagle's mission was a round-the-world mapping voyage to South America, then across the Pacific to Asia and Africa before another trip across the Atlantic to South America before returning to England. Charles' job was to record information about the geology and life at each stop. The best known of those stops was the Galapagos, a chain of Islands off the west coast of South America. There, Charles studied many unique species on the islands, including the giant tortoise and marine iguana. He also collected several specimens of different species of finches, although it would be several years of study and consultations with experts, until

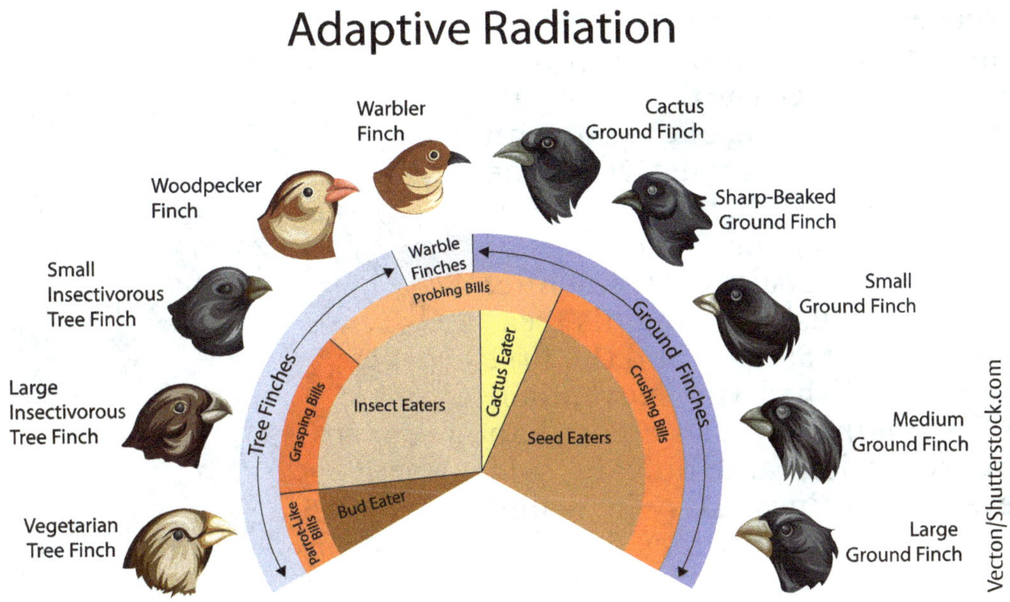

Adaptive radiation of Galapagos finches.

he realized that they were all finches. The birds all had distinctive beak shapes depending on what food source they exploited. Some were seed crushers and had short, thick bills. Others were nectar drinkers and had long, thin beaks that were good for getting inside many of the flowers on the islands.

After his five-year journey on the Beagle, Charles spend years examining his notes. He knew the Galapagos Islands were formed by volcanic action, so the animals had to have originally come from the mainland. He also realized that the locations where he found each bird corresponded with the kind of foods available. The seed crushers lived where seeds were the best available food, while the nectar drinkers lived where such flowers were abundant. After years of studying these and other specimens, Charles began to develop his theory of natural selection. He even spoke with pigeon breeders about the practices they used to ensure only desirable traits were passed down to each generation.

Charles Darwin did not want to publish his treatise right away. He knew some of his conclusions would create controversy in some quarters, so he wanted to make an air-tight case for his theory. His perfectionism kept him from publishing for 20 years. Meanwhile, he continued to correspond with other naturalists, including a young man named Alfred Russel Wallace.

While Charles came from a wealthy family and could afford to spend his time as a "gentleman scientist" without worrying about supporting himself, Wallace was not so fortunate. He had to support himself hunting fossils and selling them to collectors. In his travels looking for fossils and studying them in the Amazon Rainforest and the Malaysian Islands, Wallace made many observations that led him to independently develop the theory of evolution through natural selection. Eventually, he wrote Darwin about his ideas.

Learning that someone was working on the same theory was the final motivation for Darwin publish. However, he realized that Wallace deserved to share the credit for the discovery, so he invited Wallace to present their findings jointly. On July 1, 1858, the two presented the theory of natural selection at a meeting of the Linnean Society of London. On August 20 of the same year, they also published their findings in the *Journal of the Proceedings of the Linnean Society: Zoology*. Meanwhile, Darwin began furiously writing his treatise, *On the Origin of Species*, and the first edition was published in November, 1859. *On the Origin of Species* created an immediate sensation and quickly sold out of several printings. Darwin would go on to revise it several times over the course of his life. Unfortunately for Wallace, Darwin's fame soon eclipsed his own, though they remained friendly until Darwin's death in 1882.

Alfred Russel Wallace.

Understanding the principles of natural selection has many applications in environmental studies, including how we can selectively breed crops and livestock, understand pesticide resistance in insects, understand antibiotic resistance in infectious diseases, and how infectious diseases can gain or lose potency over time.

Evolution by natural selection is based on a number of premises such as

- Species face challenges to survive and reproduce. They must find food and avoid becoming food for others while fending off rivals for potential mates.
- Species produce more offspring that can survive.
- Individuals of the same species vary in many characteristics.
- Some characteristics favor survival and reproduction over others and those individuals who have the characteristics that are better suited for their environment are more likely to survive and reproduce than those that do not.

The last premise is often stated as "survival of the fittest" (a phrase Darwin himself did not use in his writings). But what does it mean to be "the fittest?" As we saw with the example of the polar bear and the jaguar, "fitness" is relative to the conditions of the specific environment. Traits that promote survival and reproduction in the Arctic Circle are different from those in the tropical rainforest. There is no absolute standard of "fitness" and no species is superior to another. Natural selection operates on what works best in a given environment. If an individual has traits that give it an edge in obtaining food, avoiding predators, or attracting a mate over its rivals, it will be more likely to live longer, care for its offspring, and pass its genes onto future generations. Those individuals who lack traits that favor survival and reproduction are less likely to live long enough to pass their genes along.

Natural selection operates on the existing variations of traits in a population. It does not introduce new traits. Species do not modify themselves to suit their environment. The environment acts to select those with favorable traits, called **adaptations**. This can cause the relative abundance of a given trait to shift, becoming more or less common in the population, depending on whether or not it enhances survivability and reproduction.

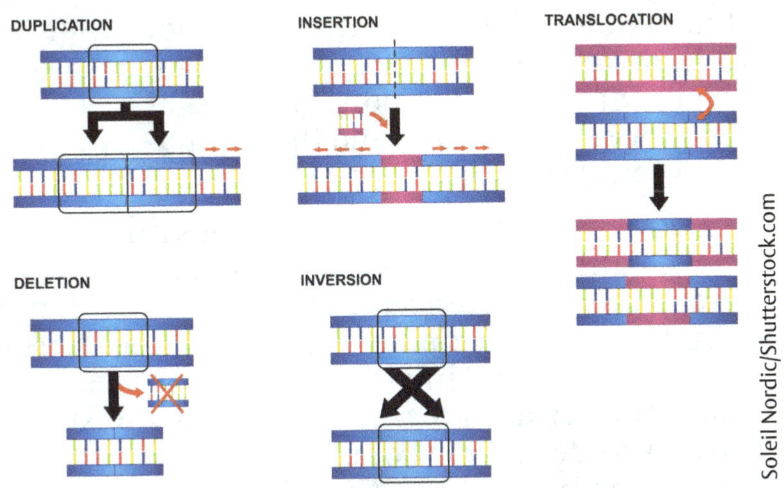

Types of genetic mutations.

New traits, however, can be introduced into a population through **mutations**, accidental changes in the DNA of organisms. Every time a cell divides, its DNA makes a copy of itself. Sometimes, mistakes are made in copying process. Perhaps an A-T pair is replaced with a C-G pair. Most mutations have no effect on the organism. Those that do have an effect, it is most often harmful. Cancer, for example, is the result of a mutation that causes cells to reproduce uncontrollably. However, a small percentage of mutations can introduce new traits that are beneficial, giving the organism a slight edge in survival or reproduction. Mutations are thus, the only way in which new traits can be introduced into a population. If a mutation has a benefit, it will be passed down to future generations and the descendants of the original organism will outcompete others and soon, more and more individuals in the population will be born with the trait.

Sexual production also has significant impacts on natural selection. Since individuals inherit half their genes from the mother and half from the father, sexual reproduction can result in novel combinations of genes faster than in organisms that reproduce asexually.

Since the traits that enhance survivability are dictated by the environment, changes in the environment may result in different changes. For example, peppered moths have two main variations: white and black. In England, the moths tended to rest on trees with light-colored bark. As a result, the white peppered moths blended in with the tree bark better than the black ones. As a result, more black pepper moths would be eaten by birds than white moths. Thus, white moths were more numerous than the black moths. However, during the Industrial Revolution, pollution darkened the tree bark, given the black peppered moths an advantage in camouflage over the white ones. Within a few generations, the black moths outnumbered the white moths. However, once pollution control standards were adopted, the trees returned to their normal color and the white moths again became more numerous.

Note that mutations and sexual reproduction affect survivability at the population level. Populations need time to adjust to changes in the environment. The peppered moth populations in England only survived the Industrial Revolution because the variation in physical traits (**phenotype**) already existed. If the environment changes too rapidly, the population may not be able to adapt in time, especially if mutations do not introduce traits that enhance survivability, and the population could go extinct.

IanRedding/Shutterstock.com

Light and dark peppered moths.

As Darwin noted with the Galapagos finches, closely related population that experience different environmental pressures will tend to diverge in their traits. The greater the difference in environmental conditions, the greater the divergence will become over time. Likewise, populations of the same species can show variation if they exist in different environments. For example, the Siberian tiger and Bengal tiger are both subspecies of the species Panthera tigris. Siberian tigers live in colder regions of northern Asia, while the Bengal tigers live in more tropical areas in southern Asia. As a result, the Siberian tiger is larger and better at retaining heat than its southern cousin.

Speciation, the emergence of new species from existing ones, can occur as populations of a species start to diverge from each other over time. Mutations and natural selection can introduce and favor different traits in different populations. Over time, enough different mutations may accumulate in the populations so that they are no longer able to interbreed and produce fertile offspring. The common path for speciation is **allopatric speciation**, when two populations are separated from each other. For example, imagine a drop-in sea level creating a temporary land bridge, enabling a population of rodents to colonize a neighboring island. This particular species of rodents can vary from very dark to very light phenotypes in color. Now, the sea level rises again, eliminating the land bridge and separating the rodents on the island with the ones on the mainland. Conditions on the island favor the very dark phenotype, while conditions on the mainland favor the very light phenotype. After a few generations, the populations will diverge, with the dark rodents being the most numerous on the island and the light rodents predominant on the mainland. Intermediate phenotypes (neither dark nor light) may disappear entirely. Meanwhile, a mutation for webbed toes emerges in the island rodents. This gives them an advantage for swimming in the shallow waters around the island, so in a few more generations, most rodents on the island have webbed toes.

Now, the sea level drops again and the land bridge reappears, enabling the two populations to interact with each other. They could still recognize each other as potential mates, however, what if a hybrid between a light, web-toed rodent and a dark, non-web-toed rodent has a disadvantage? Natural selection will favor mechanisms that avoid hybridization. Offspring of two light rodents or two dark rodents will be more

VectorMine/Shutterstock.com

Rodent color having an influence on which ones get eaten by predators.

likely to survive than offspring of a light and dark pair. Natural selection will favor the light rodents who prefer to mate with other light rodents and dark rodents who mate with dark rodents. This **reinforcement** will continue to keep the two populations reproductively isolated from each other. Over time additional mutations will continue to reinforce this isolation and soon, the two populations may no longer recognize each other as potential mates. As further generations progress, they may diverge enough to the point where, even if a dark rodent does mate with a light rodent, their offspring will not be fertile (much like a mule produced from a horse–donkey pairing). At that point, we can say that they have become two distinct species.

Kandarp/Shutterstock.com

An Indian male peacock.

How do individuals recognize potential mates? Look at peacock. At first blush, it looks like something that should not be favored by natural selection. It has bright colors and an enormous fan tail that looks ridiculous. It also creates a lot of drag and must be folded up before the peacock can fly, so it doesn't seem practical. Yet peacocks appear to love them off. They unfold their fails and strut around like they are the center of the universe. Remember, that every body part in an organism represents an investment in energy and bright colors are particularly difficult to make, so they require more energy that earth tones.

Now, look at the female of the peafowl, the penhen. She lacks the gigantic fanlike tail and is drabber in color, which is better than the peacock's bright colors for hiding from predators. So, how could natural selection favor such an appearance in the males, especially given that the female's appearance demonstrates the seemingly more practical traits exist in the species' genes? However, the peacock's strut has a purpose. He is demonstrating to potential females that he has **vigor**, good genes and health. His health is so good, in fact, that he can "waste" energy on the bright colors and fantail. He's sending out the message to females, "I have good genes and I will give you good offspring that are more likely to survive." Thus, females will preferentially mate with the peacock with the most brilliant and biggest tail over the peacock with a sad, droopy tail.

A peahen.

Blue-footed boobies engaging in a courtship display.

We observe these kinds of reproductive displays all over the animal kingdom. Colors often play this role among birds. Male cardinals are bright red, while females tend to be brown. On the other hand, both male and female blue-footed boobies, have blue feet. And as with the peacock, the blue pigment is hard to make and a good blue color is a sign of vigor. To show off the quality of their feet, the boobies engage in an elaborate courtship "dance." The male usually initiates the dance and if the female is interested, she will mimic his dance.

Other animals might use aggression as a display. Male rams will lock horns to impress the females and drive off their competitors. In many social animals, from elephant seals to chimpanzees, the strongest, dominant "alpha" male will assert control over the females and keep them all for himself, driving the beta males to the margins. However, size and aggression are not the only ways in which individuals can pass their genes on. Some do it by being sneaky. Often, when the alpha male is off feeding or otherwise distracted, the beta males will approach the females and mate with them.

Just as related populations can diverge over time, **convergent evolution** can result in different species adopting similar traits. Ichthyosaurs were aquatic reptiles that lived during the time of the dinosaurs. Even though they are separated from modern dolphins by 200 million years, they had many similar traits: a streamline body, dorsal fin, forelimbs developed into flippers, and an elongated snout. One is an extinct reptile, while the other is a living mammal, yet they have similar traits because they both were subjected to similar environmental conditions in a marine environment.

Daniel Eskridge/Shutterstock.com

Artist rendering of an ichthyosaur.

Andrea Izzotti/Shutterstock.com

Ichthyosaurs and modern dolphins display similar traits through convergent evolution.

For thousands of years, humans have unknowingly applied the principles of natural selection on our domesticated plants and animals through **artificial selection**. Artificial selection works just like natural selection, except, by deciding which individuals are allowed to breed, humans direct which traits get passed on and even enhanced. A few thousand years ago, a group of humans invited a group of wolves to share their food and live with them. Over time, these wolves evolved into the dogs we know today as our pets. Few animals have as much diversity in phenotypes as man's best friend. Through artificial selection, we have produced numerous dog breeds, from the English bulldog to the German shepherd, from the Great Dane to the chihuahua. Dogs come in a variety of shapes, big and small sizes, short hair and long hair. Each breed is the result of humans selecting for a particular set of traits and allowing only those that exhibit those traits to reproduce.

Nenilkime/Shutterstock.com

Various dog breeds are all descended from a common ancestor.

Chihuahuas were bred in Mexico to serve as "hot water bottles" for injury or illness. Their thin hair is good for the hot climate. Likewise, Saint Bernards were bred to rescue people lost in the mountains. Their thick, shaggy fur helps them retain heat in the alpine tundra and their large size enables them to pull unfortunate skiers out of the snow. Imagine a poor chihuahua sent to track lost skiers through 3 feet of snow! Likewise, collies were bred to herd sheep and retrievers were bred to, well, retrieve game after it was shot and fell into the brush or the lake.

How does artificial selection work? Say you want to breed a dog with very short legs like a dachshund. You start with your population of dogs and select those with the shortest legs and only allow them to breed. You repeat that process with each generation and over time, your population of dogs will have shorter and shorter legs.

Humans have used artificial selection on all of our domesticated animals and plants. That's why there are different horse breeds from thoroughbreds to Clydesdales and different cattle breeds such as Angus and Holsteins. Compare domesticated fruits and vegetables to their wild relations and you will see that the berries or other edible parts are always bigger. Wild carrots look almost nothing like the ones we grow in our gardens. In fact, different groups of humans have taken the same mustard species Brassica oleracea and, by selecting for different traits, turned them into brussels sprouts, cabbage, cauliflower, and broccoli!

3.2 Extinction

Extinction is the complete disappearance of a species from the Earth and no living examples still exist. According to the fossil record, the vast majority of species that have ever existed are now extinct. On average, a typical species last between 1 and 10 million years. Generally, species go extinct one at a time in what is called the **background extinction** rate. However, the fossil record indicates that there have been at least five **mass extinctions**, in which over half the species that existed went extinct in a short (geologically speaking) period of time. A mass extinction is represented in the fossil record by an abrupt disappearance of the fossils of a majority of species with many newer species appearing to take their place in layers of rock above. The biggest mass extinction, the Permian extinction, occurred 250 million years ago in which 95% of species went extinct.

The causes of mass extinctions can include volcanic eruptions, climate change, methane eruptions, and at least is attributed to an asteroid impact.

Today, biodiversity is decreasing at an alarming rate and species are going to extinct at a rate 100 to 1,000 times the background rate. As a result, many biologists believe that we are in the middle of the sixth mass extinction and the cause is us. As the human population has grown, we appropriate resources and alter the environment to suit ourselves at the expense of other species. We are causing species to go extinct through:

- Altering and destroying their habitats
- Overhunting and overharvesting species
- Pollution
- Introduction of invasive species
- Climate change

Species that have gone extinct in recent centuries as a result of human activities include the dodo, the Tasmanian wolf, and the passenger pigeon.

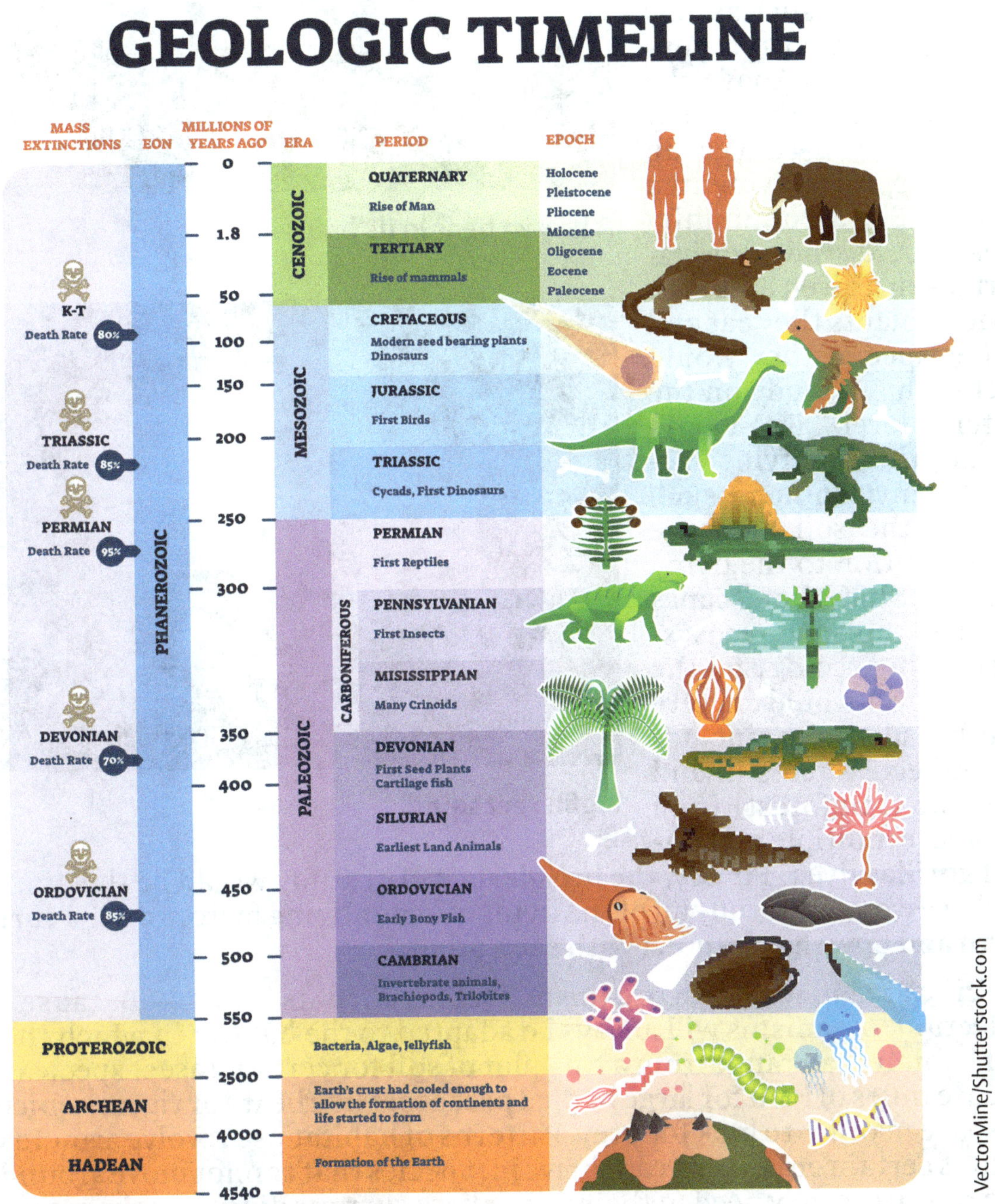

Geologic timeline with mass extinctions.

3.3 Ecology

Ecology is the study of organisms and how they interact with the environment. We can study organisms on a variety of levels. For example, we can study a single organism, such as a gorilla. We can also study a population of individuals living in a given area, such as how Diane Fossey did with a population of gorillas. Above the population is the community level. A **community** is a group of different species living in a given area. So, a community ecologist would study not just the gorillas in a given area, but the various plants they eat and any other species they would interact with. To study on the ecosystem level, we would also study the nonliving factors in the environment, such as the air, the soil, and the water in addition to the living organisms. Since landscapes encompass multiple ecosystems, studying ecology on the landscape level would involve a much broader area, including several ecosystems in and around the rainforest in a region where multiple populations of gorillas lived. Finally, the broadest level of study would be the **biosphere** level, where we would study all of the living and nonliving factors on the surface of the Earth and how they interact with each other.

Sergey Uryadnikov/Shutterstock.com

A brown bear in its habitat.

Gerald Robert Fischer/Shutterstock.com

Bigfin reef squid.

The specific area in which an organism lives is called its **habitat**. Because of natural selection, organisms will always be adapted to the habitat in which they live. Sizes of habitats can range from a tiny plot of soil for certain insect species to several square miles of land for large bodied predators like bears or tigers, which need lots of space in order to look for prey. Patterns of habitat use revolve around which conditions certain organisms can thrive in. Species that cannot move around must find the resources they need in their immediate surroundings. On the other hand, mobile species can seek out and select habitats with the best available resources for them.

Every organism also has a **niche**, its role in the community. Niches tend to revolve around feeding relationships, including what the organism eats and what could potentially eat it. A **specialist** has a very narrow niche. Specialists tend to be organisms that are very good at obtaining and exploiting one particular resource, but are less able to exploit other resources. For example, Koalas eat only eucalyptus leaves. This specialized diet limits their range as they cannot live in habitats where eucalyptus trees do not exist. It also makes them more vulnerable to changes in the environment than other organisms. Take away all eucalyptus and koalas could go extinct.

Yatra/Shutterstock.com

Koalas are specialists and have a narrow range of resources they can exploit.

In contrast, **generalists** have a much broader niche and can exploit a variety of niches. They may not be as good at exploiting a single resource as a specialist, but by being able to make use of multiple resources, they can weather changes in the environment that could limit one resource and not others. Cockroaches make great generalists. They have existed for millions of years with little change and eat almost anything. They're also resistant to many poisons and radiation. Cockroaches have thrived not only in most different types of ecosystems. They've even managed to thrive in the urban environment created by humans. This ability to exploit a variety of resources and ecosystems and weather change has enabled roaches to survive where other species have gone extinct. For this reason, many have suggested that if humans were to destroy ourselves in a nuclear holocaust, roaches would rule the world.

IrinaK/Shutterstock.com

Cockroaches are generalists and can survive in a variety of niches and exploit a lot of different resources.

3.4 Population Ecology and Reproductive Strategies

Population ecologists study the **population size**, or the number of organisms that live in a given area. They also study how populations can grow or shrink over time in response to changes in the environment. These population changes can be caused by resource loss, habitat loss, natural disasters, or the influence of other species. Changes can be drastic, such as what happened to the passenger pigeon. Once the most numerous bird in North America, the passenger pigeon is now extinct. We have writings from the seventeenth and eighteenth centuries that describe flocks of these birds so numerous that they darkened the skies for hours at a time. But apparently, they were good eating, because we hunted them down to nothing.

Different species exist in different **population densities**, the number of individuals per unit area. Some species, like colony insects, live in very high densities. On the other hand, large bodied predators that need broad territories to hunt in, like bears and tigers, tend to live in very small densities of a few individuals scattered across many square miles.

Once numerous, the passenger pigeon is now extinct.

Species also have different **population distributions** or the spatial relationships among individuals and how they are scattered. A random distribution is generally found where resources are abundant and there is little competition. Various tree species in a tree tend to have a **random distribution** with seedlings sprouting wherever seeds happen to land. On the other hand, large bodied predators tend to live in a **uniform distribution** where individuals are more or less equally spaced from one another. Finally, if resources are located in only a few scattered locations, species may exhibit a **clumped distribution**. Palm trees clustered around an oasis that is the only source of water for miles around exhibit a clumped distribution.

Sex ratio is the ratio of males to females. In a monogamous species, the ratio tends to be close to 1:1. On the other hand, in polygynous species, where a single dominant male controls the females, such as gorillas and lions, the ratio tends to skew toward a higher percentage of females to males. Finally, many hive insect species exhibit a ratio in which there is one fertile female, the queen, a small handful of males (who may be killed once mating is complete), and numerous workers who are sterile females.

Inga Linder/Shutterstock.com

A random distribution.

JONG KIAM SOON/Shutterstock.com

Large predators like tigers have a uniform distribution.

ustas7777777/Shutterstock.com

An oasis exhibits a clumped distribution.

3.5 Population Growth and Limiting Factors

Four factors influence a population's growth rate:

- Natality, or the birth rate
- Mortality, or the death rate
- Immigration of individuals coming from outside the population
- Emigration of individuals who leave the population

Natality and immigration add individuals to the populations, while mortality and emigration subtract individuals. Growth rate is then defined as the sum of all four factors and is often expressed as percentage.

Birth rates vary a lot from one species to another, but can be put into two broad categories, **r-selected species** and **K-selected species**. In r-selected species, populations focus on producing as many offspring as possible with the assumption that the vast majority of them will die before they reach maturity. They give little parental care for their offspring once they are born or the eggs are laid. For example, a mosquito might lay thousands of eggs. Over 90% of the larvae that hatch from those eggs will be eaten by predators, but the few that are fast enough or just lucky enough to avoid being eaten will reach adulthood and produce the next generation.

Mosquitos exhibit r-selected growth.

On the other hand, K-selected species tend to have a small number of offspring and focus on care for them for long period of time. In a K-selected species, the strategy is to invest effort in ensuring most of the offspring reach adulthood, rather than just flooding the zone with as many as possible and leaving their survival to chance. Most birds and mammals (including humans) are K-selected species.

Olexandr Panchenko/Shutterstock.com

Species that have few young and care for them until maturity are K-selected species.

A population that grows at a fixed percentage experiences **exponential growth**. Plotted on a graph, exponential growth starts small, but eventually skyrockets, creating a J-shaped curve. As an analogy, imagine if I promised to give you a penny on the first day of the month and give you twice that on the next day, and twice that on the next, and so on until the end of the month. So, on day one I give you one cent, on day two, two cents, on day three, four cents, day five, eight cents. Doesn't sound like much, right? But if you keep doubling it every day, but the 30th day of the month, I would have to give you over a million dollars!

ktasimar/Shutterstock.com

An exponential growth curve.

Exponential growth usually occurs with small populations where resources are abundant, competition is minimal, and there are no predators or diseases to limit growth. In other words, conditions that are ideal for growth. In the long term, exponential growth is not sustainable. If it continues, eventually, the population reaches the size where the demand for resources exceeds all available food and water. Emigration might slow exponential growth as some individuals leave the population and seek resources elsewhere, but if it continues, eventually, the resources run out and the population crashes.

Every ecosystem has a limit to as to how big a population of a species it can support, called the carrying capacity. Once the population gets above the **carrying capacity**, resources get scarce, and the population starts to die off. In general, however, there are numerous **limiting factors** that keep the population at or below the carrying capacity. In an idealized model, the J-curve would become flattened into a logistic or s-shaped curve that leveled off at the carrying capacity. In nature, however, populations are dynamic, births and deaths occur at random, and the population will fluctuate in a state of dynamic equilibrium. Think of an ecosystem as a bucket and the population being a hose filling the bucket. If water continues to fill the bucket, it will eventually overflow. Limiting factors are like holes drilled in the sides of the bucket, allowing some of the water to escape. So long as the water doesn't enter the bucket faster than it leaves, new water can be poured into the bucket indefinitely. Thus, limiting factors are what keep populations in ecosystems sustainable.

Many limiting factors are **density-dependent**. Their influence on the population size increases as the population density increases. Predators are a density-dependent limiting factor as more prey in a given area will attract more predators. Conversely, if the prey population density drops, the predators will move on and look for prey elsewhere. Disease is also a density-dependent limiting factor. The more tightly packed individuals are, the more they cough and sneeze on each other, spreading disease among the population more rapidly. The availability of food, water, space, and potential mates are also density-dependent limiting factors.

Predators like this wolf are density-dependent limiting factors.

Conversely, natural disasters such as floods and fires are **density-independent** limiting factors. The number of say, rabbits in a field has no influence on how flooded the field becomes in a heavy rainstorm.

Christian Roberts-Olsen/Shutterstock.com

Natural disasters like this forest fire are density-independent limiting factors.

Carrying capacities can change over time. If forest is clear-cut, its carrying capacity for several species will go down. Humans have used technology to increase our carrying capacity. However, we have done so by appropriating resources and habitats at the expense of other species, causing the current loss of biodiversity the world is experiencing today. Climate change is further reducing the carrying capacity in many ecosystems, driving certain species to seek new habitats by heading north or to higher elevations.

Efforts to reverse this trend, such as promoting ecotourism and protecting vulnerable habitats, are having an influence. We will discuss more of these efforts in Chapter 5.

3.6 Discussion Questions

1. Describe the differences between eukaryotes and prokaryotes.
2. How does Lamarckism differ from Darwin's theory of natural selection?
3. How is the concept of survivability dependent on the conditions in the environment?
4. How did changes in the environment affect the prevalence of different phenotypes of peppered moths?
5. How do sexual displays, such as bright colors or displays of physical prowess, aid in attracting potential mates?
6. Among horses, thoroughbreds are bred for racing, while Clydesdales are bred as draft horses. What traits have been enhanced through artificial selection for each breed to enable them to excel in their respective jobs?
7. After a mass extinction event, the surviving species tend to quickly radiate out and experience a rapid period of speciation and beginning occupying a variety of new niches. What reasons could explain this rapid speciation?
8. Which would be more vulnerable to extinction, a generalist or a specialist? Why?
9. Describe the difference between r-selected and K-selected species and give examples of each.

10. Why would an exponential growth rate be unsustainable?
11. In a certain community, wolves are preying on a population of rabbits. The rabbit population experiences a sudden expansion. How would that influence the population of wolves and how would the change in wolf population in turn affect the rabbit population?
12. Most paleontologists believe that an asteroid impact caused the End-Cretaceous mass extinction. Would this be considered a density-dependent or density-independent limiting factor? Explain.

3.7 Key Terms

Adaptation
Allopatric speciation
Artificial selection
Background extinction
Biological evolution
Biosphere
Carrying capacity
Class
Clumped distribution
Community
Convergent evolution
Density-dependent
Density-independent
Domain
Ecology
Eukaryotes
Evolution
Exponential growth
Extinction
Family
Generalist
Genus
Habitat
K-selected species
Kingdom
Limiting factors
Linnaean Classification System
Mass extinction
Mutation
Natural selection
Niche
Order
Phenotype
Phylum
Population
Population density
Population distribution
Population size
Prokaryotes
r-selected species
Random distribution
Reinforcement
Sex ratio
Specialist
Speciation
Species
Taxonomy
Uniform distribution
Vigor

Side Bar: What is a Fossil?

A fossil is an imprint in stone left behind by the remains of a dead organism. When most people think of fossils, they picture dinosaur bones, but fossils can be formed other things from the shells of invertebrates, leaves, or eggs. Sometimes even soft tissue can leave behind a fossil, though most fossils are made from the hard body parts. Trace fossils are imprints left behind by organisms, such as footprints or burrows. We've even found coprolites or fossilized dung.

A fossil forms when the remains are buried by sediments before scavengers or decomposers can scatter or destroy them. As more sediments pile on top, the pressure solidifies them into solid rock, preserving the imprints of the remains in stone.

Evidence from the fossil record uncovered indicates that life has existed on Earth for about 3.5 billion years. Life began with single cell organisms and simple structures.

Over time, these simple forms evolved into multicellular organisms with more complex structures and larger organisms. Note that complexity and larger sizes are not inevitable results of evolution. Natural selection favors whatever traits aid in survival and reproduction. Single cell organisms like bacteria still exist because they are still able to survive. Likewise, sponges are the simplest animals that exist, lacking a brain, a digestive track, or any other internal organs or even specialized tissues. They have been around for billions of years. As long as they are able to survive, there is no selective pressure to develop more complexity.

The fossil record also indicates that vast majority of species that have ever existed are extinct.

Merlin74/Shutterstock.com

Trilobite fossils.

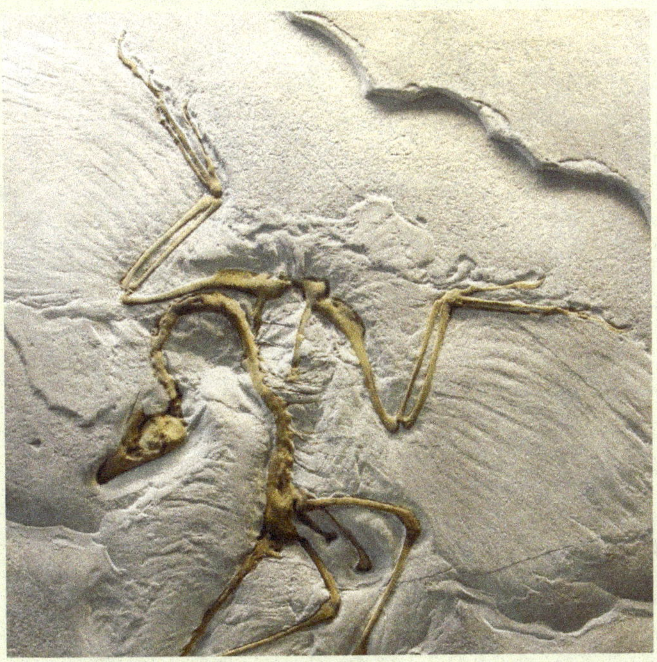

Era77/Shutterstock.com

An archaeopteryx fossil.

Chapter 4

COMMUNITY ECOLOGY AND SPECIES INTERACTIONS

Chapter Topics

4.1 Species Interactions

Sugarcane farmers in nineteenth century Hawaii had a problem. Rats, which had stowed away on ships, had invaded the islands and were attacking their crops. In 1883, they introduced the mongoose in an attempt to control the rats. Unfortunately, instead of hunting the rats, the mongooses attacked native species, include bird eggs. As a result, populations of honeycreepers and other native bird species plummeted. Why did the mongooses go after bird eggs instead of the rats? Well, part of the reason is that mongooses are opportunistic hunters who go after whatever they find easiest and eggs are easier to hunt than rats. Also, the rats were active at night while the mongooses tend to hunt during the day. So, the rats were often just not about when the mongooses were looking for a meal. The lesson here is that species interactions are complex and often difficult to predict.

There are five basic species interactions:

- Competition
- Predation
- Herbivory
- Mutualism
- Commensalism

Competition results from organisms trying to make use of the same resource. Since resources are finite, competition tends to be a strong limiting factor for populations. Competition can either be within the same species (**intraspecific competition**) or between two or more species (**interspecific competition**). Species can compete over food, water, or shelter. Intraspecific competition can also include competition over potential mates, which is how sexual displays discussed in Chapter 3 develop. In addition, plant species compete over soil nutrients, water, and sunlight and we call any plant that competes with our crops or ornamental flowers as weeds.

Rudmer Zwerver/Shutterstock.com

Competition between two male Common Redshank (*Tringa totanus*) birds, fighting in shallow water about territory borders.

Often, one species will prove to be better at exploiting a resource than another. This often happens when an invasive species enters a new area that lacks predators or diseases that would limit its growth while the native species still has to contend with these factors. This may result in the **competitive exclusion principle**, in which the losing species is driven out of the area or even to extinction.

However, competition does not always result in a winner and a loser. Sometimes, the species may engage in **resource partitioning**, in which the two or more species adopt different strategies to exploit a common resource. For example, the partitioning may be based on time, in which one species is active during the day while the other is out at night. Or, it could be spatially, in which the species focus on different areas. For example, different lizard species in the Caribbean feed on insects, but one focuses on the forest floor while another stays up the trees, looking for insects in the bark or among the leaves. Also, two species might feed on the same plant, but each has a preference for different parts of the plant. One species might eat the leaves while the other eats the stems.

Alexey Suloev/Shutterstock.com

Predators capture and kill prey species.

Predation is when one animal species captures kills and eats another. Because predators depend on the existence of prey for their survival, the two populations become intertwined. For example, coyotes may hunt rabbits in a grassland environment. If the population of rabbits rises, it will attract more coyotes to the area and the coyote population will rise. However, a rise in the coyote population may cause the rabbit population to drop as more rabbits are eaten. A drop in the number of rabbits means food will become scarce and the coyotes will either starve or move on to look for food elsewhere. With coyotes eating the rabbits, the rabbit population will rise once again, and so on.

Predation has a lot of influence on natural selection. A predator that is better at catching prey will live a longer, healthier life and be able to care for its young better, making it more likely to pass its genes on to future generations. Cheetahs are the fastest land animal on Earth. Being able to run faster than one's prey is a valuable trait to have, as are things like sharp claws and teeth to be able to strike for the kill.

Elana Erasmus/Shutterstock.com

Cheetahs are adapted to be the swiftest hunters.

There is an old joke that goes like this: Two guys are walking in the woods and see a bear coming toward them. The first guy starts to run. The second guy says, "You know you can't outrun that bear?" The first guy says, "I don't have to outrun the bear. I just have to outrun you."

Likewise, if you're a gazelle, you don't have to run faster than the cheetah. You just have to run faster than other gazelles. While cheetahs are faster gazelles, they aren't built for long distance running like gazelles are. More sprinters than marathoners, cheetahs can run very fast for short bursts. But, they only have to run long enough to catch the slowest gazelle while the faster gazelles survive to for another day. Natural selection, again, favoring those prey who can run for longer distances to escape the predator.

Natural selection can also favor other defenses against predation, such as weapons like a porcupine's quills. Many prey species have camouflage to enable them to hide from predators. Chameleons take that to another level by changing their colors to blend in with different backgrounds. Walking sticks hide from predators by looking like, well, sticks. But, the true master of disguise is the octopus. Combining color changes with some clever contortions, an octopus can disguise itself as a variety of other creatures.

Jukka Jantunen/Shutterstock.com

A porcupine's make a good defense against predation.

Brian Lasenby/Shutterstock.com

A walking stick hides from predators by looking like a stick.

Sean Xu/Shutterstock.com

The monarch butterfly displays a warning coloration that tells predators it produces a chemical that will make them sick.

Susan Schmitz/Shutterstock.com

The painted lady butterfly wards off predators by resembling the monarch butterfly.

Instead of hiding, other creatures combines defenses with standing out with warning colors. Monarch butterflies produce a chemical that makes birds sick, so their bright colors serve as a warning that says, "I am not a good meal." Other species survive by mimicking others. Several butterflies, with names like viceroy, painted lady, red admiral, queen, and soldier butterflies mimic the coloration of the monarch and benefit from being confused with them.

Parasitism also involves one species feeding at the expense of another, but unlike predators, parasites don't immediately kill their hosts. Instead, they feed or gain some other benefit from their hosts by taking a little bit at a time. Some parasites, such as tapeworms, live inside their hosts. Pathogens, disease-causing bacteria, viruses, and protists, also live inside their hosts. Many of the symptoms cause by disease, such as coughing, sneezing, and post-nasal drip, are mechanisms evolved to aid in transmitting the disease from one host to another.

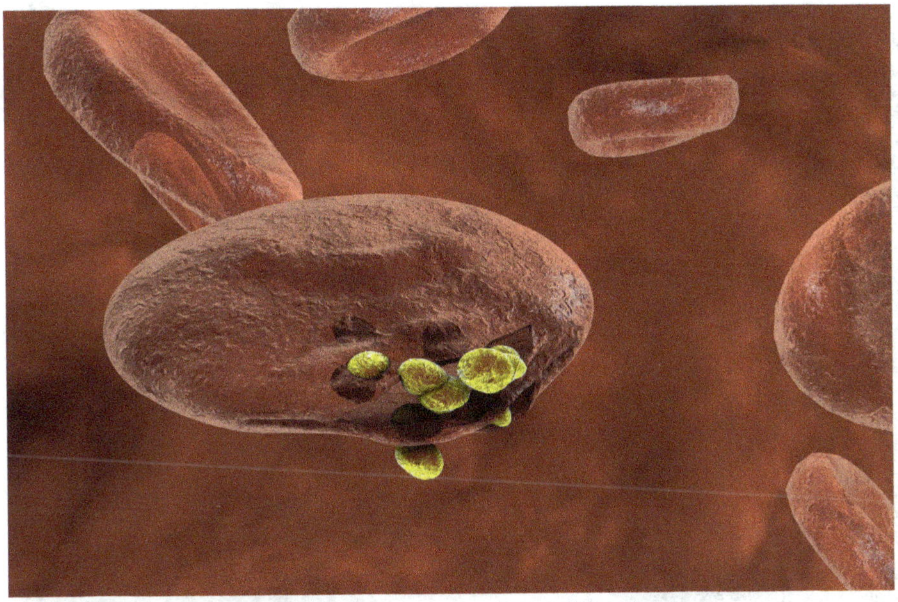

Kateryna Kon/Shutterstock.com

The malaria parasite emerging from a red blood cell.

Other parasites live on the exterior of their hosts. Lamprey eels that attach themselves to fish and drink their blood or vampire bats feeding on the blood of cattle fall into this category. Ticks, mosquitoes, and lice also fall into this pattern. Parasitic wasps exploit caterpillars by laying their eggs on the back of the caterpillar. When the eggs hatch, the larvae then eat the caterpillar.

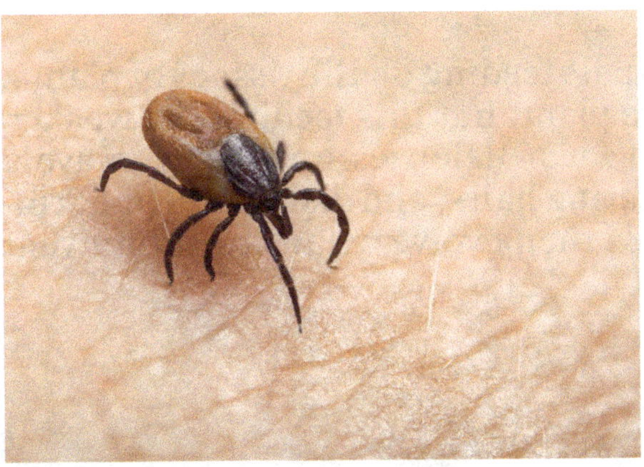

KPixMining/Shutterstock.com

A tick about to bite a host.

Meister Photos/Shutterstock.com

Caterpillar with parasitic wasp cocoons on its back.

Some parasites only encounter their hosts infrequently. Cuckoos and cowbirds have an interesting reproductive strategy. Instead of caring for their young, they seek out another bird's nest when that bird is feeding. They kick that bird's eggs out of the nest and replace them with their own. Thus, they trick that bird into care for its young.

Ernj/Shutterstock.com

The cuckoo bird will lay its eggs in another bird's nest and trick them into raising its young.

Natural selection also plays a role in parasitism, where often hosts will develop defenses against their parasites and then the parasites evolve mechanisms to get around those defenses. This kind of evolutionary arms race is often called **coevolution**.

Similar to predation, **herbivory** involves animals feeding on the tissues of plants. Many herbivorous insects don't eat the entire plant, but this feeding can weaken or impede its growth. Natural selection can favor defenses for plants to avoid being consumed, such as thorns, spines, irritating hairs, and distasteful or harmful chemicals. Naturally, plant-eating animals then evolve means to circumvent these defenses.

MattiaATH/Shutterstock.com

Caterpillar eating a leaf.

Anton-Burakov/Shutterstock.com

Plants like this rose produce thorns to ward off herbivores.

Not all interactions are exploitative. Many bird species will pick lice and other parasites out of the skin of hippos or out of the teeth of crocodiles. **Mutualism**, then, involves an interaction in which both species benefit. The birds get a source of food while the hippo gets rid of the parasites.

sirtravelalot/Shutterstock.com

Birds will pick parasites out of the hippo's skin.

Symbiosis involves mutualism in which organisms live closely together, like the bacteria in the stomachs of a cow that help it digest its food. A lichen is actually a composite organism made up of a fungus and an algae living symbiotically together. Other examples include zooxanthellae, an algae that help coral absorb food and mycorrhizea, a fungi that helps plant roots absorb nutrients from the soil.

Ivanova Tetyana/Shutterstock.com

A lichen is a composite organism made up of a fungus and an algae.

Pollination involves plants using insects, birds, or bats to help transfer the male sex cells (pollen) to the female ova, enabling the plant to reproduce. In fact, everything about a flower, the bright colors, the pleasing aroma, and the production of nectar, are evolutionary adaptions whose purpose is to attract pollinators to this mutually beneficial relationship. Some plants take this even a step further by producing a sweet-tasting container for their seeds that we call fruit. In this kind of relationship, the plant actually encourages animals to feed off them so that their seeds can be carried and dispersed away, increasing the chances that there will successfully reproduce.

Daniel Prudek/Shutterstock.com

Flowering plants depend on pollinators like the honeybee to reproduce.

Finally, **commensalism** involves an interaction that benefits one species but neither benefits nor harms the other. For example, by attaching themselves to the skin of a whale, barnacles gain mobility and avoid predators without benefiting or harming the whale.

jo Crebbin/Shutterstock.com

A gray whale covered with barnacles.

4.2 Feeding Relationships

A community consists of multiple populations of different species living in a given area. As we saw in Section 4.1, one of the most important relationships among species is feeding relationships, who is eating whom. Feeding relationships move from one **trophic level** to another, beginning with the producers. Producers use photosynthesis or chemosynthesis to build sugars. Primary consumers in turn, feed on the producers while secondary consumers feed on primary consumers. There may even be tertiary or quaternary consumers that feed on secondary and tertiary consumers, respectively. Finally, detrivores and decomposers feed on the remains and wastes of organisms, freeing nutrients for recycling in the ecosystem.

For example, in an aquatic ecosystem, phytoplankton act as the producers and are fed upon by the primary consumer, zooplankton. Fish that feed on the zooplankton make up the secondary consumer, while an eagle that swoops down and eats the fish would be tertiary consumer.

Because much of the energy in one trophic level is used for metabolic activities, not all of it is available for consumers on the next trophic level up. In fact, only about 10% of the energy in one trophic level moves up to the next. For example, if the phytoplankton in our above ecosystem has 1,000 calories of food energy available for them, there will only by 100 calories available for the primary consumers, 10 calories for the secondary consumers, and only 1 calorie for the tertiary consumers. This rule of 10% also applies to **biomass**, the total mass in an area at a particular time. If there are 1,000 kg of producer biomass, it can only support about 100 kg of primary consumer biomass, and so on. In addition, this rule of 10% applies to numbers of individual organisms as well. Generally, the prey population will be 10 times greater than the predator population that feeds on them. Any more, and the predators will overexploit the prey until they are all gone.

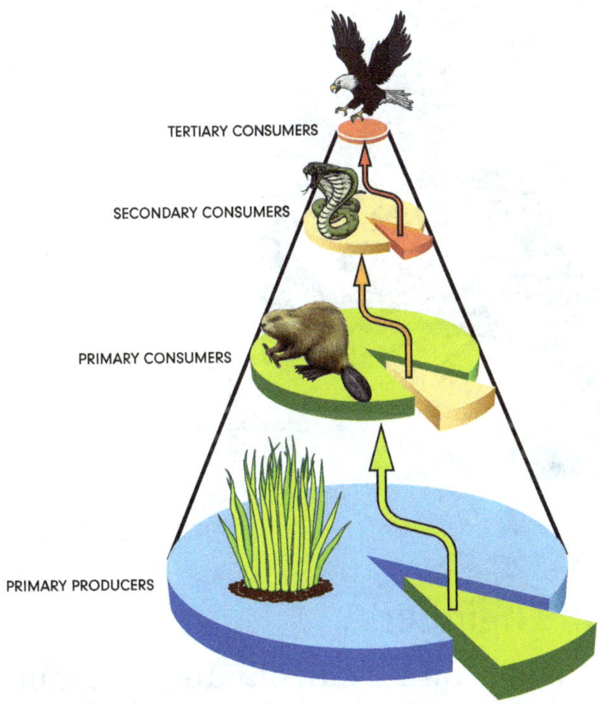

Emre Terim/Shutterstock.com

A simple food chain.

alinabel/Shutterstock.com

A simplified trophic pyramid with a polar bear as the top predator.

We can use graphical representations like **food chains** and **food webs** to illustrate feeding relationships. While often simplified representations of the feeding relationships in a given community, they are useful tools in mapping out how different populations influence each other.

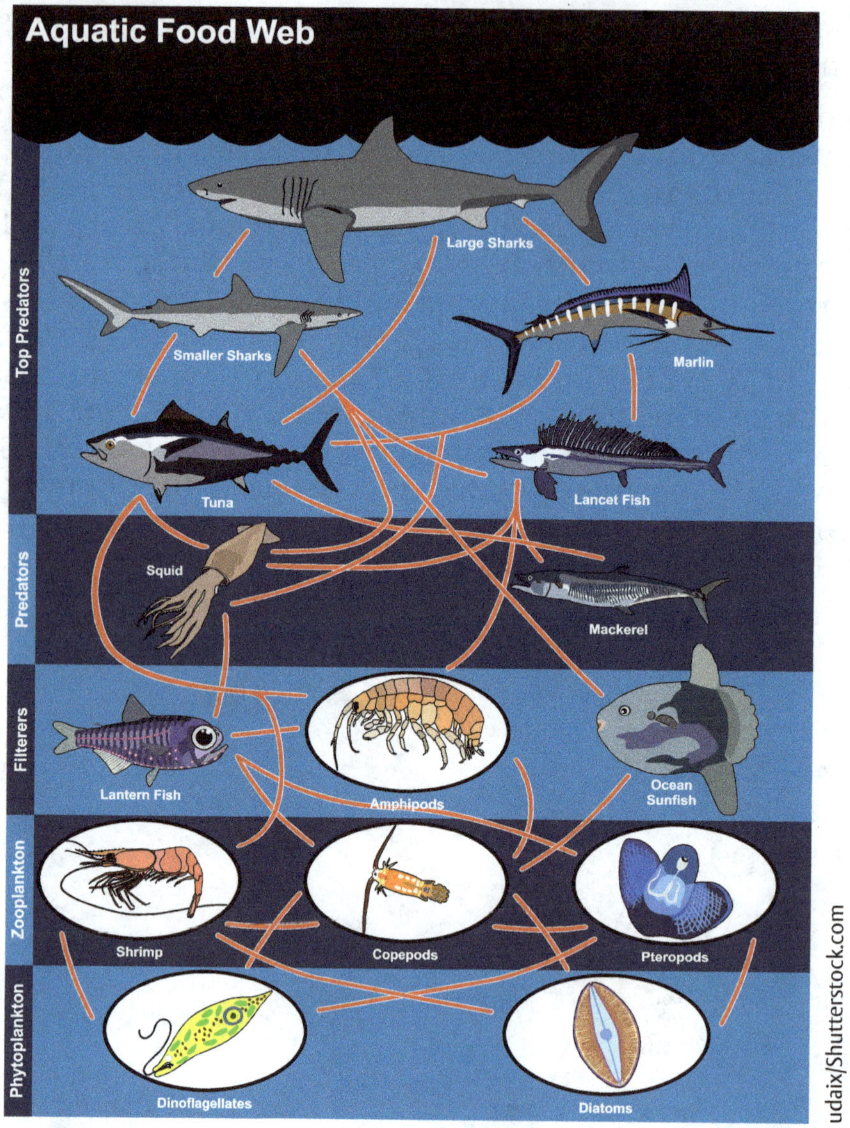

An aquatic food web.

4.3 Keystone Species and Ecosystem Engineers

Keystone species have a disproportional impact on its surrounding environment. These can include top predators, decomposers who help recycle nutrients, and ecosystem engineers. For example, in the Northeastern United States, people eliminated top predators like mountain lions and wolves. This resulted in an explosion in the deer population. After the deer overgrazed their forest habitats, they migrated out to farms and began eating the crops. Now, deer populations have to be managed through hunting. Without annual deer hunting to reduce the population in some places, thousands more deer could potentially starve as their food source becomes scarce through overgrazing. This could result in a **trophic cascade**, in which the entire ecosystem could collapse.

The elimination of keystone species like the mountain lion from some ecosystems has permanently altered the communities.

Ecosystem engineers physically alter their ecosystems. For example, prairie dogs dig tunnels that form prairie dog "towns" that can stretch for hundreds of acres. Other species, from black-footed ferrets, burrowing owls, and snakes then make use of these towns for their own habitats. Also, elephants roam across the plains of Africa eating lots of vegetation and creating depressions that fill with water that could be the only source of drinking water for many species doing the dry season. As they migrate, the elephants also deposit lots of organic fertilizer in their wake. As the elephant population has gone down, the natural grasses have been replaced with scrubby vegetation, converting the savannas into scrub forests that are less palatable to many species. Beavers build dams that create ponds of standing water that can benefit other species. These dams trap sediments, allowing cleaner water to flow through, which is ideal for salmon spawning downstream. Beaver dams also create wetlands, which filter nutrient runoff and other pollutants.

Beavers are ecosystem engineers.

4.4 Community Change Disturbance

Often, ecosystems experience a **disturbance**, in which there is a sudden and drastic change to the ecosystem. Some can be as localized as a tree fall creating a gap in the forest. Others can be widespread and caused by natural disasters like hurricanes and fires while others are human-caused, like clean-cutting a forest or strip mining an area. Some may be seasonal or cyclical, like prairie fires or insect outbreaks.

A community that remains stable after a disturbance with little significant change shows **resistance**. On the other hand, if the community is altered by the disturbance but is able to return to its original state, it shows **resilience**. A severe disturbance may destroy nearly all the species in a community, in which case, the community may rebuild itself through a series of stages called **succession**. Succession may involve several stages until it reaches the final, **climax community**. First succession begins with the **pioneer species**, which could be grasses or even lichens, depending on the severity of the disturbance.

jflin98/Shutterstock.com

Primary succession begins with mosses and lichens growing on rocks.

If a disturbance not only removes all the species in a forest ecosystem and strips all the soil away down to bare rock, succession begins with **primary succession** as lichens colonize the rocks. The lichens produce acids that break down the rocks and start forming the basis of new soil. Once a thin layer of soil is built up, it switches to **secondary succession** where grasses and forbs take root. These low-lying plants are eventually replaced by shrubs and fast-growing trees like poplars. These are in turn replaced by shade-intolerant trees like evergreens. Finally, the evergreen trees actually shade themselves out and are replaced by shade-tolerant hardwood trees like maple or oak.

Matthew R McClure/Shutterstock.com

Secondary succession in a pine forest after a devastating fire.

Note that if a disturbance removes the plant life but leaves the soil in place (such as clear-cutting a forest), succession can begin with secondary succession with the need for primary succession. Note that the above example is an idealized model. In the field, different ecosystems might go through different stages, where local conditions could inhibit or even block profession from one stage to another.

Sometimes, the community may be permanently altered, causing a **regime shift** in which the new climax community is very different from the previous community. Regime shifts can be caused by the loss of a keystone species, introduction of an invasive species, or climate change. For example, the conversion of savanna ecosystems into scrub forests due to the reduction in elephant populations is a regime shift. Also, the loss of wolves in many parts of North America have also caused regime shifts as prey populations have exploded.

4.5 Restoration Ecology

Restoration ecology is the study and practice of restoring natural areas that have been altered by human activities. The goal of restoration ecology is to restore a functioning ecosystem in the area and return it to some semblance of its previous conditions. Often, restoration efforts are mandated by law, for example, federal law requires companies to post a bond and have a restoration plan in place before receiving a mining permit. Forest managers will require replanting of trees as companies harvest timber.

Large areas of prairie grasslands in the Midwest have been converted to farmland. In Illinois, workers are replanting tallgrass prairie vegetation to restore a 1,000-acre area near Chicago to a presettlement condition. In Florida, a series of canals, dams, and levees were constructed in an effort to control flooding. Unfortunately, these efforts actually made flooding worse in Southern Florida. In addition, wetlands in the Everglades dried up, causing wading bird populations to drop by 90%. Now, people are removing the canals, dams, and levees to restore the natural flow of water.

Nancy Bauer/Shutterstock.com

Prairie restoration in the Midwest is converting farmland back into grassland.

4.6 Terrestrial Biomes

While every community and ecosystem is unique, we can organize them into large regional complexes based on similar patterns of vegetation and function called **biomes**. Climate dictates what kind of biome can be found in a given region and the two most important factors in climate are average annual precipitation rate and average annual temperature. For example, a hot and very rainy climate will produce a tropical rainforest, a hot and dry climate will produce a desert, while a cold and dry climate will produce a tundra.

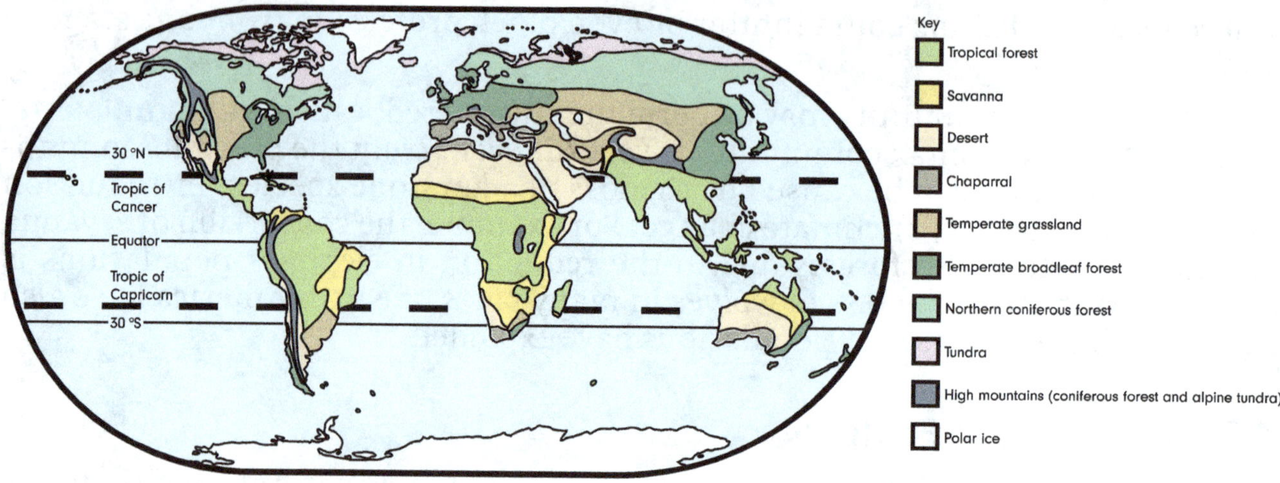

Emre Terim/Shutterstock.com

The world's major biomes.

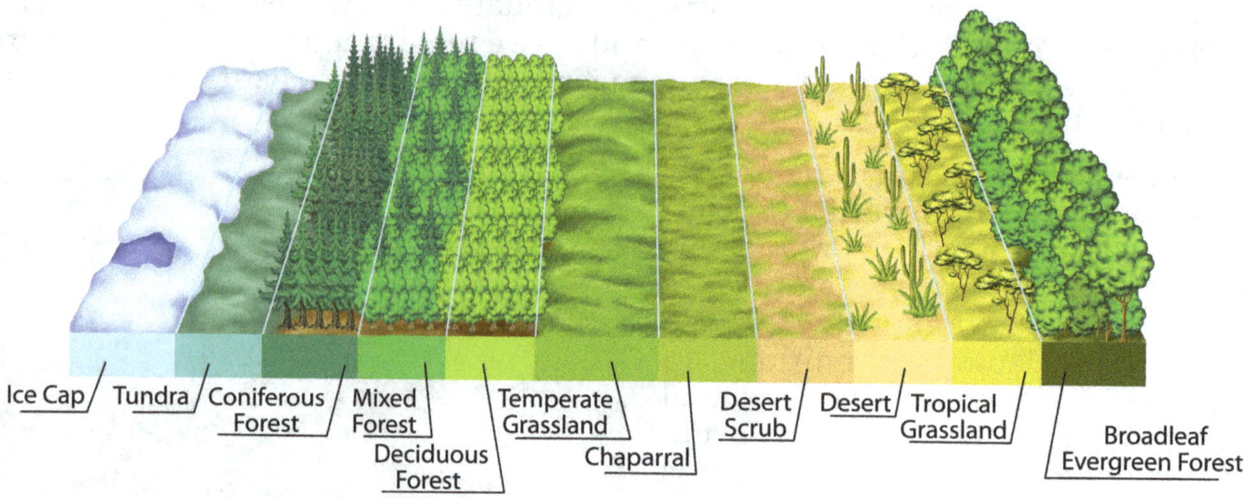

stihii/Shutterstock.com

Latitudinal distribution of biomes.

Temperate deciduous forests feature trees that shed their leaves in the fall and go dormant in the winter months. Deciduous trees include hardwoods like maple, beech, or oak. They are found in Eastern North America and portions of Europe. They are characterized with steady, year-round rain and seasonal temperatures, with cold winters and warm summers.

Jeff Holcombe/Shutterstock.com

A temperate deciduous forest in Vermont.

Moving into the interior of large continents like North America and

Eurasia, we find areas that receive too little rain to support forests. **Temperate grasslands**, also called steppe or prairie, predominate in the Midwestern planes states and nations such as Ukraine. Temperate grasslands receive less annual rainfall than forest biomes and have more extreme differences in seasonal temperatures. Winters are very cold and summers are very hot. Animals found in the prairies include large grazers like bison or pronghorn sheep, burrowing animals like prairie dogs, and ground-nesting birds. Soils in temperate grasslands are rich in nutrients and most of them have been converted in farmland.

HelloRF Zcool/Shutterstock.com

Grassland in Mongolia.

Temperate rainforests occupy most of the Pacific Northwest and parts of Japan. As the name in implies, they receive a lot of rain. If you've ever heard someone says that it rains nine months out of the year in Seattle, Washington (or if you live in Seattle), you get a good picture of how much rain they receive. Like other temperate zones, they have seasonal variations in temperature. Trees are mostly coniferous like the giant redwoods of the Pacific Northwest. Soils are high in fertility, but the heavy rains make them vulnerable to erosion and landslides if the trees are removed, especially in hilly or mountainous regions.

Zhukova Valentyna/Shutterstock.com

Sol Duc rainforest at Olympic National Park, Oregon Coast.

Moving to the equator, we find **tropical rainforests**. These are found around the globe in South America, Africa, India, and Indonesia. They receive steady, year-round rain and warm temperatures throughout the entire year. They lack the four seasons found in temperature zones. Tropical rainforests are characterized by dark, lush interiors with the highest biodiversity found anywhere in the world. While tropical rainforests are densely packed with trees, the variety of life means that the density of individual species is very low. Soils in tropical rainforests are acidic and poor in nutrients and organic material. The hot, wet conditions promote rapid decomposition and the rains wash the nutrients away quickly. So, if you're looking for nitrogen and phosphorus, you won't find much in the soil. Instead, the trees are adapted to hold onto their nutrients. As a result of the low-nutrient content, soils in tropical rainforests make for poor farmland and require large amounts of fertilizer.

South America, Africa, Asia, and Australia are also home to **tropical dry forests** also known as tropical deciduous forests. Like in temperate deciduous forests, trees in tropical dry forests also shed their leaves and go dormant, but they do that when the climate turns dry instead of cold. Tropical dry forests have only two seasons, the dry season and the rainy season. During the dry season, they receive almost no rain, but get heavy, monsoon rains during the rainy seasons. Like tropical rainforests, tropical dry forests have warm temperatures throughout the year.

Africa, Australia, and Asia are also home to **savannas**, which are tropical grasslands interspersed with low-density trees like acacias. Like tropical dry forests, savannas have a dry season that receive virtually no rain and a rainy season, although they receive less rain overall than tropical dry forests. Animals in savannas include large grazers like buffalo, gazelle, and elephants as well as large predators like lions, hyenas, and cheetahs.

RPBaiao/Shutterstock.com

Tropical rainforest in the Amazon.

Naypong Studio/Shutterstock.com

Macaws are one of many species of birds found in tropical rainforests.

Vladislav T. Jirousek/Shutterstock.com

Tropical dry forest in Madagascar.

Deserts are defined as regions that receive less than 25 cm of rain a year. While many deserts are hot, there are also cold deserts. Some deserts like the Sahara are barren with very little vegetation while others like the Sonoran have lots of vegetation. Either way, plants in deserts are adapted to retain most of their water. Soils are high in mineral content but very low in organics. Because there is very little water in a desert, they often do not retain much heat after sunset. As a result, many deserts can have extreme differences in temperature between day and night. It could be blazing hot during the day but drop below freezing at night. Animals in deserts tend to nocturnal, in which they burrow in the sand to avoid the heat of the day or nomadic, moving from one water source to another.

Regions that receive little rain throughout the year but have a consistently cold climate are called **tundra**. Large tundra regions are found in Canada, Scandinavia, and Russia. Also, small tundra biomes, called alpine tundra, can be found at the tops of tall mountains throughout the world, even in tropical regions. Tundra have no trees with mostly low-lying vegetation and lichens. Underneath the soil, there is a layer called permafrost, where the soil is frozen throughout the entire year because the heat never penetrates deep enough to thaw it. Polar bears, caribou, and wolves are among the animals common to the tundra.

Simon Dannhauer/Shutterstock.com

Savanna in Tanzania.

Anton Foltin/Shutterstock.com

The Sonoran Desert near Phoenix, Arizona.

Andrei Stepanov/Shutterstock.com

Arctic tundra in Siberia, Russian.

Pi-Lens/Shutterstock.com

Boreal forest in the Yukon Territory, Canada.

saulty72/Shutterstock.com

Boreal forests are important habitats for many vulnerable species, such as the caribou.

Boreal forests or **taiga** are found in northern climates in Alaska, Canada, Scandinavia, and Russia. Like tundra, they are cold, but receive more rain. They have long, cold winters and short, mild summers. They are dominated by evergreen trees with animals like elk, wolves, and brown bears. Soils in boreal forests are acidic and poor in nutrients.

Finally, **chaparrals** are found in small patches in places like Southern California, the Mediterranean, and Chile. Usually, the climate is influenced heavily by nearby oceans. Chaparrals have warm, dry summers and mild, wet winters. The dominant vegetation is evergreen shrubs found in dense thickets. The dry summers make wildfires frequent. After the fires clear much of the vegetation, when the rains come in the fall, they often trigger mudslides, especially in mountainous chaparrals like in California.

Tim Gray/Shutterstock.com

A dry chaparral in Southern California.

Henrik Lehnerer/Shutterstock.com

Firefighters putting out a brush fire in Ventura, California.

4.7 Discussion Questions

1. Why was the mongoose ineffective in controlling rats in Hawaii?
2. What might be factors that would result in intraspecific competition?
3. Many grazing animals gather in large herds. What might be a survival advantage for a prey species to gather together in large numbers on an open plane?
4. In comparing the eyes of many predators and prey species, biologists find that predators tend to have eyes that are closer together, giving them binocular vision with good depth perception while prey species have eyes that are widely spaced, giving them a wider field of vision. What would be the advantages for each type of organism's vision?
5. A simple food chain involves kelp, sea urchins, and sea otters. The sea urchins feed on kelp while the sea otters prey on the sea urchins. Describe the effect on the community if the sea otter was removed.
6. What factors contribute to the soils in tropical rainforests being poor in nutrients?
7. Why do tropical rainforests have high levels of biodiversity?
8. What climate conditions do deserts and tundra have in common? In what ways are they different?
9. What kinds of adaptations would desert plants need in order to survive?
10. Why are plants in chaparrals vulnerable to annual brush fires?

4.8 Key Terms

Biomass
Biome
Boreal forest
Chaparral
Climax community
Coevolution
Commensalism
Competition
Competitive exclusion principle
Desert
Disturbance
Ecosystem engineers
Food chain
Food web
Herbivory
Interspecific competition
Intraspecific competition
Keystone species
Mutualism
Parasitism
Pioneer species
Pollination
Predation
Primary succession
Regime shift
Resilience
Resistance
Resource partitioning
Restoration ecology
Savanna
Secondary succession
Succession
Symbiosis
Taiga
Temperate deciduous forest
Temperate grassland
Temperate rainforest
Trophic cascade
Trophic level
Tropical dry forest
Tropical rainforest
Tundra

Chapter 5

BIODIVERSITY AND CONSERVATION

Chapter Topics

5.1 Scope and Benefits of Biodiversity
5.2 Extinction and Causes of Biodiversity Loss
5.3 Conservation Biology
5.4 Efforts to Conserve Species and Habitats
5.5 Discussion Questions
5.6 Key Terms

In the past 40 years, 95% of all the rhinos have been lost worldwide. It is estimated that there are about 30,000 rhinos living in the wild today, down from 500,000 at the beginning of the twentieth century. There are five species of rhinos: Black, White, Greater One-Horned, Javan, and Sumatran. The black-and-white rhinos are found in Africa, while the other three species are found in Asia. The table below summarizes the numbers of each rhino species and their classification under the Convention on International Trade in Endangered Species (CITES) and The International Union for Conservation of Nature (IUCN).

Emma Geary/Shutterstock.com

The black rhinoceros.

Species	Numbers in the wild	Classification
Black	5,000	Critically endangered
White	20,000	Nearly threatened
Javan	65	Critically endangered
Sumatran	100	Critically endangered
Greater One-Horned	3,550	Vulnerable

The causes of the rhino's decline include habitat loss and poaching. While the trade in rhino horns have been banned under the CITES treaty since 1977, poachers still hunt these animals every year, mostly for their horns. Many cultures use powdered rhino horn in a variety of folk medicines, although there is no scientific evidence that it causes any benefit. The rhino has also lost a lot of its territory as more and more land has been converted to agriculture.

Elephants have also become endangered due to habitat loss and poaching. Although the trade in elephant ivory has been banned under the CITES treaty in 1975 for Asian elephants and 1989 for African elephants, poaching of both is on the rise. The IUCN estimates that 111,000 African elephants, a fifth of the total population, were killed between 2006 and 2015. Despite the global ban, there are still places in the world, including the United Kingdom, Thailand, Japan, and parts of the United States where the sale of elephant ivory is still legal.

Tigers are also listed as endangered by the IUCN with two subspecies, the Malayan and Sumatra, listed as critically endangered. The global population of tigers is estimated to be 3,900, down from 100,000 a century ago. Once again, the biggest threats are the loss of habitat to farming and poaching for sale of the body parts for trophies and use in folk remedies.

Despite laws and treaties intended to ban the poaching and the illegal trade in these animals, many countries lack the resources or the will to enforce these laws. In many cases, the penalties are light compared to the profits.

5.1 Scope and Benefits of Biodiversity

Biodiversity, the variety of life across all levels of biological organization, from genes to ecosystems, is in decline worldwide. Despite increased efforts, many species are in danger of going extinct. This has implications not only for ecosystems, but also for ourselves as a species.

The number and variety of species is **species diversity**. The number of species in a given area is **species richness**. As a rule, species richness is highest at the equator and goes down as one heads toward the poles. This is due to larger amount of geographic area, abundant solar energy, stability of the climate, and the lack of glaciation, which can disrupt ecosystems. The more the variety of niches, the higher the biodiversity. Forests, being more structurally diverse, tend to have higher levels of biodiversity than grasslands, deserts, or tundra.

Evenness or relative abundance is the number of species compared to other species in a given area. As we noted in Chapter 4, one way in which relative abundance varies is by feeding relationships, with prey species being more numerous than predators.

The differences in the DNA among species are called **genetic diversity**. The reason that dogs come in so many shapes and sizes is because, as species, dogs have a staggering amount of genetic diversity. However, the situation is different when we look at individual breeds. Pressure among breeders to conform to idealized versions of breeds has resulted in many breeds having a low genetic diversity, putting purebred dogs at risk for disorders such as heart ailments, hip dysplasia, and epilepsy. Populations with low genetic diversity are vulnerable to **inbreeding depression**, where genetically similar parents produce inferior offspring. They can also be less able to weather changes in the environment, since they lack a variety of traits to call upon that could help the population survive.

Ricardo Reitmeyer/Shutterstock.com

David Osborn/Shutterstock.com

Dr Ajay Kumar Singh/Shutterstock.com

The North American bison, elephant seal, and cheetah are all species that have low genetic diversity.

Examples of species with low genetic diversity include the elephant seal, cheetah, and American bison. The reason for the low genetic diversity is that these species passed through a **bottleneck**, in which the majority of population was killed in some catastrophe. The result was a small breeding population. Even if the number of individuals may have returned to previous levels, the small surviving population means that the new population was bred from a limited gene pool. While the causes of many bottlenecks are unknown, in the case of the American bison, the bottleneck occurred in the nineteenth century when they were hunted nearly to extinction.

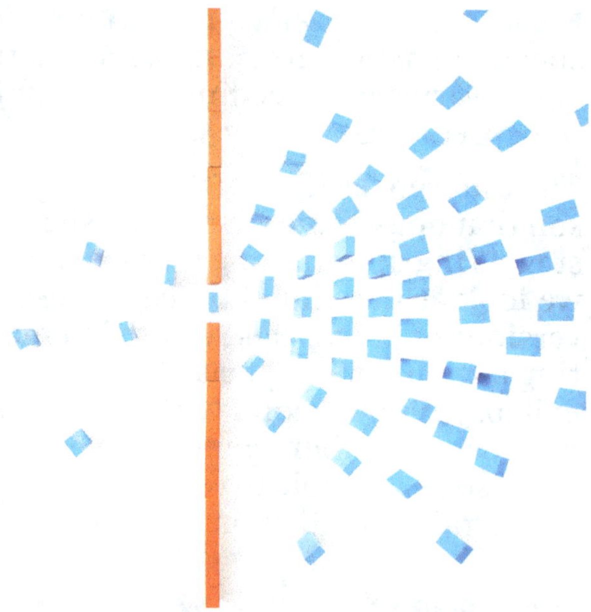

Kaesler Media/Shutterstock.com

A bottleneck occurs as a result of an event in which only a small number of individuals survive.

Ecosystem diversity refers to the number and variety of ecosystems in a given landscape. The greater the ecosystem diversity, the greater the number of potential niches that can be exploited by different species.

In terms of the overall relative abundance of species, arthropods have the most species. In fact, there are more insect species than any other group of organisms. There are more insect species than all other animal species combined. Among vertebrates, fish make up half of all species. Though we often think of our current era as the Age of Mammals, in fact, mammals are among the groups that have the least number of species.

Overall, there are about 1.8 million species that have been classified in the world today. Estimates of the actual number of species range from between 3 and 100 million, with around 10 million being the most widely accepted estimate. Our knowledge of species is limited for a number of reasons. Many species, such as bacteria and protists, are microscopic and we haven't examined enough of the microscopic world. Also, many organisms look very similar and may be mistaken for belonging to the same species, but actually are two or

Don Mammoser/Shutterstock.com

The Sumatran orangutan is one of the three recognized species of living orangutan.

more species. For example, we once thought there was only one species of orangutan, when in fact there may be as many as three. Biologists originally classified the forest and plains elephants of Africa as a single species, but many are starting to conclude that they in fact are two.

Ethan Daniels/Shutterstock.com

Biodiversity provides numerous benefits for us. Wild species may serve as a source of genetic traits that our domesticated crops and animals may need, or they may be potential new sources of food or medicine. For example, in the nineteenth century, Ireland was hit with a devastating fungal infection call the blight. The blight devastated the potato crops on the island as none of the strains of potatoes grown there had any resistance to the disease. Farmers went bankrupt and lost their lands and millions of people faced famine. This triggered a massive migration as people left Ireland and moved to places like Great Britain or the United States in search of a better life. Wild plant species could hold the key to resisting the next blight that could be introduced to our crops through crossbreeding or genetic engineering. This could prevent another disastrous famine like the one that hit Ireland.

chrisdorney/Shutterstock.com

The Irish potato blight devastated farming on the island.

In the event that one of our key domestic food sources does become threatened with a devastating famine, wild species could be potential replacements. For example, the maca is a root vegetable similar to a radish that's found in Latin America. In South America, animals such as the capybara (the world's largest rodent) and the vicuna (a relative of the llama) are easily ranched and could serve as alternatives to many of our current sources of meat.

Kevin Xu Photography/Shutterstock.com

The capybara.

abriendomundo/Shutterstock.com

The vicuna.

About half of the major pharmaceuticals have been discovered from wild species. Captopril, a drug used to control high blood pressure comes from the venom of the Brazilian pit viper. Taxol, a key anticancer agent is produced from the Pacific yew tree. The velvet bean plant is the source of L-dopa, a drug used to suppress the symptoms of Parkinson's disease. Quinine, one of the most commonly used antimalarial agents is made from the yellow cinchona plant. These are just a few examples of important, life-saving drugs that come from wild species. Imagine the treatments that could be lost forever, never discovered as we lose species without having the chance to even examine their potential.

Fabio Maffei/Shutterstock.com

The Brazilian pit viper's venom is used to make a drug to control high blood pressure.

Lavandina/Shutterstock.com

The Pacific yew tree produces an anticancer drug, Taxol.

Chatchai.wa/Shutterstock.com

The velvet bean produces L-Dopa, which controls the symptoms of Parkinson's disease.

LIKIT SUPASAI/Shutterstock.com

Dried cinchona being prepared to produce quinine, an antimalarial drug.

Biodiversity also keep ecosystems stable and sustainable. This enables ecosystems to perform many of the ecosystem services we discussed in Chapter 1, including

- Sources of fuel, food, and shelter
- Breaking down air and water pollution
- Decomposing waste
- Climate stabilization
- Pollination
- Providing habitats for predators that control pests and disease vectors
- Cultural and aesthetic benefits

bsd/Shutterstock.com

Biodiversity also helps ecosystem maintain resilience in the face of disr-uptive events or adapt to change.As we noted in Chapter 4, the reduction in elephant herds in Africa has resulted in savannas being converted into scrub forests. The removal of keystone species or ecosystem engineers can result in the character of an ecosystem changing drastically.

Biodiversity can also help support economics by providing opportunities for outdoor recreation. People don't spend their time and money camping or hunting in an area that looks like a barren wasteland. Outdoor activities are popular for the wildlife they can see. Many communities depend on the money from different kinds of ecotourism such as whale watching, scuba diving in coral reefs, or taking tours of the rainforest. Interaction with nature also gives many people a feeling of peace and psychological well-being, reducing any forms of stress and anxiety.

5.2 Extinction and Causes of Biodiversity Loss

As human alter the environment to suit themselves, they also affect the way other species interact with the environment, creating "winners" and "losers." Often, humans alter the ecosystem to produce a simpler environment, with lower levels of biodiversity, more open spaces, and more pollution. As a result, some species made it difficult to find the resources they need. On the other hand, other species may thrive. Most of the winners tend to be generalists, who can exploit a variety of resources, while specialist, who depend on a more limited range of resources, find it more difficult to cope. The table below summarizes the losers and winners.

Losers	Winners
Specialists	Generalists
Confined to a limited geographic range	Widespread, global in their distribution
Need dense habitats like old growth forests	Able to exploit open habitats in the early stages of succession
Need conditions to remain stable	Able to adapt to rapidly changing conditions
Large and take a long time to reproduce, often K-selected	Small-bodied and reproduce quickly, often r-selected
Top predators	Low on the food chain, scavengers
Require large habitats to hunt for food	Can thrive in small habitats
Island species	Mainland species

So, as rule, the large-bodied, top predators such as bears and tigers are often the biggest losers, while the small scavengers such as rats and roaches are the winners when humans alter the environment.

There are a number of ways in which biodiversity loss is measured. For example, the United Nations Environment Program (UNEP) has maintained the Living Planet Index since 1970. This measures the decline in over biodiversity in the world. Using 1970 as the baseline, the most recent report was published in 2018 and indicates a decline of 60% between 1970 and 2014. Tropical regions saw the most decline in those years among terrestrials areas. In aquatic systems, freshwater ecosystems saw the most decline.

Another report is the Red List, which tracks species most threatened with extinction. The most recent report of the Red List was published in 2017 and found the following percentages of species are endangered:

- 21% of mammal species
- 13% of bird species
- 20% of reptile species
- 32% of amphibian species
- 14% of fish species

Extinction is the complete disappearance of a species and occurs when the last member in the world dies. A localized form of extinction is called **extirpation**, in which species have lost much of the original territory and are now only found in more limited locations. For example, the black rhino in Africa has lost most of its traditional ranges and is now found mostly in small protected areas. In order to protect the black rhino from poachers, many conservationists have taken to removing the feature most often prized by poachers, its horn. Note the rhino's horn is made of the same material as your fingernails, so cutting it off doesn't cause it any pain, though it probably doesn't like it.

Extinction is a natural process, as new species emerge, older species will often fade away. Examination of the fossil record indicates that about 1 out of every 1 to 10 mammal and marine species will go extinct each year. However, as we noted in Chapter 4, there have been at least five mass extinctions in the fossil record in which at least half of the species in existence go extinct. These are summarized in the table below:

Extinction Event	Timeline	Cause	Impact
End Ordovician	444 million years ago	Climate change and drop in sea levels	85% of marine species extinct
Late Devonian	375 million years ago	Unknown	75% of species lost
End Permian	251 million years ago	Possibly volcanism	90% of all species lost
End Triassic	200 million years ago	Unknown	80% of species extinct, set the stage for the age of the dinosaurs
End Cretaceous	66 million years ago	Possible asteroid impact	75% of species wiped out, including the nonavian dinosaurs

The End Cretaceous extinction was likely caused by an asteroid impact, which ended the reign of the nonavian dinosaurs.

The current extinction rate is between 100 and 1,000 times the normal background rate and species are going extinct at a rapid rate putting us in the Sixth Mass Extinction. While modern industrialization has accelerated this extinction rate, humans are actually very good at wiping out species even with primitive, Stone Age technology. The fossil record indicates that every time humans moved into a new region, there is a drop in the number of large-bodied animals. Humans likely caused or at least contributed to the extinction of creatures such as the mammoth, the woolly rhinoceros, and the elephant bird.

Today, the major causes of the decline in biodiversity are

- Habitat loss
- Pollution
- Overhunting
- Invasive species
- Climate change

By bar, the greatest threat to biodiversity is habitat loss. Ways in which humans have altered habitats include converting the land into farms, pasture, and suburban sprawl. Clearing forests and damning rivers have also contributed to the loss of biodiversity by destroying important niches and resources. Often, habitat loss begins as a piecemeal process in which one continuous habitat is fragmented into smaller patches. **Habitat fragmentation** is a particular threat to large-bodied predators, such as bears and tigers, which need vast territories in order to hunt for food. In the United States, habitat fragmentation has also resulted in an increase in human–animal encounters. Bears that lose their territories often enter suburban areas to scavenge for food. Often, they end up digging through trash cans or otherwise becoming a nuisance. In some parts of the country, mountain lions and coyotes are also showing in suburban communities, often posing a threat to humans or pets found in the yard. Deer that lose their habitats descend on farmland and feed on crops and have a high risk of getting hit by traffic on the highway.

Habitat loss affects every biome, with wetlands having the most impact. Many wetlands have been drains in order to build homes and businesses. This loss of wetlands often increases problems with fertilizer runoff and flooding. Overall,

Jay Ondreicka/Shutterstock.com

Loss or fragmentation of habitats is the leading cause of the decline in biodiversity.

habitat loss is responsible for about 80% of the decline in population of threatened birds and mammals.

Pollution also affects biodiversity in many ways. Air pollution contributes to acid rain, which damages forests and lowers the pH of lakes, making them less habitable. Water pollution directly impacts fish and amphibians. Agricultural runoff damages food webs and causes eutrophication, which leads to the formation of hypoxic dead zones in aquatic ecosystems. Persistent synthetic organic compounds and toxic heavy metals directly poison organisms. Plastic debris poses a choking, entanglement, and drowning threats to organisms. Thermal pollution can trigger massive fish kills. Even noise and light pollution can cause stress, affecting behavior and production of organisms.

GroanGarbu/Shutterstock.com

Pollution causing a fish kill.

Rich Carey/Shutterstock.com

Plastic debris is a threat to many aquatic species.

Overharvesting and overhunting often threaten specific species. Poachers hunt elephants, tigers, and rhinos for trophies and use in folk remedies. Rare trees like teak and mahogany are being cut down rapidly. Gorillas and other primates are hunted for their meat. Sharks are harvested to make shark fin soup. Whales are also still hunted for meat. Worldwide, fish stocks in many locations are on the verge of collapse due to overfishing. Despite laws to prevent poaching, in many parts of the world, enforcement is lax and the penalties are light.

Svetlana Foote/Shutterstock.com

Poaching threatens African elephants.

Invasive species pose a threat to native species. While most species that are introduced into a new environment do not succeed in establishing themselves, many are able to thrive. Often, in the new environment, the predators and other limiting factors that kept the species in check in the old environment are not present in the new one. As a result, the invaders are able to proliferate and outcompete native species that still have to contend with many limiting factors. Examples of invasive species include:

Jeff Caughey/Shutterstock.com

Zebra mussels often stick to boat propellers, causing damage.

Zebra Mussel: Introduced in the Great Lakes in the 1980s when a ship from Eurasia dumped its ballast water into Lake St. Clair, these tiny mussels have outcompeted native mussels and clams in the Great Lakes and have since moved into the Mississippi River system. Not only have they caused native mussels and clams to decline, they cause millions of dollars in damage by clogging intake pipes for water treatment plants and factories.

Mongoose: As noted in Chapter 4, the mongoose was introduced into Hawaii to control rats that were threatening sugarcane crops. Instead, the mongoose has been feeding on Hawaiian bird eggs, causing many bird populations to drop.

Frank Fichtmueller/Shutterstock.com

The Javan mongoose has devastated bird populations in Hawaii.

Kudzu: Brought into to the American Southeast from China to combat erosion, this vine has spread all over the South. The warmer climate prevents kudzu from dying back as it does in China and it engulfs whole areas, replacing native plants and causing property damage.

Gypsy Moth: Introduced in Massachusetts in an attempt to create a silk industry to compete with Chinese silkworm silk. The effect failed and the moth spread across the country. Every few years, outbreaks of the gypsy moth defoliate trees all over the United States.

Luke P Ferguson/Shutterstock.com

An abandoned building covered in kudzu.

Safwan Abd Rahman/Shutterstock.com

Gypsy moth caterpillar eating a leaf.

European Starling: A wealthy drug manufacturer and Shakespeare enthusiast decided to release every bird mentioned in Shakespeare's plays into New York's Central Park. The starling, mentioned in Henry IV, was the most successful in spreading across the United States. Within 75 years of being released, the starling became one of the most numerous birds in the United States, displacing many native birds.

Emerald Ash Borer: Traveling into the United States inside wooden pallets from China, this insect first appeared in Michigan in 2002. By 2010, it had spread to 12 states and into Canada, where it has killed millions of ash trees.

arjma/Shutterstock.com

The European starling.

Herman Wong HM/Shutterstock.com

The emerald ash borer.

Rat: Historically, this rodent has followed people wherever they have gone. Often stowing away on ships, they have even colonized many remote islands. Not only do they threaten crops, they also carry diseases such as the Hanta virus and Bubonic plague.

Brown Tree Snake: Another animal that has stowed away on ships, this snake traveled from Asia to the island of Guam where it has devastated bird populations.

JaklZdenek/Shutterstock.com

Norway rat.

Firepac/Shutterstock.com

The brown tree snake has threatened bird species on the island of Guam.

Rabbit: Two dozen rabbits were introduced into Australia in order to promote sport shooting. They soon multiplied to millions and have overgrazed countless acres of grassland and farmland. Today, Australia attempts to control the rabbit population with the rabbit hemorrhagic disease virus.

Feral Pig: Feral pigs have become a problem on many islands. In particular, they have invaded the Galapagos where they feed on the eggs and young of the giant tortoise.

Houshmand Rabbani/Shutterstock.com

Rabbits have stripped grasslands of vegetation in Australia.

Ivan M Munoz/Shutterstock.com

Feral pigs.

Climate change is also having an impact on species. As conditions shift, many species are shifting their ranges up in elevation or higher in latitude. This puts species that already live on the margins at particular risk. For example, the loss of Arctic sea ice has made it difficult for polar bears to hunt for prey during the summer months.

Jan Martin Will/Shutterstock.com

Polar bears are threatened due to climate change and the loss of Arctic sea ice.

Some species are threatened by multiple sources. For example, the monarch butterfly has declined in population due to a combination of pesticides killing the milkweed that they depend on and deforestation of the winter habitat in Mexico. Colony collapse disorder, which has been devastating bee populations, is likely caused by a combination of pesticides and parasitic mites.

Worldwide, one of the biggest warning signs has been the collapse of amphibian populations. Amphibian populations have plummeted due to a combination of pesticides, habitat loss, climate change, invasive species, and disease. Chemical pollution in particular has caused amphibians to become deformed as they mature from the tadpole stage. Often, the adult amphibians are missing limbs, have extra limbs such as three hind legs, or have reproductive abnormalities, such as becoming hermaphroditic.

Charlotte Bleijenberg/Shutterstock.com

The golden mantella frog of Madagascar is critically endangered.

5.3 Conservation Biology

The study of the loss of biodiversity and research into efforts to prevent or reverse it is **conservation biology**. Conservation biologists engage in field research to study the behavior of species and look for causes of species of decline. Tools of conservation biology include collecting field samples, attaching radio trackers to map migratory patterns, or setting up camera traps to record species activity in an area.

Of particular concern for conservation biology is the **minimum viable population size**. This is the smallest a population can get before it runs into problems such as inbreeding depression and genetic abnormalities. Too much of a decline in population and the species could lose too much genetic variation and end up like the cheetah or American bison today. Even if their numbers rebound, they may continue to have problems with abnormalities or even sterility.

5.4 Efforts to Conserve Species and Habitats

In the United States, the primary legal means to protect species is the **Endangered Species Act (ESA)**. Passed in 1973, this law forbids the killing and trading in species that have been designated as endangered or threatened. It also protects the habitats of endangered or threatened species. The ESA is administered by the US Fish and Wildlife Service (FWS) and the US National Oceanic and Atmospheric Administration (NOAA). The FWS maintains the list of **endangered** and **threatened species**. As of 2019, 1,274 species were listed as endangered and 386 were listed as threatened.

rokopix/Shutterstock.com

For a species to be listed as endangered, the FWS must determine that the species is likely to go extinct without intervention. The lesser standard of threatened means that the species is in decline and may possibly become endangered in the future.

The ESA has been successful in many cases. A combination of habitat protection, captive breeding programs, and the banning of DDT has enabled three birds – the bald eagle, the peregrine falcon, the brown pelican – recover to the point where they have been removed from the endangered or threatened list.

Collins93/Shutterstock.com

Ray Hennessy/Shutterstock.com

Three species of birds, the bald eagle, the peregrine falcon, and the brown pelican have recovered to where they could be removed from the endangered/threatened lists.

Unfortunately, federal authorization for the ESA expired in 1992 and it has been limping along with a year to year appropriation. As a result, the FWS has twice had to impose a moratorium on new applications. A species only receives protection from ESA after it has been added to either the endangered or threatened list. This requires individuals or groups to file a petition to the FWS along with data justifying its inclusion. Due to lack of funding, the FWS has developed such a backlog that they have twice had to announce that they could not receive any more petitions until they have cleared their current backlog. This has resulted in many species not receiving protection that they may have needed.

Political opposition to ESA is often based on a perception that the law imperils jobs and commercial development. In practice, however, few development projects have been halted by ESA. Often, development will go on if the landowner agrees to adopt a habitat conservation plan in which part of the habitat may be harmed if the overall conditions of the habitat are improved. The government may also adopt a safe harbor agreement in which the government agrees not to pursue additional actions against the landowner if the landowner agrees to assist in the species' recovery.

One of the biggest controversies regarding ESA happened in the late 1980s when the FWS set aside a large area of old growth forest as a protect habitat for the northern spotted owl. Naturally, the timber industry objected and demanded that their economic interests and the jobs of timber workers take precedence. Eventually, a compromise plan was reached, called the Northwest Forest Plan, which allowed some timber harvesting while still protecting much of the habitat.

The Northern spotted owl.

More recently, a conflict among federal officials, cattle ranchers, and the oil and gas industry arose about the greater sage grouse and whether it should be listed as endangered. As with the northern spotted owl, a compromise was reached to protect the grouse while still allowing ranchers and drillers access to the land. The federal government determined that this voluntary program would protect the grouse adequately.

Tom Reichner/Shutterstock.com

The greater sage grouse.

On the international level, species are protected by the **Convention on International Trade in Endangered Species of Wild Fauna and Flora (CITES)**. Adopted in 1973, this treaty bans the international trade in body parts of protected species. Currently, 183 nations have adopted CITES and abide by its principles. Parties of CITES meet every three years and vote on whether new species should be included in the list of protected species or if new rules should be adopted.

The International Union for Conservation of Natura (IUCN) maintains the Red List of Threatened Species. The IUCN maintains eight categories of species:

The first two are for extinct or near-extinct species.

- Extinct: Completely gone from the world.
- Extinct in the wild: No longer existing in the wild, but individuals still exist in captivity.

The next three are for species considered by the IUCN as threatened. As most to least serious, they are

- Critically endangered: Extremely high risk of extinction
- Endangered: Very likely to become extinct in the near future
- Vulnerable: Likely to become endangered unless conditions improved.

The final three categories for those species of lower risk. They are

- Near threatened: May become threatened in the future, but does not currently meet the criteria for threatened.
- Conservation dependent: Dependent on conservation efforts to avoid becoming threatened.
- Least concern: Not considered the focus of conservation.

In addition, IUCN can also categorize species as Data Deficient (not enough information is available to evaluate the status of the species) or Not Evaluated (Not yet assessed by the IUCN).

One of the most successful efforts to help species recover has been the **captive breeding program**, where the young or eggs of an endangered species are taken and raised in a protected environment with the intention of returning them to the wild when they are mature enough to have a better chance of survival. For example, black rhinos are raised in a preserve in South Africa and then released to the wild. In the Galapagos Islands, where giant tortoise species are threatened by feral pigs and rats, the eggs are taken to the Charles Darwin Conservation Center. There, the eggs are hatched and the young are raised until they are about 5 years old. At that age, they are too big for the rats or the pigs to threaten them so they can then be released into the wild.

In 1992, the population of the California condor, the largest bird in North America, had dwindled down to 22 individuals in the wild. Captive breeding programs have brought their numbers up to nearly 300 in the wild and about 170 still in captivity. In these programs, condor chicks are cared for using hand puppets designed to look like adult condors. The puppets ensure the condors will seek out other condors in the wild and mate with them instead of bonding and following humans around. Although hunting was originally the biggest threat to the California condor, today the biggest problem is lead shot. As carrion eaters, these birds feed on recently dead animal carcasses. If the carcass was killed by a hunter using lead shot, the condor will swallow the shot as it tears chunks of flesh from the bones. This can give the condors lead poisoning. In 2019, the state of California banned the use of lead shot in order to protect the condor.

FRAYN/Shutterstock.com

The California condor has been critically endangered, but captive breeding programs have given it another chance.

Ian Kennedy/Shutterstock.com

The Galapagos tortoise.

One proposal to help species recover has been **cloning**, where the DNA taken from a donor organism is placed inside an egg and then implanted inside a surrogate for gestation. The result is a genetic twin of the donor organism. The first mammal to be cloned was Dolly the sheep in 1996. Since then, many other mammals, such as monkeys and cats, have been successfully cloned.

Not only has cloning been proposed to help rebuild endangered populations, but some have suggested that it could be used to resurrect extinct species such as the Tasmanian tiger or the dodo if viable samples of the DNA could be found. Sadly, for fans of the Jurassic Park movies, it is very unlikely that dinosaurs could be cloned. In the movie, scientists obtained dinosaur DNA from mosquitoes trapped in amber than had bit a dinosaur just before getting stuck in tree sap. However, DNA degrades over time and the likelihood of finding dinosaur DNA inside the gut of a fossilized mosquito that is intact enough for cloning is practically zero. Which is probably for the best, since cloning dinosaurs never seems to work out pretty well for the people in the movies.

On the other hand, woolly mammoths have only been extinct for about 10,000 years and it is possible that viable DNA could be obtained. In fact, biologists have already sequenced the mammoth's genome. So, we could either clone a mammoth or perhaps alter an elephant's genome to create something that looked like a mammoth.

However, there are limitations to cloning. For one, since cloning produces a genetic twin of the donor, even if we can produce large numbers of an organism, they would be low in genetic diversity, putting them at risk for inbreeding depression. In addition, creating large populations of species may not be a good idea if we don't have sufficient habitats for them. As appealing as seeing a live herd of woolly mammoths might be, they'll just end up as zoo specimens if there isn't sufficient tundra set aside for them to inhabit.

Reimar/Shutterstock.com

Is cloning the wooly mammoth a good idea?

Oftentimes, the larger and more well-known species receive the most attention in gaining public support for conservation efforts. However, protecting the habitat of one species often results in it acting as an **umbrella species**, extending protection to other species. For example, elephants have an enormous range and protecting their habitat also benefits other species that make use of the same habitat. In addition, many conservation organizations will select a popular, charismatic species as a **flagship species** to promote conservation of endangered species in general. For example, the World Wildlife Fund has the giant panda as its symbol because people like pandas and may be more likely to donate money to protect them instead of say, an endangered worm.

clkraus/Shutterstock.com

The World Wildlife Fund uses the giant panda as its symbol for fundraising.

Some locations around the world have been designated as biodiversity **hotspots** because they contain many **endemic species**, that is, species that are found nowhere else in the world. For example, the island of Madagascar has been designed as a hotspot because of the many unique species found there, from lemurs to chameleons. To be designated a hotspot, a region must have 150 unique plant species and have already lost 70% of its habitat area to human activity.

Hajakely/Shutterstock.com

The Island of Madagascar is the only place in the world where the ring-tailed lemur is found.

Many other areas have been designated as parks or preserves, making them protected areas. Areas of the ocean are also being made into marine protected areas. Unfortunately, due to lack of funding, many of these areas are not adequately patrolled or protected. Often, these **paper parks** (existing only on paper) are still subjected to illegal poaching and logging.

Another method found to be successful in promoting conservation is enlisting the local people in **community-based conservation**. These efforts include giving the native people of the region authority over wildlife protection. Getting them involved and giving them ownership in exchange for jobs, healthcare, or education is effective in recruiting local support. This is especially importation in countries that have a history of suffering colonialism and might be wary of westerners coming in and trying to take control of their lands again. It also provides economic support for people who might otherwise turn to poaching as a means of supporting their families.

5.5 Discussion Questions

1. What factors might explain why species diversity is highest in tropical rainforests and lowest near the poles?
2. What practices might alleviate the problems associated with inbreeding depression in many dog breeds?
3. Describe the five major causes of biodiversity loss.
4. What invasive species are found in your region and what effects are they having in species diversity?
5. Describe the relationship between biodiversity and food security? Pharmaceuticals?
6. Is cloning a viable means for preventing species extinction? What limitations might hinder using cloning to restore lost biodiversity?
7. What is a captive breeding program? How have they helped prevent the loss of biodiversity?

8. Describe some successes of the ESA.
9. How has the banning of DDT affected populations of birds of prey?
10. Why is poaching still a problem despite laws against it?
11. How does habitat loss contribute to loss of biodiversity?
12. How does protection for an umbrella species also protect other species that may be threatened?
13. What is a hotspot? How does a location become designated as one?

5.6 Key Terms

Biodiversity
Bottleneck
Captive breeding
Cloning
Community-based conservation
Conservation biology
Convention on International Trade in Endangered Species of Wild Fauna and Flora
Ecosystem diversity
Endangered species
Endangered Species Act
Endemic species
Evenness or relative abundance
Extirpation
Flagship species
Genetic diversity
Habitat fragmentation
Hotspot
Inbreeding depression
Minimum viable population size
Paper park
Poaching
Relative distribution
Species diversity
Species Richness
Threatened species
Umbrella species

Chapter 6

LAND MANAGEMENT AND FORESTRY

Chapter Topics

The largest rainforest in the world, the Amazon, performs many important ecosystem services. It regulates climate, produces oxygen, and contains some of the highest levels of biodiversity in the world. Unfortunately, since 1970, nearly a quarter of the Amazon's land area has been lost due to clear-cutting. Historically, Brazil has pushed into the rainforest to relieve crowding in its coastal cities. Demand for cattle pastureland as well as farmland for sugar, soy, palm oil, and corn has also driven efforts to clear cut the rainforest. As the rain forest shrinks, we are experiencing an irreversible loss of biodiversity. The effect of this loss of biodiversity is impossible to measure as we still do not have a full accounting of all the species that exist in the rainforest.

The Amazon rainforest is also a major sink for carbon dioxide. So, as the trees are cut down, global warming is expected to increase.

In addition, indigenous people are losing their land and are being put at risk to exposure to diseases they have no resistance to.

To combat deforestation, Brazil strengthened its forest code, requiring landowners to conserve 80% of the land as forest and other rural landowners conserve 20% of their land as forest. The demand for sustainably grown forest products has increased, putting market pressures on producers to adopt more sustainable practice. As a result of these and other efforts, deforestation in the Amazon began to drop sharply in 2004. Unfortunately, this trend has begun to reverse in recent years as changes in government policy have again encouraged deforestation. Clear-cutting combined with wildfires resulted in more land area of the rainforest being lost in 2019 than any year since 1970.

Toa55/Shutterstock.com

The fires in the Amazon rainforest raise concerns for serious loss of biodiversity.

6.1 Forest Ecosystems

A **forest** is defined as any ecosystem with a high density of trees. As noted in Chapter 4 (Section 4.6), forest biomes can be found in many climates from tropical rainforests, temperate deciduous forests, and boreal forests. In addition, within each forest biome, there can be variations of forest ecosystems based on differences in soil conditions, temperature, and rainfall. Different forest types can be classified by the predominant tree species, such as oak-hickory, maple-spruce-beech, or elm-ash-cottonwood, to name just a few forest types found in North America. Altogether, forests cover about 31% of the Earth's land surface.

Compared to other biomes, forests have more structural complexity and more niches to be exploited by many species. As a result, they have higher levels of biodiversity and productivity compared to grassland, desert, or tundra biomes. Vertically, forests can be divided into several levels:

- **Canopy**: The upper level of the treetops or crown of the forest
- **Subcanopy**: The middle forest of the tree forest level
- **Understory**: A shaded region below the subcanopy consisting of shrubs and short trees
- **Forest floor**: The ground layer consisting of ground-cover plants

SaveJungle/Shutterstock.com

Forest layers.

Dead and dying trees that are still standing are called **snags**, which are often homes to insects and in turn, provide a food source and nesting cavities for birds. When a dead tree falls, it becomes a fallen log and creates a **treefall gap** which lets in more sunlight. This promotes the growth of younger trees.

Forests in general perform many ecosystem services. In addition to having high levels of diversity and productivity, the roots of trees stabilize soil and reduce erosion. In addition, their deep roots draw up minerals from deep soil layers, making them usable for ecosystems. Leaf litter also slows runoff and increase water infiltration, enable groundwater sources to be replenished. Forests also provide many recreational opportunities, such as hiking, camping, hunting, and nature photography. They act as important carbon sinks and sources of oxygen. Trees and tree products also produce many natural resources including lumber, paper products, food, medicines, dyes, and fibers. Finally, many people gain a sense of peace of mind from the aesthetic beauty of forests.

6.2 Causes and Impacts of Deforestation

As with many extraction industries, the harvesting of timber has benefited from advances in technology. Trees are no longer cut down by rugged lumberjacks wielding axes. Instead, they are cut down, gathered, and loaded onto trucks using heavy machinery. **Deforestation**, the cutting down of forests faster than the trees can regrow, results in loss of biodiversity, soil erosion, and desertification. The Food and Agriculture Organization (FAO) of the United Nations has conducted satellite surveys that have indicated that the world is losing forest cover by about 3.3 million hectares a year, even taking regrowth into account.

Peter Bocklandt/ Shutterstock.com

Old-growth forests are more structurally diverse and have more niches than secondary forests.

Deforestation drove the growth of the United States. Prior to 1620, nearly all the land west of the Mississippi River was covered with forest ecosystems. In the 1800s, as Americans spread westward, rapid deforestation destroyed most of the old-growth forests. Today, most of the few remaining **primary or old-growth forests** are found scattered around the Pacific Northwest and the Rocky Mountains. Nearly all of the old-growth forests in the Eastern United States is gone.

The good news is that North America has been experiencing a net regrowth in forest cover. Today, there is more forested land in the United States than there was a century ago. However, these new, **secondary forests**, are less ecologically valuable than old-growth forests. Secondary forests have less structural complexity and fewer niches to be exploited by species. Often times, they have a different species composition and nutrient balance than the old growth forests they replaced.

There are still many hectares of old growth forests still uncut in developing countries. However, many of these forests are being cleared at an alarming rate. Demand for new agricultural land, new human settlements, and fuelwood drives a lot of this deforestation. In addition, many developing countries are carrying huge debts that they owe to organizations like the World Bank and are desperate for economic growth. In order to raise needed funds, they sell off the rights to timber in the forest through the collection of **concession fees**. So, while North America and Europe are experiencing a regrowth of forests, this often comes at the expense of forests in developing countries as timber harvested in South America, Asia, and Africa is sold to wealthier nations. The short-term economic benefits of collecting concession fees leave these nations with long-term environmental problems.

Brazil has experienced the most deforestation in South America, as forestland is converted into soybean farms and cattle ranchland. However, the soils in the Amazon are acidic and poor in nutrients and the heavy rains leach away fertilizers. It doesn't take long for farmers to exhaust the soils in one area, then move on to clear another section of the forest. This pushes the edge of the rainforest further and back, making it difficult for the forest to reclaim the lost land.

In Southeast Asian countries, like Borneo, much of the rainforest land has been cleared and replaced with palm oil plantations. Palm oil is used as a biofuel and is a major component in snack foods and other processed foods. Over the past century, Borneo has lost over half its forest cover and is currently losing forest at a rate of 1.3 million hectares a year.

Rich Carey/Shutterstock.com

Deforestation in Malaysia.

Rich Carey/Shutterstock.com

A palm oil plantation.

Solutions to the deforestation problem are emerging. Conservation groups are promoting community-based conservation (Chapter 5) giving indigenous people control and ownership over the land. Other organizations raise money to pay the concession fees. Many governments of developing countries do not care if anyone actually exercises the timber rights they purchase, only that they collect the money from the fees. So, groups are buying the timber rights and putting the land into protected, private reserves. Other international organizations are promoting debt-for-nature swaps, where nations can get portions of their debt forgiven in exchange for putting their forests into protected reserves. The *Reducing Emissions from Deforestation and Forest Degradation* (REDD+) program where developed nations can purchase carbon credits to offset their carbon dioxide emissions by paying developing nations to preserve their forests. Other programs are designed to steer palm oil plantations toward already cleared land instead of clearing the remaining forests. Research is also being conducted on developing oils and biofuels from algae that could replace palm oil altogether.

6.3 Forestry and Forest Management

Forestry, the practice of managing forests, often involves a careful balancing act between two competing interests: the demand for timber products vs. the need to maintain forests as viable ecosystems. Timber is a renewable resource, but it is not an inexhaustible one. If the trees are cut down faster than they can regrow, we can lose them as a resource, much like how the inhabitants of Rapa Nui ran out of trees as they cut them all down to make log rollers. Forestry is an example of **resource management**, the practice of managing the harvesting and the use of renewable resources with an eye toward sustainable. Many resource management strategies focus on determining the **maximum sustainable yield**, the largest amount of a resource that can be extracted and still allow for sustainable regrowth.

The goal of maximum sustainable yield is often to maintain a tree population at about half the carrying capacity. This is the population size which promotes the most rapid regrowth. However, this still means a population of trees that is significantly smaller than it would be naturally. This results in a forest with fewer niches and resources for species to exploit. Alternative to maximum sustainable yield is an **ecosystem-based management** approach. In this case, the idea is to manage a forest in a way that mirrors how ecosystems behavior naturally and minimize the impact on the ecosystem and its ability to provide ecosystem services. This means protecting particularly sensitive areas, protecting areas that provide important services, and consider how these impacts affect life on the landscape level.

Forest ecosystems are complex and our understanding of them is imperfect. We don't always know how our activities will affect the entire community. In many cases, ecosystem-based management involves a practice of **adaptive management** or a trial-and-error strategy. Forest managers may try several different strategies and study which ones are more effective than others. Then, they will adjust their practices as they collect and analyze data, doing more of what works while halting practices that do not.

Mariusz Szczygiel/Shutterstock.com

A tree plantation.

While timber harvesting does still occur on federal and state land, most of the timber harvesting in the United States today occurs on private lands. The timber industry has been shifting emphasis toward **tree plantation's** maximum sustainable yield.

Tree plantations often contain **monocultures**, land devoted to growing a single plant species. They also tend to be **even-aged**, in which all the trees are planted at the same time and then harvest at the same time. These stands are allowed to grow for a certain period called the **rotation time** and then they are cut down and replaced with seedlings. This makes tree plantations very different from natural forests that consist of a variety of tree species at different ages. Lacking in biodiversity, even-aged stands have fewer niches and resources that other organisms can make use of. They are also more vulnerable to insect and disease outbreaks. Planting a monoculture is akin to opening up an all-you-can-eat buffet for the pest species that prefer these trees.

To more closely mimic natural forests, some plantations use **uneven-aged stands**, with a mix of ages and species. This produces a system with more structural diversity and resources that can support other species.

8H/Shutterstock.com

Clear-cutting.

The most efficient method of harvesting timber is **clear-cutting**, the removal of all the trees in a given area. This also causes the most ecological damage, mars the beauty of a forest, and is politically unpopular. Clear-cutting increases erosion and degrades many of the microclimates in the area by exposing them to more sunlight than they are adapted to.

Other approaches include:

- **Seed-tree approach:** Some seed-producing trees are left behind to reseed the areas that have been cleared.
- **Shelterwood approach:** Some mature trees are left to shelter the newly planted seedlings.
- **Selection systems:** Maintaining an uneven-aged stand by cutting down trees at differing ages.

Timber that is harvest in a sustainable manner can be certified by the organizations like Forest Stewardship Council (FSC). Sustainable forest certification can be granted if the FSC judges that it was harvested according to their criteria and can then be marketed as sustainable.

One aspect of forest management that has changed in recent decades relates to fire. Originally, forest managers would extinguish all fires as fast as possible. However, we soon learned that fire plays an important role in forest ecosystems. Over time, dead leaves and twigs accumulate on the forest floor. This dead material dries out and increases the **fuel load** of the forest. When there is a period burning on the forest floor, the brush is destroyed, leaving the living trees alone. However, if the fuel load continues to build up year after year, a fire could reach high enough to ignite the tops of the living trees, changing a ground fire into a crown fire. Once this happens, the fire becomes devastating and destroy healthy ecosystems, spread rapidly, and possibly threaten people's homes. Periodic burns also free up nutrients and return them to the soil so that they can be taken up by the living trees. Finally, there are certain species of pine trees whose pinecones form with a waxy coating. These cones cannot open and release their seeds until they've been scorched by flames and the wax melted. These **fire climax systems** need a period of burning in order to complete their reproduction cycle.

As a result of these discoveries, today, forest managers will conduct periodic **prescribed burns** in order to reduce the fuel load of the forest. These low-level fires are set purposefully so the brush can be cleared out in a controlled manner that doesn't threaten private property or people's lives. Wildfires are still controlled if they threaten private property, endanger the public, or threaten the functioning of the ecosystem.

Nashepard/Shutterstock.com

A prescribed burn in South Carolina.

In the Amazon, a recent spate of drier weather has made the forest more vulnerable to devastating fires and the government of Brazil has been criticized for its response to the fires. Many ecologists think the deforestation of the Amazon has contributed to these fire outbreaks. Loss of tree species reduces transpiration and less water is released into the air. This in turn reduces precipitation in nearby areas.

In North America, climate change has resulted in a recent outbreak of the pine bark beetle. Billions of conifer trees have been killed by the beetle, leaving many snags that increase the risk of a wildfire. Recent milder winters in parts of the country have allowed the pine bark beetle to overwinter at a more northern latitude than normal and warmer winters have increased their feeding. Even-aged plantation monocultures are particularly vulnerable to the pine bark beetle.

6.4 Federal and State Land Management

The federal government has several agencies that manage federally owned land. Much of that land is in the western part of the country. This is a result of the settlement pattern of the United States. After the original 13 colonies split from Great Britain and formed the United States, the new Congress pass laws like the General Land Ordinances that federalized the land west of the established states and encouraged people to move westward. Much of the land in the eastern United States fell into private ownership while the federal government-maintained ownership of a lot of land in the west. The major federal land management agencies and the type of land they oversee.

- National Forest Service: Manages the federal forests
- Bureau of Land Management: Manages federal lands that are open for grazing by ranch animals
- National Park Service: Operates the national parks and monuments
- Bureau of Indian Affairs: Oversees the Native American Reservations
- Fish and Wildlife Service: Manages wildlife refuges

The Forest Service sells timber rights to areas of the National Forests. About 2% of the national forest land area ends up harvested each year. The timber rights are sold below market value and as a result, taxpayers are subsidizing the harvest of timber on these lands. In addition, the Forest Service pays for the construction of roads providing accusing to the timber, providing another taxpayer subsidy.

6.5 Parks and Protected Areas

In the nineteenth century, the United States created the first **national parks**. The goal was to preserve parts of the American west for recreation and nature appreciation. Yellowstone National Park was the first park created in 1872. Other parks soon followed including Sequoia, Kings Canyon, Yosemite, Mount Rainier, and Crater Lake. Today, national parks include many famous landscapes like the Grand Canyon as well as famous battlefields like Gettysburg. In addition, the National Park Service also maintains national monuments like the Lincoln Memorial and the Vietnam Veterans Wall in Washington, DC and Independence Hall in Philadelphia.

shaferaphoto/Shutterstock.com

Yosemite National Park.

KateShort/Shutterstock.com

The Gettysburg Battlefield.

Wildlife refuges are established to provide havens for wildlife. These are administered by the U.S. Fish and Wildlife Service. While they are off-limits for development, many recreational activities, such as hunting, fishing, and photography are permitted in wildlife refuges.

In 1964, Congress passed the Wilderness Act and created another category of protected lands: **wilderness areas**. These are land designated within existing federal lands, such as wildlife refuges and national parks. These lands have the strictest protections where only low-impact activities like hiking and photography are permitted.

Robert Mutch/Shutterstock.com

The Klamath National Wildlife Refuge.

In addition to federal lands, each state has land management agencies. Nationwide there are about 7,000 state parks as well as state forests and state game lands. The rules on what activities are permitted on these lands can vary from one state to another. There are also **private land trusts** that are maintained as protected areas by nongovernmental organizations. Examples include Big Sur in California and Jackson Hole in Wyoming.

Travelling Thilo/Shutterstock.com

Big Sur, California.

Worldwide, protected areas and parks have grown dramatically since 1970. Unfortunately, due to lack of resources and funding, many of these are referred to as **paper parks**, in that they only exist on paper. Despite laws meant to protect these lands and the wildlife in them, they are often still plagued by illegal poach, logging, and other extraction activities.

Michal Stipek/Shutterstock.com

Entlebuch, the first biosphere reserve in Switzerland.

The United Nations maintains protected areas in countries like India called **biosphere reserves**. These are lands judged to have exceptional biodiversity and are maintained for the benefit of local residents. Each biosphere reserve is designed around three concentric zones. In the middle is the core zone, which contains the lands in most need of biodiversity protection. The core zone receives the highest level of protection. Around the core zone is the buffer zone, where limited development, ecotourism, and educational programs are permitted. Finally, the outermost zone is the transition zone where local residents practice sustainable agriculture.

The United Nations also designates **world heritage sites**, these are lands that cross national boundaries or have exceptional natural or cultural values. Examples of these include the Sangha Trinational reserve, which encompasses land in Cameroon, Congo, and the Central African Republic and serves as a protective haven for species including chimpanzees and the western lowland gorilla. Other sites designated as world heritage sites include Easter Island (Rapa Nui), the Statue of Liberty, Chichén Itzá, the Grand Canyon, the Everglades National Park, and that Galapagos Islands.

Alison Hancock/Shutterstock.com

Elena Berd/Shutterstock.com

WitR/Shutterstock.com

Stonehenge, Machu Picchu, and the Great Pyramids are examples of world heritage sites.

6.6 Habitat Fragmentation

Expansion of human settlements and development often results in dividing up the landscape into small, unconnected patches. These patches are often too small to support large-bodied predators and other animals that need wide territories. Research on the size and nature of these patches of forests have found that the size and distance from the continuous forest have a big influence on the biodiversity of these "islands" of forests created by fragmentation. As rule, smaller patches have the greatest losses in biodiversity while those closest to the continuous forest are most likely to be recolonized by organisms. Corridors of forestland connecting these patches can also assist in recolonization.

Fragmentation also increases the impact of edge effects. For example, species that are adapted for a dark, lush interior often have trouble adapting to being on the edge where they are exposed to more sunlight and dryer conditions. They are also more vulnerable to predators and diseases that favor open areas.

Global warming can often worsen problems associated with fragmentation. As temperatures rise, species may be forced to shift their ranges up in elevation or further north. This may make current protected areas no longer suitable. Without corridors connecting them to new habitats, these species may not be able to make the transition. Also, species that are already high in elevation or living in the Arctic region may not have any place they can migrate to.

6.7 Discussion Questions

1. Why are old growth or primary forests more diversity than secondary forests?
2. What are concession fees and why do many developing nations need the revenue from them?
3. How as the expansion of palm oil plantations affected biodiversity in Borneo?
4. What is maximum sustainable yield and how is it used to manage timber harvests?
5. How do tree plantations differ from natural forests?
6. Describe the different methods for harvesting timber in plantations such as even-aged stands, seed-tree approach, shelterwood approach, and selection systems.
7. What is the difference between a monoculture and a polyculture?
8. How do prescribed burns reduce the risk of a devastating forest fire?
9. Describe the different federal land management agencies and what kinds of lands they oversee.
10. Describe the difference between national parks, wildlife refuges, and wilderness areas and how each protects against the loss of biodiversity and deforestation.
11. What is a private land trust and how does it differ from public lands?
12. What is a biosphere reserve and how designed?
13. What is a world heritage site?
14. How is global warming expected to worsen the problems of habitat fragmentation?

6.8 Key Terms

- Adaptive management
- Biosphere reserve
- Canopy
- Clear-cutting
- Concession fee
- Deforestation
- Ecosystem-based management
- Even-aged stand
- Fire climax system
- Forest
- Forest floor
- Forestry
- Fuel load
- Maximum sustainable yield
- Monoculture
- National park
- Paper park
- Prescribed burn
- Primary or old-growth forest
- Private land trust
- Resource management
- Rotation time
- Secondary forest
- Seed-tree approach
- Selection system
- Shelterwood approach
- Snag
- Subcanopy
- Tree plantation
- Treefall gap
- Understory
- Uneven-aged Stands
- Wilderness area
- Wildlife refuge
- World heritage site

Chapter 7

AQUATIC ECOSYSTEMS

Chapter Topics

The Great Barrier Reef off the coast of Australia is the largest reef system in the world. Formed from the exoskeletons of millions of corals, it is home to staggering amounts of biodiversity. The many niches in the reef provide opportunities for numerous fish, shellfish, and sea stars to search for food and avoid predators.

In addition, the reef serves to protect the coastlines from erosion due to wave action and tropical storms. It also helps recycle nutrient and provides a source of nitrogen for aquatic food chains. Finally, the reef assists in the fixing of carbon and nitrogen.

However, the Great Barrier Reef is under stress. Climate change leads to a condition called coral bleaching, where the corals turn white and eventually die. Ocean acidification from increased absorption of carbon dioxide also threatens the coral by reducing their shells. About 50% of the coral population has been lost in the past 50 years.

Runoff from the land produces dead zones and outbreaks of harmful species like crown-of-thorns sea stars. Plastic litter and other debris also threaten the aquatic life through choking and entanglement. Finally, illegal fishing and poaching also have a significant impact on reef populations.

Edward Haylan/Shutterstock.com

Aerial photo of the Great Barrier Reef.

7.1 Distribution and Importance of Water

No substance is more critical for life as we know it. It is the medium used to transport nutrients and the solvent that facilitates biochemical reactions that power life. Water is so crucial for life that when astronomers look for any possible signs of life on other worlds, whether on Mars, one of Jupiter's moons, or one of the many exoplanets that have been discovered, the first thing they look for is whether liquid water is possible.

About 71% of the Earth's surface is covered with water. The vast majority of it (97.5%) is **salt water** in the oceans. The 2.5%, that is, **freshwater** is divided between frozen ice caps and glaciers (79%), groundwater (20%), and surface water (1%).

Water is a renewable resource. The hydrological cycle constantly renews water through the processes of evaporation, transpiration, condensation, and precipitation. As water evaporates/transpires into the air, it rises and condenses into clouds. Water falls back to the ground precipitation, such us rain, snow, hail, or sleet. Precipitation then either enters the ground through infiltration or runs along the surface, collecting into streams which merge into rivers. These rivers then carry the water to a lake or the oceans. **Groundwater** also makes its way to the oceans or returns to the surface through **springs**. In addition, ice caps and glaciers melt in the spring, renewing many ground and **surface water** sources. This movement of water through an interconnected web renews water resources.

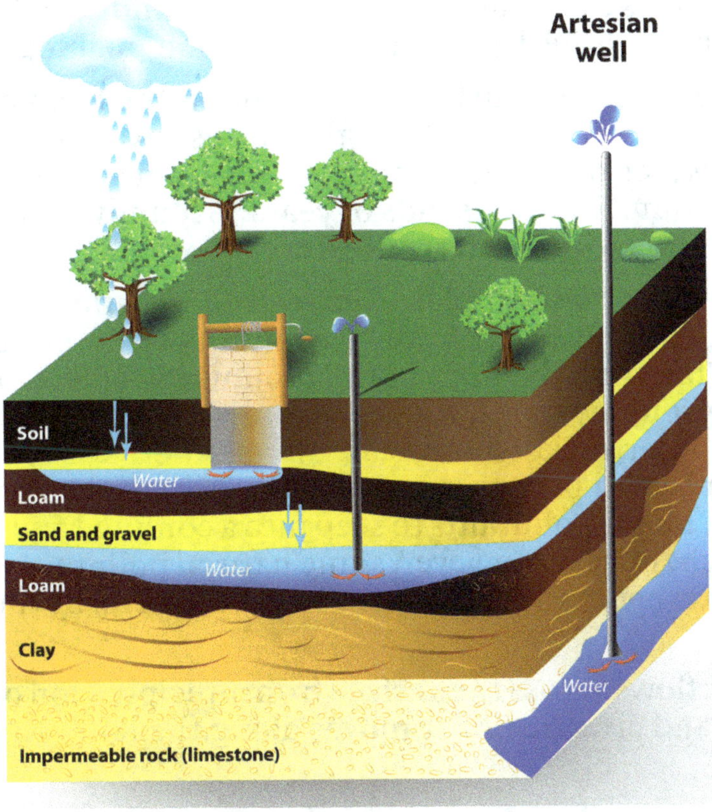

Designua/Shutterstock.com

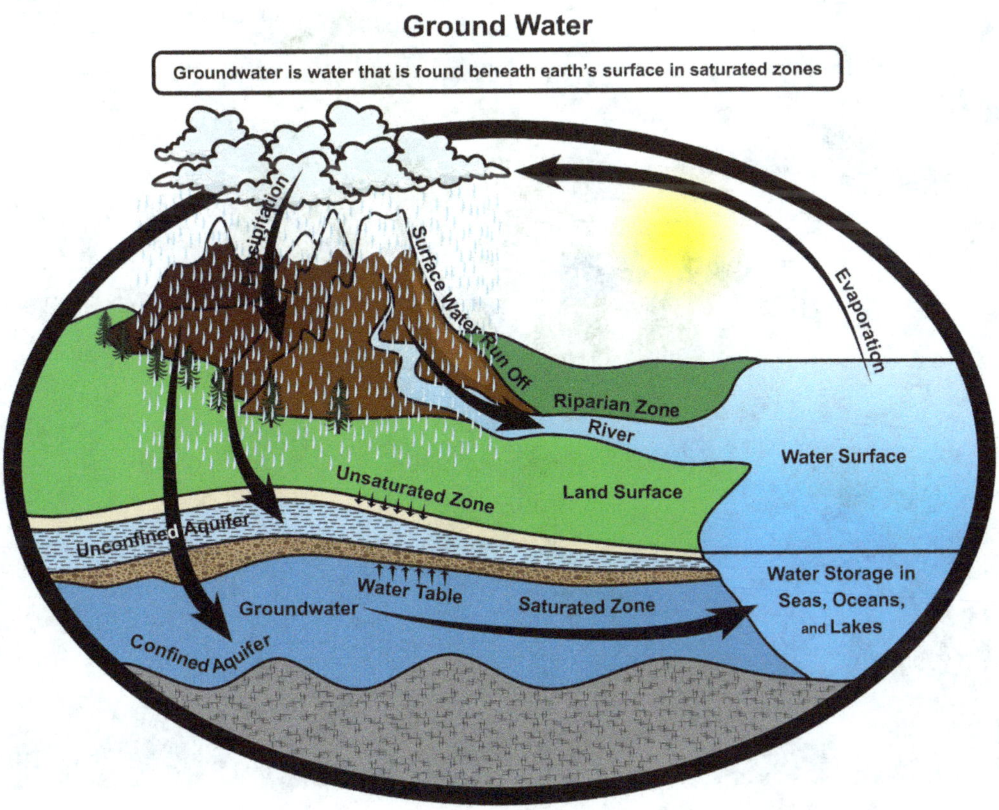

udaix/Shutterstock.com

A cross section of aquifers.

However, water sources can be depleted if they are withdrawn faster than they can be renewed. They can also be degraded with pollution, making them usable.

While surface water is on the surface of the land in the form of rivers and lakes, groundwater flows underground through porous layers of rock or soil called **aquifers**. Groundwater flows much more slowly than surface water, sometimes remaining underground for thousands of years. Groundwater that's been underground that long is sometimes referred to as **fossil water**.

The **water table** marks the boundary between the aquifer and the porous layer about it that is mostly filled with air. Groundwater enters the aquifer at the **recharge zone**. This could be high in the mountains where snowmelt provides the source for the groundwater. A **confined aquifer** is trapped between to impermeable layers of soil or rock, whereas an **unconfined aquifer** sits on top of an impermeable layer with a porous layer above it. Because the upper impermeable layer makes it difficult for pollutants to seep into a confined aquifer, the water may have less harmful chemicals. However, some natural pollutants such as arsenic or sulfides could be present in the rock layer and get taken up by the water in the confined aquifer.

Runoff, water that flows across the surface from precipitation or melting ice, converges into rivers and streams. As it moves along the land, runoff could pick up

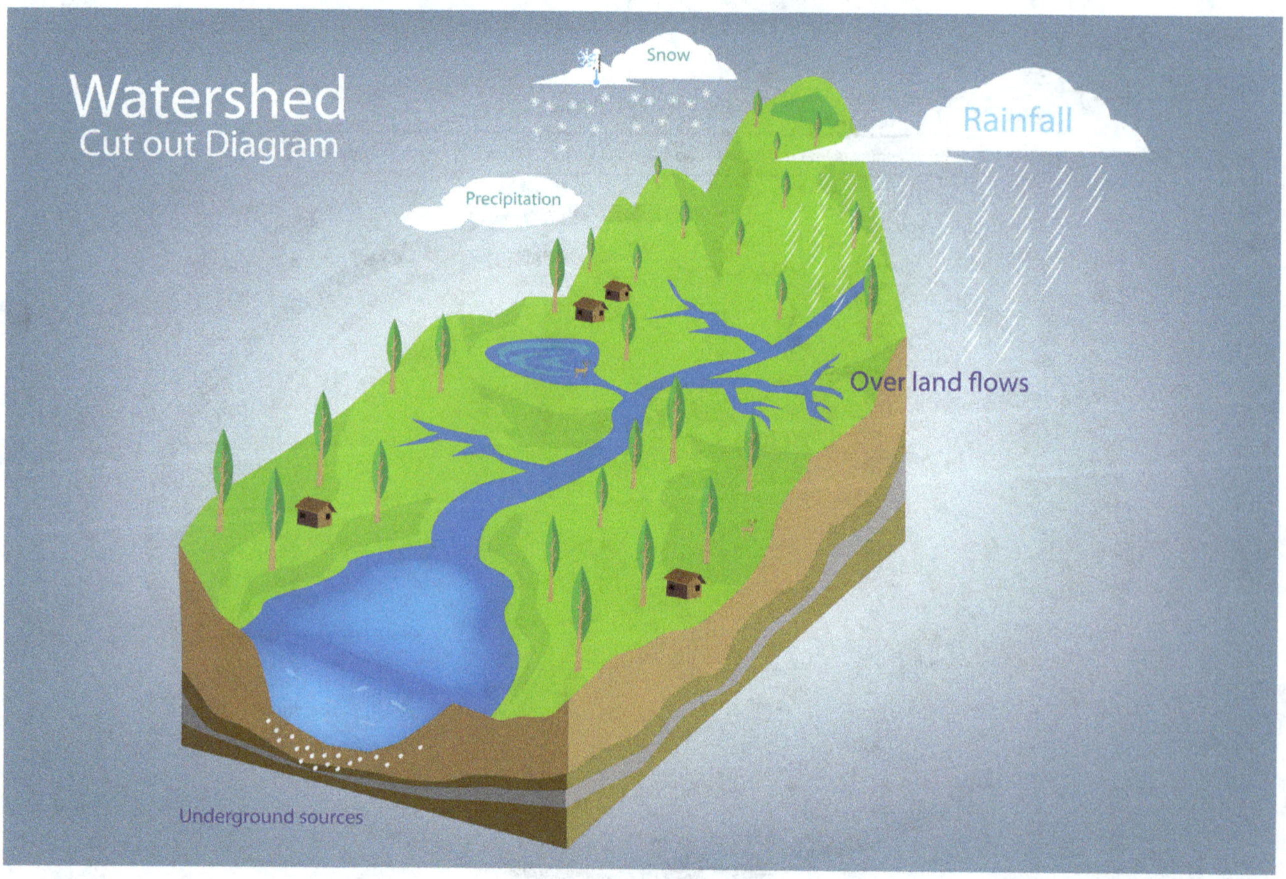

Crystal Eye Studio/Shutterstock.com

A watershed.

a variety of pollutants, such as fertilizers, pesticides, motor oil, or road salt. All of the land area in which the runoff is drained into a single river and its tributaries is called a **watershed**. As rivers flow, they erode and shape the landscape, carving valleys and canyons. Many rivers periodically flood their banks and when they do, they deposit fresh sediments onto the **floodplain**. Annual floodplains are areas of land that get flooded every year. For example, before the Aswan Dam was built, the Nile River would flood its banks every spring, leaving nutrient-rich sediments behind. These sediments made it possible for the Egyptians to grow crops in an otherwise arid land. Other floodplains may be expected to flood less frequently, only once a decade or century on average. Sediments left behind in a floodplain can also support **riparian forests** along the river banks, making them areas of high productivity.

FMPortella/Shutterstock.com

A riparian forest along the shores of a river in Canada.

7.2 Freshwater Aquatic Ecosystems

Bodies of standing water, **lakes and ponds**, can be divided into different zones. For example, the **littoral zone** makes up the shallow edge around the lake. The littoral zone supports aquatic plants such as cattails and lily pads. These plants provide habitats and food sources for insects, which in turn attract predators such as frogs and birds. This makes littoral zones of high productivity. The area along the bottom of the lake, the **benthic zone**, is inhabited by scavengers like catfish and crayfish. The **limnetic zone** is the area of the open water that receives enough sunlight for microscopic organisms called phytoplankton to

Valerii_M/Shutterstock.com

A lake.

perform photosynthesis. The phytoplankton add dissolved oxygen to the water and form the basis of a food chain. In between the limnetic zone and the benthic zone is the **profundal zone**, which does not receive enough sunlight for photosynthesis.

As rivers carry nutrients into lakes and ponds, their biochemistry can change. An **oligotrophic lake** has high levels of dissolved oxygen and low concentrations of nutrients. Over time, it may become a **eutrophic lake**, which has low oxygen content and high concentrations of nutrients. A eutrophic lake may have fewer predatory fish and more scavengers and may be dominated by plants and algae. If this process continues, the lake may eventually fill in completely with sediments and disappear.

Any area with shallow standing water is called a **wetland**. Although the term swamp is often used interchangeably with wetland, a **swamp** is specifically a wetland in a forest ecosystem. An open wetland or **freshwater marsh** has lots of plants such as cattails and bulrushes. Ponds that are covered with matts of thick floating plants are called **bogs**. As bogs fill in, the plants may start to decompose into peat, a material that is often used as a fuel source of heating and cooking in many parts of the world. Some wetlands, such as **vernal pools**, are seasonal, filling with water in the spring rains and drying out in the summer months.

Wetlands perform many ecosystem surfaces. They filter pollutants and absorb nutrients, thus reducing eutrophication in coast areas. They also slow runoff and reduce flooding. Many wetlands have been impacted by humans. Many have been drained for agriculture or for human settlements. Washington, DC, and New Orleans, LA, are two cities that have been built upon drained wetlands. The land New Orleans sits on, however, is still very soft and spongy and is slowly sinking, making the oldest parts of the city below sea level. This makes flooding in New Orleans after events like Hurricane Katrina worse as the water rushes into the "bowl" formed by the sinking land. Other wetlands have lost water by being diverted for irrigation or dried out by dams that prevent the water from reaching it.

Abbie Warnock-Matthews/Shutterstock.com

A swampland.

Wirestock Images/Shutterstock.com

A freshwater marsh.

Luis Jurado/Shutterstock.com

Vernal pools.

Lucas T. Jahn/Shutterstock.com

Bogs.

7.3 Coastal and Marine Ecosystems

While there are four named oceans (Atlantic, Pacific, Indian, and Arctic), they form one continuous body covering the majority of the Earth's surface. The oceans consist of 96.5% water and the remaining percentage is made up of salt and other minerals. These salts came from the land as rivers carried them into the oceans. As water evaporated, the salts remained, raising the salinity over millions of years. The waters of the oceans constantly flow in a series of currents, riverlike flows of water driving by temperature differences, wind, gravity, and the outlines of the continents.

One important current is the North Atlantic Current, which begins with the Gulf Stream in the Gulf of Mexico and brings warm water across the eastern seaboard of North America and then crosses the Atlantic toward Europe. The warm waters of the North Atlantic Current influence the climate of Europe, giving a much warmer and milder climate than it would have based on its latitude. Without the North Atlantic Current, Western Europe would be much colder than it is today.

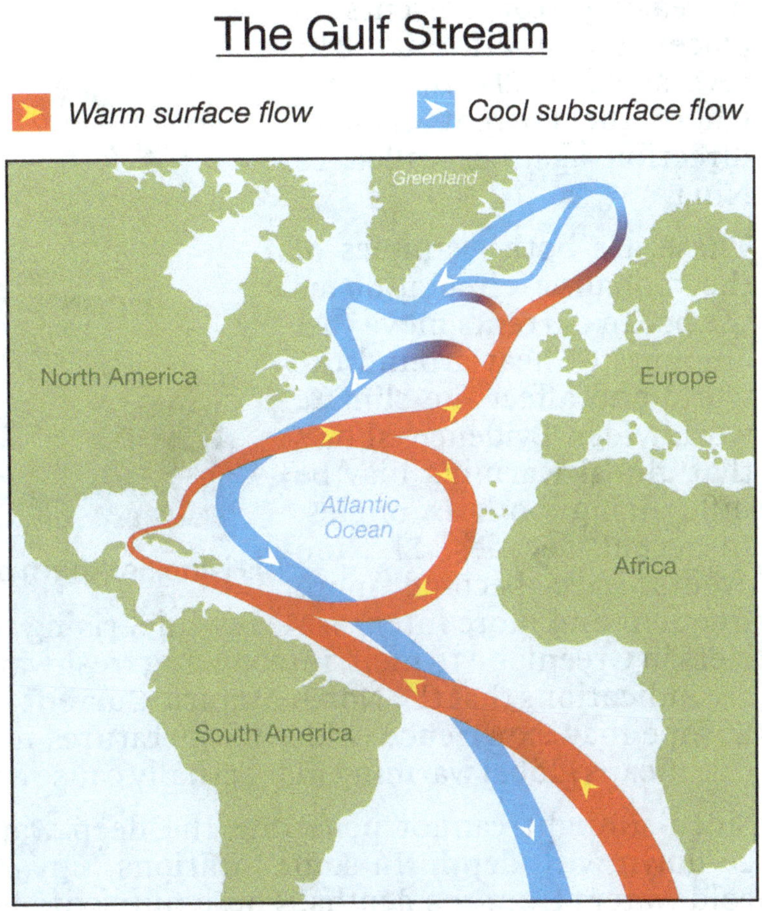

Peter Hermes Furian/Shutterstock.com

The Gulf Stream and North Atlantic Current.

The Equatorial Current across the Pacific Ocean also influences climate. A cold water upwelling off the coast of Peru brings nutrients to the surface, forming the basis of an ecosystem and a highly productive fishery. Prevailing winds drive a warm water current that moves across the Pacific, causing warming temperatures in the Malay Archipelago. This warm water evaporates and then returns as precipitation, giving these island very rainy conditions.

However, every 2 to 10 years, a drop in air pressure causes the current to reverse in what is call an **El Nino** event. This shuts down the upwelling on the west coast of South America and the fish go away. The warm water around the islands in the South Pacific shifts to the middle of the Pacific Ocean and the rains move with them. El Nino events thus can cause drier conditions in Indonesia while increasing storm events in places like California. The reverse of an El Nino event, where the current changes direction again is called La Nina.

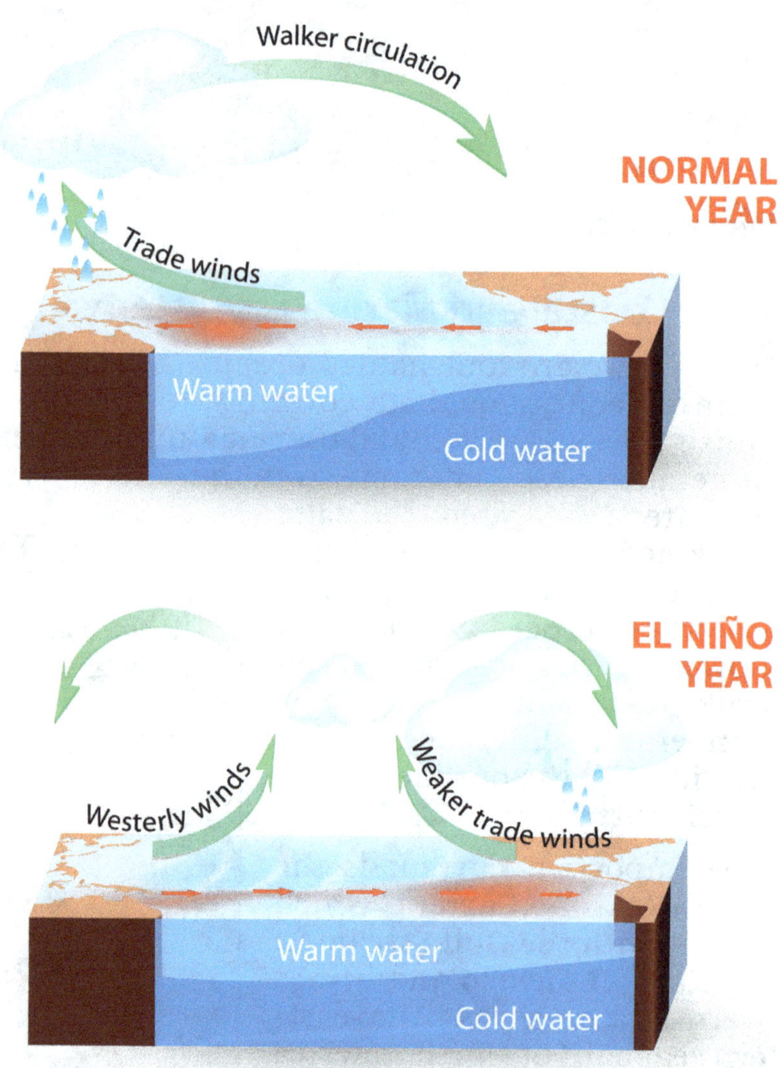

Designua/Shutterstock.com

El Nino Southern oscillation.

These are both examples of thermohaline circulation as the ocean currents move vast amounts of heat around the world and affect the climate worldwide. Evidence shows that global warming may be influencing both examples. In recent decades, El Nino events have become more frequent and more intense. Meanwhile, rising temperatures are causing the glaciers in Greenland to melt, introducing freshwater into the North Atlantic. There are indications that the North Atlantic Current is slowing down. If this continues, Europe may experience colder temperatures in the next century. Paradoxically, this means global warming may actually cause a colder climate in this region.

Since sunlight cannot penetrate the deep waters of the ocean, temperatures go down with depth. In some locations, **upwellings** bring deep, nutrient-rich cold waters from the depths. These nutrients feed phytoplankton that form the basis of productive fisheries like the one off the coast of Peru. In other locations,

Upwellings.

downwellings transport warm, oxygen-rich water to the depths, providing dissolved oxygen for benthic species in the ocean depths. Downwellings also carry carbon dioxide to the seafloor, which form carbonate minerals with the sediments. These minerals are eventually incorporated into the Earth's crust, making the oceans an important sink for atmospheric carbon dioxide.

The floor of the oceans has many rugged features. The **continental shelf** forms areas of relatively shallow water along the coastlines. The shelves drop-off steeply at the shelf-slope break and the continental slope angles down toward the ocean depths. In addition, the oceans have underwater ridges formed by the separation of tectonic plates, such as the mid-Atlantic Ridge, the longest mountain range in the world. In other locations, one tectonic plate is sliding underneath another, forming deep undersea trenches like the Marianas Trench. Often, at these subduction zones, magma is forced up through the crust, forming chains of volcanic islands like Japan and the Philippines.

In terms of marine and coastal ecosystems, these regions vary greatly in productivity and the kinds of species they support. Unlike terrestrial biomes, which are defined by the predominate plant species, marine ecosystems are defined by the animal species they support. Conditions in marine ecosystems vary based on characteristics such as temperature, salinity, topography, sunlight, and available nutrients.

Every day, the gravitational pull of the moon and (to a lesser extent) the sun causes the sea level to rise and fall as the **tides**. **Intertidal zones** span the area of land between the high-tide and low-tide levels. As a result, they are filled with species

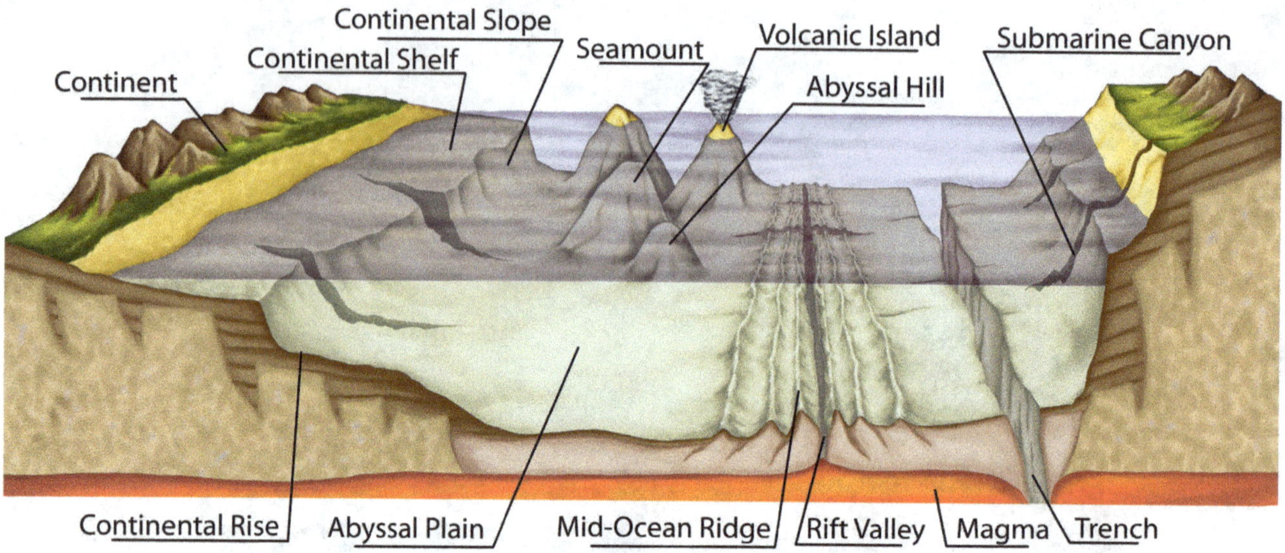

stihii/Shutterstock.com

The Ocean floor.

adapted to spending part of their time above the water and part below their water. During low tide, they are exposed to sunlight while at high tide, they are submerged. In particular, rocky intertidal zones have high levels of biodiversity due to variations in temperature, salinity, and moisture.

Denise Lett/Shutterstock.com

Intertidal zone in California.

Estuaries include bays and other bodies where freshwater from rivers mixes with the salt water of the oceans. Often, the topography of the land shield estuaries from the heavy waves of the ocean, providing relatively calm waters. This enables them to provide habitats for plants, which in turn form the basis of a food chain for shellfish, fish, and shorebirds.

Salt Marshes are wetlands along the coast where the tides flow over a silty substrate and are called **salt marshes**. As the tides rise and fall, water flows in and out of tidal creeks that form channels in these marshes. Like freshwater wetlands, salt marshes provide habitats for many species and filter pollutants

Delmas Lehman/Shutterstock.com

The Chesapeake Bay at sunset, the largest estuary in the United States.

such as nutrient runoffs. They also help reduce erosion online the coastlines.

In tropical coastal communities, forests of mangrove trees grow. **Mangroves** have some roots that grow up to absorb oxygen and others that go down to home in the sediments. These roots give the appearance that trees are standing on stilts. Many aquatic organisms including fish, shellfish, and snakes live among the roots, while birds nest and feed among the trees. Mangroves also protect the shoreline from erosion and the damage from heavy surf and tropical storms.

Kelp forests grow near the coasts of temperate climates. Kelp is a type of brown algae that growth from the sediments into long stalks that can be up to 300 feet tall. Kelp forests protect the shoreline from heavy surf. They also provide habitats and food for species such as sea urchins. Today, kelp forests are under stress as the kelp is harvested for food. Kelp is also used to strengthen paper and as a thickener in cosmetics.

Over thousands of years, coral can build up masses forming **coral reefs**. While they are related to jellies, coral do not float freely in the water. Instead, they root themselves in the sea floor and secret shells made of calcium carbonate, using their tentacles to capture food. Reefs made from generation after generation of coral skeletons can be extensions of a shoreline or as a barrier reef that runs parallel to

John Wesolowski/Shutterstock.com

A salt marsh.

Oliver Osvald/Shutterstock.com

A mangrove forest.

Ethan Daniels/Shutterstock.com

A kelp forest.

the shoreline. Sometimes, they can form an atoll, a ring around a submerged island. Colonies of coral can form reefs that provide numerous niches for other species. Many reefs have tremendous levels of biodiversity as many species use reefs as places to look for food or hide from predators.

Coral has a symbiotic relationship with a species of algae called zooxanthellae, which help them absorb nutrients. Many corals, however, have become afflicted with a condition called "**coral bleaching**," where the zooxanthellae turn white, weaken, and eventually die. This deprives coral of many of the nutrients they need. The causes of coral bleaching include rising ocean temperatures and certain pollutants. In addition, ocean acidification, caused by the absorption of carbon dioxide from atmosphere, can reduce the calcium carbonate, inhibiting the corals' ability to survive.

In the open ocean, sunlight can only penetrate the uppermost layer called the **photic zone**. There, microscopic organisms called **phytoplankton** perform photosynthesis. These in turn are eaten by microscopic animals called **zooplankton**. Both the phytoplankton and zooplankton are fed upon by fish and baleen whales.

In the **pelagic zone**, the region between the photic zone, and the ocean floor, species include scavengers that feed on detritus that sinks from above. In addition, predators like angler fish that use bioluminescent lures

Irina Markova/Shutterstock.com

A coral reef.

Sabangvideo/Shutterstock.com

Coral bleaching.

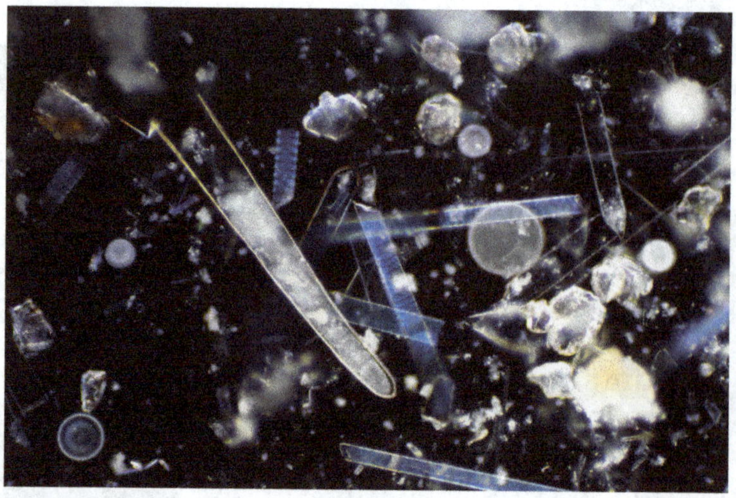

Choksawatdikorn/Shutterstock.com

Planktons are organisms that just drift in the currents.

to capture prey live in the dark, deep waters of the ocean. **Benthic habitats** are found on the ocean floor. Even though sunlight cannot reach the ocean floor, hydrothermal vents or "smokers" release heat from the Earth's mantle. Instead of photosynthesis, species of bacteria perform chemosynthesis and use hydrogen sulfide to build sugars. These form the basis of ecosystems that feed organisms such as tube worms.

7.4 Ocean Fisheries and Marine Protected Areas

Like all other harvesting industries, fishing has become a high-tech business, using sonar, GPS, and other technologies to locate, track, and harvest schools of fish. Despite these technologies, the total catch of wild species has actually leveled off since 1998. We are not catching any more fish than we did 20 years ago, despite increased effort. Ships are traveling further distances and staying at sea longer to catch the same amount of fish. Aquaculture has made up the difference between the current catch level and the rising demand for fish, shrimp, and other marine species. Some fish stocks, like the cod off the coast of Canada and New England, have already collapsed. Others may collapse in the coming decades based on current rates of capture.

Industrialized fishing uses a variety of techniques to maximize harvest:

- **Purse seining** uses a bowl-shaped net to surround entire schools of fish and capture them.
- **Drift netting** uses long nets supported by floats to capture passing fish.
- **Longline fishing** is similar to fishing with a rod and reel, except instead of using a single baited hook, it uses a long, floated line with hundreds or even thousands of baited hooks.
- **Trawling** drags a weighted net through the pelagic zone or along the sea floor to capture fish or shrimp.

All of these methods have a problem with **bycatch**, the capture of nontarget species. Purse seining and drift netting capture species such as dolphins, sea turtles, and seals that get caught in the nets and drowning. Longline fishing harms sharks, seagulls, turtles, albatrosses, and other species that go after the baited hooks. Bottom trawling for benthic species is particularly damaging, destroying entire benthic committees.

Recent studies indicate that the oceans contain only one-tenth of the large-bodied species of fish and sharks as they did a century ago. In many cases, fishing fleets are capturing smaller fish that would have been rejected decades ago as the big fish are no longer present. In addition, many are "fishing down the food chain" by harvesting the prey species of the fish that were once harvested.

Fishing managers have employed many techniques to study and protect fish harvests. These include population surveys and catch monitoring. Often, fisheries are managed according to the principles of maximum sustainable yield. Despite these efforts, many fish populations continue to decline. As a result, many managers are switching to more ecosystem-based management strategies.

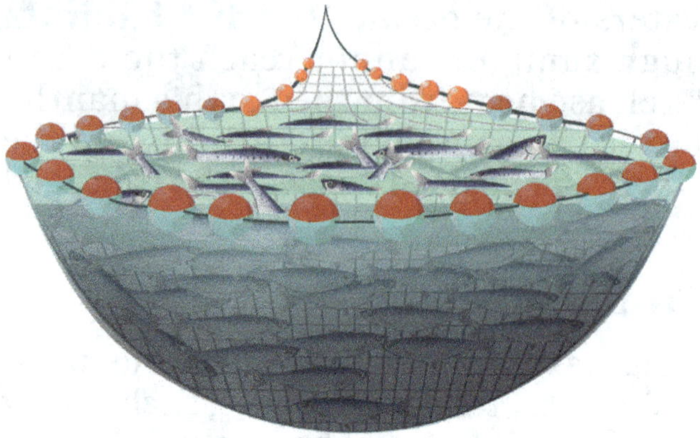

Purse seining.

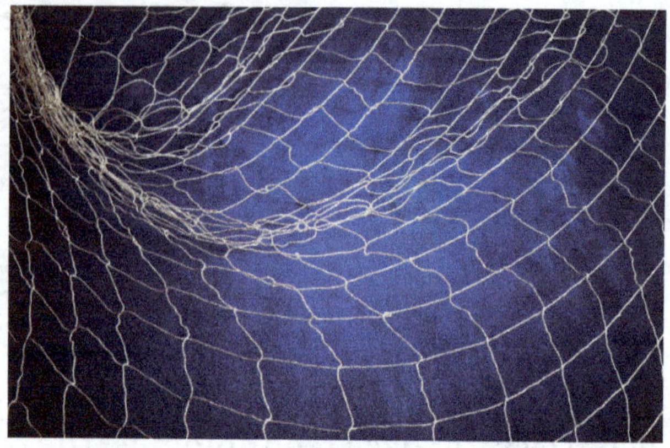

Drift nets.

Longlines.

Trawler net.

Some areas of the oceans have been established as **marine-protected areas** (MPAs). In MPAs, certain activities, such as oil drilling are restricted to protect the fish populations. Fishing and other harvesting practices are permitted in MPAs, however. On the other hand, fishing is entirely prohibited in **marine reserves**. While many commercial and recreational fishers do not like marine reserves, they have been demonstrated to help fish populations recover. Studies have shown that marine reserves result in increases in species density, total biomass, and the number of organisms in the reserves. In addition, as fish in marine reserves grow, they spread out into other areas, resulting in increases in both species number and size in areas where fishing is permitted. Marine reserves are therefore said to "seed the seas" by providing a protected area where fish can breed and grow to maturity before being captured.

7.5 Whaling

Historically, whales have been hunted for their meat as well as for their oil to fuel lamps. The discovery of kerosene (and later, electric lighting) made whale oil obsolete. Today, nations practice commercial **whaling** for the meat.

In 1946, 10 nations gathered in Washington, DC, and signed the International Convention for the Regulation of Whaling (ICRW). This treaty regulates the practice of whaling through the International Whaling Commission (IWC). Today, 89 nations are signatories to the ICRW.

As with other harvesting activities, whaling has gone high-tech, using high-power harpoon cannons and whaling ships that have bring the whale carcass on board and process it before it returns to harbor.

Due to declines in whale populations, the IWC adopted a near-moratorium of whaling in 1982. The moratorium has two major exemptions. For one, indigenous people in Greenland, Canada, and Alaska are permitted to hunt whales using traditional methods. The other exception is whaling for scientific purposes, such as studying its stomach contents to learn about its effect on other species. Two countries, Japan and Iceland, have continued commercial whaling under the guise of scientific studies. In addition, though Norway is a signatory to the ICRW, it objects to the moratorium and continues to engage in commercial whaling. The Russian Federation also objects to the moratorium, but has not resumed commercial whaling. In 2019, Japan withdrew from the IWC and has resumed commercial whaling.

Hunting has devastated many whale species, three of which were driven nearly to extinction. The largest mammal in the world, the blue whale, has been reduced to between 10,000 and 25,000 in population and is listed as endangered. The humpback whale's population had been hunted down to about 1,5000, although since the 1982 moratorium, it has rebounded to about 15,000. The most impacted whale species, however, is the North American right whale. Studies indicate that there may be less than 500 of these whales left. Today, the primary species of whales that are hunted include minke whales, belugas, narwhals, and pilot whales, which are some of the smallest species of whales. In addition, other species of whales killed include gray whales, sei whales, fin whales, bowhead whales, Bryde's whales, sperm whales, and humpback whales.

wdeon/Shutterstock.com

A Japanese whaling ship.

wdeon/Shutterstock.com

A whaling harpoon.

LouieLea/Shutterstock.com

Freshly caught minke whale for sale at a Butcher in Nuuk, Greenland. This is a seasonal delicacy, strictly regulated by a government quota system.

Nico Faramaz/Shutterstock.com

Humpback whale rising out of the water.

7.6 Discussion Questions

1. How is climate change and water pollution putting the Great Barrier Reef and other coral reefs at risk?
2. Describe the components of the hydrologic cycle and how it serves to renew water resources.
3. What is the difference between a confined aquifer and an unconfined aquifer?
4. How does annual flooding affect the productivity of riparian communities and farmland along the river banks?
5. Describe the different zones of a lake and what kinds of life are found in each one?
6. What is the difference between an oligotrophic lake and a eutrophic lake?
7. What are the ecosystem services provided by wetlands?
8. How does the North Atlantic Current affect the climate of Western Europe?
9. What is an El Nino event and how does it affect climate?
10. Where does the energy for ecosystems on the ocean floor come from and how do organisms store it in the form of sugars?
11. Describe the four methods of industrial fishing and the environmental impacts of each.
12. How do marine reserves "seed the seas," both within the protected areas and in areas where fishing is permitted?
13. What were the exceptions to the 1982 near-moratorium on whaling and what countries have continued whaling since then?

7.7 Key Terms

Aquifer
Benthic habitats
Benthic zone
Bog
Bycatch
Confined aquifer
Continental shelf
Coral bleaching
Coral reef
Downwellings
Drift netting
El Nino
Estuary
Eutrophic lake
Floodplain
Fossil water
Freshwater
Freshwater marsh
Groundwater
Intertidal zone
Kelp forest
Lakes and ponds
Limnetic zone
Littoral zone
Longline fishing
Mangroves
Marine-protected areas
Marine reserves
Oligotrophic lake
Pelagic zone
Photic zone
Phytoplankton
Profundal zone
Purse seining
Recharge zone
Riparian forest
Runoff
Salt marsh
Salt water
Spring
Surface water
Swamp
Tides
Trawling
Upwellings
Vernal pool
Water table
Watershed
Wetland
Whaling
Unconfined aquifer
Zooplankton

Chapter 8

ECONOMICS AND ENVIRONMENTAL POLICY

Chapter Topics

In 2015, a pediatrician in Flint, MI noticed a problem with a number of her patients. After a few tests, she determined that they had elevated lead blood levels. Lead is a toxic heavy metal that damages the central nervous system. The effects of lead poisoning can include permanent learning disabilities, hyperactivity, and memory loss.

Flint, a majority African American community with 40% of its residents below the poverty level, is a small city near Detroit. In addition, the elevated lead levels, residents complained about the quality of their drinking water, stating it was cloudy and brown. Eventually, other health problems began to plague the residents, including outbreaks of *Escherichia coli* and Legionnaire's disease.

A year earlier, the city's emergency manager had made the decision to switch the city's water supply from the Detroit water system to the Flint River. It was sold as a cost-savings measure. Experts warned that the water in the Flint River had serious problems, including a low pH. Despite these warnings, the city did not treat the water to raise the pH. Even though lead solder had been banned in plumbing by the early 1980s, Flint, like many communities with an aging infrastructure, still had a lot of plumbing that predated the ban. This low pH, caustic water began eroding the lead solder in the piping, resulting in elevated lead levels in the drinking water.

At first, city and state officials denied there was a problem with the water. Mean-while, General Motors stopped using water from the Flint River in its manufacturing, citing concerns about its corrosivity. Eventually, studies confirmed the residents' fears. The water in Flint had elevated lead levels. By January 2016, both Governor Snyder and President Obama declared a state of emergency in Flint. The residents were told not to drink their tap water and instead, they received shipments of bottled water for drinking and cooking.

Linda Parton/Shutterstock.com

FLINT, MICHIGAN January 23, 2016: Bottled Water Distribution by National Guard at Fire Station 6, in Downtown Flint.

Meanwhile, Flint switched back to the Detroit water supply, but the damage to the city's plumbing was already done. The state and the federal government have spent millions of dollars replacing the old plumbing with new pipes that do not use lead solder and compensating the residents of Flint. Several state and local officials were ultimately charged with violating the state's Safe Drinking Water Act and negligence of duty.

As of 2019, workers are still replacing the city's plumbing and many residents are still dependent on bottled water. The tragedy of Flint is that it was completely avoidable. The city and state officials could have avoided this by not switching to the Flint River or by adequately treating it. What was supposed to save the city money ultimately cost much more to fix the problem, in addition to the health effects it inflicted on the residents of Flint. It's an illustration of the importance of using good science and long-term thinking in making policy decisions that can affect the quality of life of citizens.

Linda Parton/Shutterstock.com

The Flint River in downtown Flint.

8.1 Economic Principles and the Environment

People convert limited resources into goods and services that they can exchange with each other through the social system of an **economy**. A **good** is defined as any material product while a **service** is any work performed on behalf of another. We study the means by which people exchange goods and services through the academic discipline of **economics**.

While economic systems vary greatly from country to country and time to time, there are three main categories:

- **Barter economy**: This is the oldest kind of economic system in which people directly exchange goods and services directly, without the use of money. If I have chickens and you have cows, we can agree to exchange some of the eggs produced by my chickens for some of the milk produced by your cows. Barter economies, however, has a difficulty in determining what is a fair exchange. How does one decide how many eggs a gallon of milk is worth?

Barter.

Nik Symkin/Shutterstock.com

- **Market-based economy**: In this kind of economy, buyers and sellers use money to exchange goods and services. In a market-based economy, money is used a medium of exchange. The value of a particular good or service is determined by what the buyer is willing to pay and the supply, the amount of the goods produced, will be determined by the demand, the number of people who want to buy it.
- **Centrally planned economy**: In this kind of economy, a central authority such as a government, makes decisions on how much of a good or service is produced and how many resources should be allocated toward providing it.

While most economies throughout the world reflect market-based principles to some degree or another, the reality is, there are not many which could be considered purely market-based. Likewise, there are very few that could be considered purely centrally planned, such as North Korea. Most economies, including the United States, are mixed economies, combing both market-based principles and **central planning**. For example, while virtually all manufacturing in the United States is market-based, services such as public education, police and fire protection, and national defense are established through central planning.

All economies interact with the environment by making use of natural resources and then discharging waste back into the environment. However, most economic theories do not take these interactions into account. Instead, they focus only on the human element. Businesses produce goods and services by hiring and paying workers. The workers in turn spend their wages purchasing goods and services from the companies that produce them. The source of natural resources and whether or not they are renewable is often overlooked, as are the many ecosystem services that these businesses depend on.

The Scottish philosopher Adam Smith is regarded as the father of modern economic theory. His theory, now called **classical economics**, states that if individuals are free to pursue their economic self-interest, society will benefit through competition. The market is guided by what he described as the "invisible hand," which corrects things to eliminate goods and services that are shoddy or too expensive. For example, Apple revolutionized the cell phone industry with the introduction of the iPhone. The company developed this product because they were motivated by a desire to make a profit. In response, other companies got into the smartphone business and Google developed their Android operating system for use by these companies in their phones. Under the theories of classical economics, the competition between Apple and Google will motivate them to improve their respective operating systems to increase their quality and usability, while keeping the costs down. If either company produced a product of low quality or sold it at a price higher than their customers are willing to pay, they would not be able to sell it. Eventually, they will either withdraw the product or go out business. To avoid that, they will strive to produce products that are better and cheaper than their competition.

Matt Ledwinka/Shutterstock.com

Statue of Adam Smith in Edinburgh in front of St. Giles Cathedral at Parliament Square.

Neoclassical economics is a refinement of classical economics, which digs deeper into the interactions between buyers and sellers. As a rule, sellers want to sell their product at as high a price as possible so that they can make more money. On the other hand, buyers want to pay as little as possible. Somewhere in the middle, they reach a compromise at the "correct" price. In addition, the amount of products produced, the supply, is balanced by the amount people wish to buy, the demand. As demand rises, producers will step in to increase the supply. If the demand drops, producers will curtail production and produce less of the product.

Economic theory also includes a **cost–benefit analysis**, in which people weigh the cost associated with an action against its benefits. If the benefits outweigh the costs, people will generally pursue the action. For example, a person may wish to purchase a Ferrari, but unless they have enough income to make the payments, they unlikely to be able to buy one. When it comes to environmental policy, however, a cost–benefit analysis can be difficult to perform. Costs associated with policies may be easy to quantify; however, it can be much harder to quantify the benefits. For example, when considering implementing a new program for reducing air pollution emissions, a company can easily calculate the cost of purchasing and installing the new equipment. But how does one put a dollar value on cleaner air? One could conduct epidemiological surveys and how the cleaner air might reduce the prevalence of certain health conditions, but that leaves a lot of uncertainty as every case is unique. This uncertainty can make building a case in favor of such a policy.

Olivier Le Moal/Shutterstock.com

Weighing costs and benefits.

Standard economic models often assume that natural resources are inexhaustible. However, we know that this is not the case. Fossil fuels, for example, are nonrenewable and it's only a question of when, not if, they will run out. We've also seen that fish stocks, pollinators, and timber products can be depleted if not managed properly. Even inexhaustible resources like water can be degraded with pollutants.

Economic models also only take into account internal costs. **External costs**, those borne by someone other than the buyer or the seller, are discounted. For example, let's say a power plant pollutes the local groundwater which is used as a source of drinking water for a local community. This causes the residents to experience certain health problems. These ailments would be an external cost, as these people would have to deal with them even though they do not gain the benefits of the economic activity of the power plant. In addition to health impacts, external costs can also include aesthetic damage, loss of property values, and depletion of resources that others depend on. The 2010 Deepwater Horizon imposed all of these costs on people in the Gulf of Mexico whose livelihood depending on the tourism and fishing industry by contaminating the water and beaches with oil.

This can also result in the **free-rider** effect, in which one party gains the benefit of an action without incurring the same costs as others. For example, a government agency might put restrictions on commercial fishing in a given area in order to help the fish stocks recover. Most of the commercial fishers abide by these restrictions, however, one refuses to. They gain the benefit of an improved fish population without making the sacrifice that others are making.

Another issue that standard economic models such as neoclassical economics often overlook is the long-term impacts of environmental impacts such as pollution. Many environmental problems take years to become apparent, whereas most businesses are focused primarily on short-term profits. This can lead to **discounting** of these impacts in favor of immediate profits. This is especially true for issues like ozone depletion and global warming, which can take decades or even centuries to unfold. Therefore, motivating people to make short-term sacrifices or changes in behavior in order to avoid problems that may not even affect them in their lifetime can be difficult.

Neoclassical economics also always assumes that all growth is good. However, just as an exponential growth rate can overgraze ecosystem resources, too much rapid growth can overtax natural resources. If the demand for timber or fish exceeds the ability to produce these resources, it could result in a market failure, where the market collapses due to lack of sustainability. Think of the story of the Lorax again, where the Once-ler harvested trees recklessly by expanding his business faster than the trees could go, demonstrating that this excessive growth is an unsustainable business model.

Because of these issues, many people advocate for adopting a more environmental economic model, where issues like external costs and discounts are taken into account. For example, regulations that require polluters to reduce their emissions are designed to internalize the external costs by making them just part of the cost of doing business. A more radical step might be to abandon the growth model altogether and go to a steady-state economy, the neither grows nor shrinks, but maintains a dynamic equilibrium much like populations in an ecosystem do. Advocates of the steady-state model argue that quality of life can still improve through advances in technology and efficiency, even if it does not continuously produce more overall wealth.

8.2 The Value of Ecosystem Services

How does one put a value on ecosystem services and biodiversity? Efforts to protect endangered species and wildlife preserves still enjoy popular public support, so clearly, people do value these things, even if they do not receive a direct benefit from them. There are many ways in which such **nonmarket values** can be defined.

Dudarev Mikhail/Shutterstock.com

Use value.

For example, we can apply a **use value** in which people place a value on something because they can directly use it. Diving in the Great Barrier Reef, rock climbing, and whale watching are all examples of use values.

Lara Zanarini/Shutterstock.com

Existence value.

In some cases, we can apply an **existence value** or intrinsic value but stating that an ecosystem or species adds value to the world simply by its existence. The bald eagle is a majestic bird, so majestic that the Founding Fathers of the United States adopted it as the symbol of the country, even though Benjamin Franklin wanted the turkey instead. The African elephant, the Siberian tiger, the giant panda, and the black rhino are similarly magnificent creatures. Even though they may never gain an economic benefit from these creatures, many people feel that the world would be a less wondrous place without them in it. So, support for protecting these endangered species remains strong.

Lillac/Shutterstock.com

Option value.

Then there are **option values**, in which decide not to extract a resource so that we may make use of them in the future. In other words, we leave some trees in place so that we will have trees (and timber products) in the future.

Alexey Suloev/Shutterstock.com

Aesthetic value.

Aesthetic values are values we place because certain ecosystems are unique and awe-inspiring. People would strongly object to a proposal to strip mine in the Grand Canyon or clear-cutting Yosemite National Park. We put the beauty and natural wonder of these places ahead of what resources could be extracted from them.

Some places could be preserved for the **scientific value**, so that we can learn more about life and how it interacts with the world from them. Likewise, we would want to preserve some ecosystem for their **educational value** for use as teaching tools.

Sathachak/Shutterstock.com

Scientific value.

Finally, we might want to preserve certain places for their historic or **cultural value.** Such locations might be preserved because they are part of our shared heritage as humanity. The National Park Service preserves many key battlefields from the Civil

War such as Antietam, Gettysburg, and Vicksburg. Archeological sites like Rapa Nui and the Lascaux cave paintings also fall into this category as they tell us something about the people who once lived in these places and left their mark there.

Bob Pool/Shutterstock.com

Cultural value.

There have been studies that are trying to put an actual monetary value on ecosystem services. One such study was conducted in 2014 and examined the following 17 ecosystem services:

1. Storing water supplies
2. Regulating water flow
3. Genetic resources
4. Soil formation
5. Pollination
6. Controlling erosion
7. Controlling pests
8. Reducing disturbance
9. Raw materials
10. Habitats
11. Regulating climate
12. Purifying air
13. Treating waste and filtering runoff
14. Cycling nutrients
15. Providing food
16. Recreational opportunities
17. Cultural and aesthetic values

The study examined how much money people spent to visit national parks, how views of nature increase property values, how much it would cost to restore these services or replace them with technology. The figure they came up with was $148 trillion. This is greater than the total dollar value of all goods and services produced by people. In other words, there is literally not that much money in the entire world.

8.3 Environmental Policy Making

If you've ever had to engage with any organization, you will inevitably encounter its policies. A **policy** in a formalized set of instructions intended to provide uniformity in decision-making and resolve difficulties. Policies are important for consistency. After all, you wouldn't want to call your credit card company and have the customer service representative make up a random answer on the spot. You want to know that you will get the same result every time a particular issue comes up. **Public policy**, the policy adopted by the government, covers a wide area and we going to focus on environmental policy.

Governments often intervene whenever there is a market failure. For example, it may adopt policies to eliminate free riders or internalize external costs. Or they may be adopted to address what economist Garrett Hardin described as "**tragedy of the commons**." Imagine if an area of land is open for grazing. Anyone can bring as many animals as they want onto this land to graze and no one is responsible for maintaining it. This is what Hardin described as a **commons resource**. Without any restrictions on the use of the resource and no requirements that anyone pays the cost of maintaining it, it will inevitably become depleted. The tragedy of the commons is another example of a market failure that may result in government intervention to put restrictions on its use or charge taxes or usage fees in order to pay the cost of maintaining it.

There are three main ways in which governments can intervene to correct market failures. One would be tax harmful activities in order to discourage them. For example, the government may decide to levy an excise tax on tobacco to make cigarettes more expensive and thereby, discourage people from purchasing them.

Another way would be put laws and regulations in place restricting certain activities, including penalties imposed in the case of noncompliance. For example, the government might require a factory to pretreat its wastewater to remove pollutants before discharging it into the river. If the business does not comply, they could be subject fines or other penalties. Or, the government could decide that a pesticide is too hazardous and ban its use, such as when Dichlorodiphenyltrichloroethane (DDT) was banned in the United States.

The third way governments could intervene is through economic incentives. This could be through direct subsidies, such as paying farmers to plant trees along the edges of streams to capture runoff. Or they could be in the form of tax incentives, such as allowing businesses to write off the cost of installing a new air pollution control system from their corporate income taxes.

Ideally, environmental policy makers should make their decisions based on accurate information. It is therefore crucial that sound scientific principles be adopted when conducting research. For example, before the Environmental Protection Agency (EPA) considers new regulations on a pesticide, it may review the scientific literature relating to its use and its effects on wildlife and human health. It may even commission new studies if there is not enough data and analysis available. The scientists conducting this research could then advise the EPA on the possible consequences of using the pesticide as well as the effect new regulations might have.

In a representative democracy, citizens also have an opportunity to weigh in on policy decisions through voting, lobbying, or making campaign contributions to politicians that advocate for policies they favor. Industry and other private organizations can also influence policy decisions through lobbying and making campaign contributions. These interests may conflict, especially if citizen groups advocate for a policy that may cost industry money in order to comply with any new rules.

Because many environmental policies may cause certain interests to have to make a sacrifice for the benefit of others, there is often strong resistance to new regulations. Companies, think tanks, and other organizations spend large sums of money attempting to influence policy makers. As a result, moneyed interests can sometimes have a disproportionate influence on the policy-making process. This is especially true for environmental issues that take a long time to become apparent or where there may be some uncertainty in the scientific consensus.

In addition, many politicians will ignore the scientific consensus for ideological reasons. Three areas where this is a common issue despite the overwhelming scientific evidence are global warming, vaccine safety, and evolution. Politicians may also suppress scientific studies that do not support the conclusions they want or they may appoint unqualified individuals to head agencies in order to reduce their effectiveness.

8.4 History of U.S. Environmental Policy

The United States has long been regarded as a bellwether for environmental policy. A lot of innovative policies have been adopted first in the United States. These policies have served as models for other countries. In addition, because of its economic and military might, the United States exerts a strong influence on many other nations.

The foundation of the U.S. government is the Constitution, which establishes three branches of government, each with its own set of powers and duties. The first is the legislative branch in the form of Congress, which approves the federal budget and writes legislation. The second branch is the executive branch, headed by the president. The executive branch carries out and enforces the laws written by Congress through various administrative agencies. The third and final branch is the judiciary, represented by the federal court system with the Supreme Court as the highest court in the country. The judicial branch interprets laws and resolves any disputes pertaining to them.

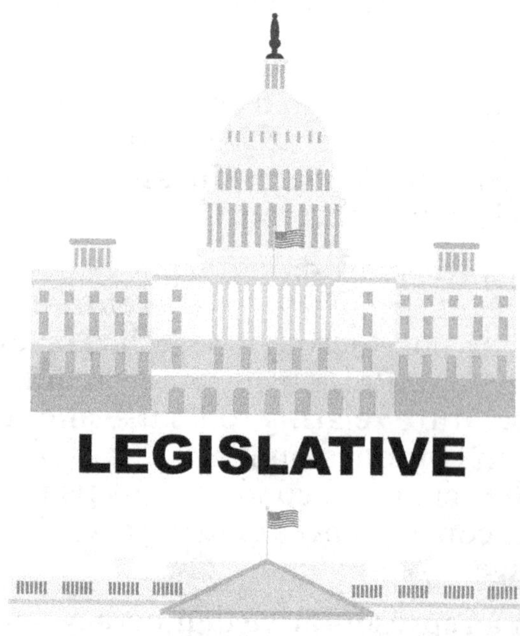

JPL Designs/Shutterstock.com

Under the U.S. legal system, there are also three main types of laws. The first is **statutory or legislative law**. These are laws that are written and approved by Congress. Congress cannot pass any law that violates the principles of the Constitution. The second category is **regulatory law**. These are rules that are adopted by the administrative agencies in the executive branch to implement the laws passed by Congress and to provide more details as to what standards should be followed. For example, Congress may pass a law directing the EPA to set maximum contaminant levels (MCLs) for various pollutants in public drinking water. The EPA then would review the scientific studies to determine things like which pollutants are of highest concern for public health, what level of an MCL would be protective of public health for each pollutant, and what level of reduction is both technically and economically feasible. No administrative agency can adopt regulations that are outside of the authority granted to it by Congress.

The final category of law is **common law** or the law of precedent. These are laws established by the courts, which are then binding on similar cases in the future. For example, the Supreme Court could determine that an administrative agency has exceeded its authority granted by Congress or that a law passed by Congress violates the Constitution. In which case, the law would be nullified and no longer enforceable.

State governments mirror the federal government in that each state has its own constitution and three branches of government. The state legislature, governor, and state court system taking on the same roles as the U.S. Congress, president, and federal court system, respectively. Under the system of federalism in the United States not state can pass a law that violates the principles of the U.S. Constitution.

The history of U.S. environmental policy can be said to have come in three different waves. The **first wave** began shortly after the adoption of the U.S. Constitution and ran until the end of the nineteenth century. Most of the laws passed during this period encouraged westward expansion and settlement and the extraction of resources. These included the General Land Ordinances of 1785 and 1787. These laws put most of the land west of the original 13 colonies under federal control and made

Everett Historical/Shutterstock.com

Nebraska homesteaders circa: 1887.

them available for settlement. These laws allowed for millions of acres to be claimed by citizens, railroad companies, timber companies, and mining companies with little restriction. These policies relived crowding in Eastern cities and propelled the settlement of the west. They also resulted in the displacement of Native American tribes who were forced off their traditional lands. These laws reflect the attitude of **manifest destiny**, in which the United States would be able to expand its borders indefinitely and that the land was an inexhaustible resource.

Everett Historical/Shutterstock.com

Wagon train west!

By the late nineteenth century, however, people began to realize that the land was not inexhaustible and that the boundless frontier was closing. Federal policy began to shift toward the conservation and preservation of wildlife and land. During this **second wave**, Yellowstone was established as the first national park in 1872. The first national forests were created in 1891 and the first wildlife refuge was established in 1903. The second wave continued through the 1930s with the establishment of soil conservation services during the infamous Dust Bowl. The final law passed under the second was the U.S. Wilderness Act of 1964.

Everett Historical/Shutterstock.com

Young Teddy Roosevelt. An avid hunter, Roosevelt became pro-active in land conservation during his presidency.

By the end of World War II, the United States emerged as a global superpower and entered into its golden age of manufacturing. By the 1950s, American manufacturing dominated the world and advances in technology were improving the quality of life for millions of Americans. However, people also began to notice the effects of pollution on our air and water. For example, the Cuyahoga River in Ohio, which flows into Lake Erie, became so polluted with industrial chemicals that it caught fire several times in the 1950s and 1960s. The Great Lakes were dangerously polluted, with Lake Erie all but declared dead by 1970. Incidents like Times Beach and Love Canal, where people's homes were contaminated by toxic chemicals, began to mobilize public sentiment in favor of reducing pollution. This became the basis of the **third wave**.

In 1962, Rachel Carson published her book, *Silent Spring*. This became one of the first popular works to discuss the effects many synthetic chemicals were having on public health and the environment. In particular, she focused on how the pesticide DDT was causing the eggshells of birds of prey such as eagles and hawks. These eggs would then collapse before a chick could form, thus imperiling the birds' ability to reproduce. *Silent Spring* was a major sensation and along with incidents like the Cuyahoga River fires, galvanized public opinion in favor of government intervention.

Solodov Aleksei/Shutterstock.com

Rachel Carson. Her book, *Silent Spring*, influenced public opinion on pollution issues.

During this time period, legal scholar Joseph Sax developed the **public trust doctrine**, which states that the environment and natural resources belong to all citizens and should be protected by the government for the benefit of all. In addition, the first **Earth Day** was declared by Wisconsin Governor Gaylord Nelson.

Several federal laws were passed during this period, including:

Ron Dale/Shutterstock.com

- **The Clean Air Act (1963 and amended in 1970 and 1990):** Set standards for air quality and emissions and allows citizens to sue parties for violating these standards.
- **National Environmental Policy Act (NEPA, 1970):** Created the Council on Environmental Quality which advises the White House on environmental research and requires **environmental impact statements** (EIS) for any federal action that may significantly impact the environment.

- **Endangered Species Act (ESA, 1973):** Protects endangered and threatened species and their habitats.
- **Safe Drinking Water Act (1974):** Directs the EPA to set standards for public drinking water suppliers.
- **Resource Conservation and Recovery Act (RCRA, 1976):** Requires landfills to be permitted and follow regulatory standards and requires "cradle-to-grave" tracking of hazardous materials.
- **Toxic Substances and Control Act (TSCA, 1976):** Requires the EPA to monitor and regulate hazardous chemicals.
- **Clean Water Act (1977):** Regulates the discharge of pollutants into the waters of the country.
- **Soil and Water Conservation Act (1977):** Requires the Department of Agriculture to survive soil quality throughout the country and develop soil conservation plans.
- **Comprehension Environmental Response, Compensation, and Liability Act (CERLCA, 1980):** Established the "Superfund" to prioritize and clean up areas that have been contaminated with hazardous chemicals.

Of particular note is the EIS requirement in NEPA. This requires that the federal government prepare and publish an EIS for any federally funded action, such as a highway construction project that may significantly impact the environment. An EIS is a summary of what the expected impacts on the environment will be and it is required to be made available to public with a comment period. While NEPA does not require that the federal government to mitigate these impacts, it must respond to all public comments. This does incentivize federal agencies to work to reduce some of these impacts in order to avoid a long and complicated public comment process.

Finally, in 1970, then-President Nixon created the EPA, establishing for the first time a single federal agency tasked with setting environmental policy for the entire country. Prior to this, most pollution issues were regarded as the responsibility of state governments and were usually handled by the state departments of health. These health departments were often not equipped with the resources and expertise to handle many environmental issues such as air and water pollution. And of course, pollutants do not respect state lines, so pollutants generated in one state can affect the quality of life for residents in a neighboring state. After the establishment of the EPA, the states soon followed suit, creating state-level environmental protection agencies.

Il.studio/Shutterstock.com

The EPA's mandate is to set standards for air and water pollution, monitor and regulate new chemicals being brought to the market, fund research, and provide public education on environmental issues. In practice, while the EPA will establish regulations for the country, in most cases, it turns enforcement of these rules over the state agencies. The federal government provides funding to states for enforcement and the EPA can takeover direct enforcement if it determines if the state's enforcement is inadequate.

Since 1980, some of these laws, such as the Clean Air and Clean Water Acts have been amended, but there has been very little major legislation adopted since the passage of CERCLA. Many politicians and industry lobbyists have worked to weaken or even repeal many of the rules adopted under these third wave laws. This is based on the assumption that many of the regulations are an economic burden to the industry. While many long-established policies, such as the ban on leaded gasoline and DDT, remain in place, the regulatory process has not been keeping up with new advances. For example, thousands of new chemicals are being brought to the market every year and the EPA lacks the resources to thoroughly review all of them. In Chapter 5, we have seen the U.S. Fish Wildlife Service has had difficulty keeping up with the demand to add new species to the endangered and threatened list. Also, new technologies like hydraulic fracturing pose new challenges for which environmental policy makers are not prepared.

There are three main approaches to environmental policy:

1. Lawsuits
2. Command-and-control
3. Economic incentives

The earliest means for addressing environmental issues was through filing a **lawsuit**. For example, if residents were being harmed by a polluter, they could sue and ask the court to issue an injunction ordering the company to clean up the problem or pay a fine. Sometimes, the company would find it was just cheaper to pay the fine than reduce their emissions. Other times, judges were reluctant to issue rulings that hinder business. In most states, judges are elected, which means they have periodically run for reelection. A judge who ruled against a company that was also a major employer in the region could find himself facing a well-financed opponent. Finally, large corporations have deeper pockets than most citizens and can afford more expensive legal

RomanR/Shutterstock.com

representation. Often, they could bury the plaintiff in motions and appeals until they could no longer afford to continue the lawsuit.

EtiAmmos/Shutterstock.com

Because of the limitations of lawsuits, the federal government began adopting **command-and-control** strategies during the third wave. Under command-and-control, an administrative agency such as the EPA sets limits on emissions or prohibits certain actions with penalties associated with noncompliance. Compared to lawsuits, command-and-control is more proactive. It seeks to prevent pollution from becoming a problem rather than restricting it after it has already caused harm. Most of the reductions in air and water pollution that were achieved during the third wave happened under command-and-control.

Many people, however, dislike command-and-control because it is the heavy hand of government. They argue that it stifles freedom and innovation. As a result, **economic incentives** have been adopted in many cases since the early 1980s. Economic incentives can take a variety of forms.

One example is a **subsidy**, in which the government gives money, either directly or through tax incentives to promote certain activities. Farmers receiving money to reduce erosion is an example of a subsidy. Subsidies, however, can also promote more environmental impacts. For example, the federal government subsidizes the timber industry by paying the roads that provide access to timber in the National Forests. In terms of energy sources, fossil fuels have long received the lion's share of federal subsidies compared to renewable sources. These subsidies

Dominik Bruhn/Shutterstock.com

not only give fossil fuels a competitive advantage over renewable sources, they also keep the real costs of these fuels artificially low. The average consumer does not pay the full cost at the gas pump, for example, not realizing that their taxes are helping to make up the difference.

Another example is the **green tax**, which is a tax designed to discourage a harmful activity. A tax on carbon dioxide emissions is an example of a green tax. The idea behind a carbon tax is that companies can either reduce their carbon dioxide emissions or pay the tax. Money collected from the tax can then be spent on other efforts to reduce greenhouse gas emissions. While carbon taxes are popular in other countries, they have only been adopted in a few states and efforts to pass a national carbon tax were blocked in Congress. Green taxes could also be collected to reimburse people from the environmental effects of certain activities. For example, the state of Pennsylvania charges an environmental impact fee on natural gas drilling. This money is used to compensate communities whose drinking water supplies have been polluted by the gas drilling.

Since the Clean Air Act Amendment of 1990, a new form of economic incentive has emerged, that of **emissions trading or cap-and-trade programs**. Under an emissions trading scheme, the federal government sells the right to industry to emit a certain amount of a pollutant through credits. However, the government only makes a limited amount of credits available, establishing an overall cap on emissions. If a new power plant or factory needs emissions credits, they need to purchase them from a company that has more than they need (the "trade" part of the program). Over time, the government can lower the cap, requiring businesses to phase in new reductions. The first cap-and-trade program was established for sulfur dioxide, one of the major contributors to acid precipitation. It has been successful in reducing overall emissions of sulfur dioxide and thus reducing the impact of acid rain.

Since then, the European Union and some U.S. states have adopted emissions trading programs for carbon dioxide as part of the effort to reduce greenhouse gas emissions.

Other economic incentive programs can include pay-as-you-throw trash collection in which residents are charged by the amount of garbage they produce. Another example would be providing tax incentives for installing rooftop solar panels or buy high-efficiency appliances.

Ecolabeling is the practice by which companies advertise how "green" their products are. Certified organic produce is a means by which consumers can purchase food in which no chemical pesticides or synthetic fertilizers have been used in growing them. The federal Energy Star program provides consumers with information about how much an appliance will cost them on their electric bill, given them the opportunity to make an informed choice as to how much energy efficiency they are willing to pay.

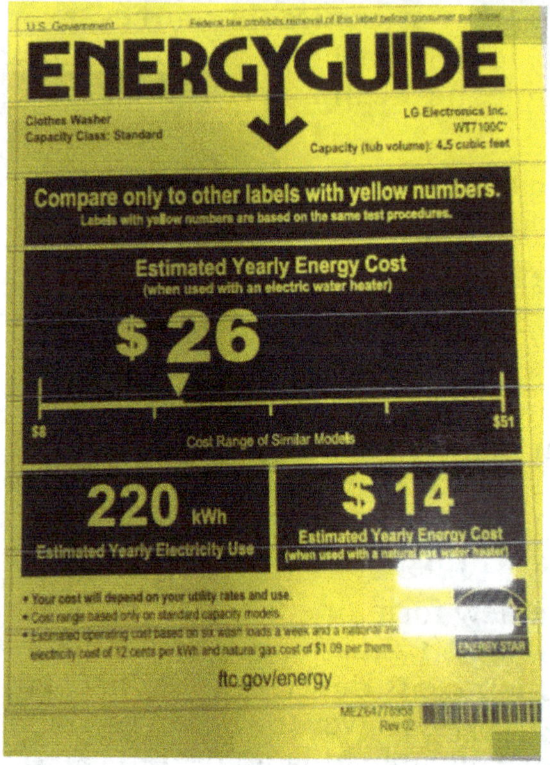

Source: Dennis Wilson

The federal Energy Star program provides consumers with information about the costs associated with the energy consumption of appliances.

ducu59us/Shutterstock.com

Unfortunately, many companies tend to exaggerate how "green" their practices are and can criticize what is called **greenwashing**. Companies that engage in greenwashing may just engage in PR programs and don't actually deliver on any positive environmental impact.

Tamixes/Shutterstock.com

8.5 International Environmental Policy

carlos castilla/Shutterstock.com

Pollution does not stop at the borders between countries and with increasing **globalization**, in which nations are becoming more interconnected with technology and trade, there is a need for international agreements to address these problems. Acid precipitation, stratospheric ozone depletion, and global warming are all examples of problems that have international implications and require global solutions. While international agreements have made significant progress in reducing the first two problems, global warming remains a pressing issue. Over the past few decades, nations have gathered together in several conferences in attempts to find a lasting solution.

As a rule, international law is weaker than the laws of nations. Countries jealously guard their sovereignty and are often unwilling to cede too much power to international bodies. International law can be described as either customary law or conventional law. **Customary law** comes from the informal practices nations have developed to promote cooperation and good relations. On the other hand, **conventional law** takes the form of written treaties or conventions, formal agreements among nations.

While the president of the United States may sign treaties, they must be ratified by the Senate before they become law. While the United States has often taken the lead in drafted many environmental treaties, since the 1990s, it has become increasingly difficult to get them ratified in the Senate. Listed in the following table are several key treaties relating to the environment. Note that none of the treaties signed by various presidents since the 1990s were ever ratified. In particular, while the Obama administration signed the Paris Agreement on climate change, one of the first actions of the Trump administration was to announce plans to withdraw from it. In November 2019, the Trump administration then began the formal process of withdrawing from the Paris Agreement.

Treaty	Year It Came Into Force	Nations That Have Ratified It	Status in the United States
CITES: Convention on International Trade in Endangered Species of Wild Fauna and Flora	1975	175	Ratified

Treaty	Year It Came Into Force	Nations That Have Ratified It	Status in the United States
Ramsar Convention on Wetlands of International Importance	1975	159	Ratified
Montreal Protocol, of the Vienna Convention for the Protection of the Ozone Layer	1989	196	Ratified
Basel Convention on the Control of Transboundary Movements of Hazardous Wastes and Their Disposal	1992	172	Signed but not ratified
Convention on Biological Diversity	1993	168	Signed but not ratified
Stockholm Convention on Persistent Organic Pollutants	2004	152	Signed but not ratified
Paris Agreement, of the United Nations (UN) Framework Convention on Climate Change	2016	197	Signed but not ratified; the United States has withdrawn

There are several international organizations that can influence environmental agreements. The most significant is the **United Nations** (UN). Formed at the end of World War II, the UN provides a forum for nations to discuss and (hopefully) resolves issues peacefully. In addition, the UN sponsors conferences and publishes research on a variety of global issues, such as human rights and the environment.

Osugi/Shutterstock.com

Kristi Blokhin/Shutterstock.com

The **World Bank** is the largest lender of funds for developing countries. It provides funding for a variety of infrastructure projects, including roads, dams, and irrigation. The purpose of the World Bank is to benefit low-income populations. However, it provides funds for many projections that are unsustainable.

The **World Trade Organization** (WTO) is a sort of international court for resolving trade disputes among nations. It can assess penalties against countries that it finds are engaging in unfair trade practices. Unfortunately, it has a history of ruling that environmental regulations are unfair barriers to trade. For example, when the United States adopted cleaner gasoline regulations in 1995, Brazil and Venezuela filed a complaint with the WTO, alleging unfair trade practices. The WTO agreed and forced the United States to weaken those standards.

EQRoy/Shutterstock.com

In addition to these international governmental organizations, **nongovernmental organizations** (NGOs) can also influence international environmental policy. Some, like Greenpeace and Population Convention, take an active role in influencing policy through lobbying and protest. Others, like the Audubon Society and the Sierra Club mostly avoid politics and instead focus on educational and conservation projects.

One of the goals of global agreements and organizations is to promote **sustainable development**, which the UN defines as *"meets the needs of the present without sacrificing the ability of future generations to meet their own needs."*

While development and environmental protection have often been seen as conflicting interests, sustainable development aims to make them work together to improve the quality of life of people throughout the world in a manner that doesn't deplete natural resources faster than they can be replenished. The UN has established a "triple bottom line" of environmental, social, and economic goals for sustainable development.

In 2015, nations adopted a set of sustainable development goals to meet the triple bottom line. Many of these goals have 2030 as the target date for achieving them.

The following table summarizes the sustainable development goals, including the environmental, social, and economic goals for developing nations.

United Nations Sustainable Development Goals
• End poverty
• End hunger and promote sustainable agriculture
• Promote health and well-being for all
• Ensure quality education for all
• Achieve gender equality
• Ensure water and sanitation for all
• Ensure access to affordable, reliable, and sustainable energy sources
• Promote sustainable economic growth and employment
• Build resilient infrastructure, promote sustainable industry, and promote innovation
• Reduce inequality within and among nations
• Make cities safe and sustainable
• Ensure sustainable consumption and production
• Combat climate change and its impacts
• Conserve marine resources
• Protect and restore terrestrial ecosystems and halt biodiversity loss
• Promote peaceful, inclusive, and just institutions
• Renew partnerships for sustainable development

AlexLMX/Shutterstock.com

D-Krab/Shutterstock.com

8.6 Discussion Questions

1. What were the causes and impacts of the Flint water crisis?
2. Describe the three main kinds of economies and how they differ from one another.
3. How is the "invisible hand" supposed to produce quality goods and services at affordable prices?
4. What are external costs and how do events like the Deepwater Horizon spill impose them on people?
5. What is the "tragedy of the commons" and how do societies avoid it?
6. How do statutory law, regulatory law, and common law differ from one another?
7. What were the primary focus of the first and second waves of U.S. environmental policy?
8. What events raised public awareness about the problems associated with pollution?
9. What problems did people attempting to use lawsuits to address environmental problems run into?
10. What is a cap-and-trade program and how was it used to reduce sulfur dioxide emissions?
11. What is greenwashing?
12. What role has the United Nations played in international environmental policy? The World Trade organization?

8.7 Key Terms

Aesthetic value
Barter economy
Cap and trade program
Centrally planned economy
Central planning
Classical economics
Command-and-control
Common law
Commons resource
Conventional law
Cost–benefit analysis
Cultural value
Customary law
Discounting
Earth Day
Ecolabeling
Economic incentives
Economics
Economy
Educational value
Emissions trading or cap-and-trade programs
Environmental Impact Statement
Existence value
External cost
First wave
Free rider
Globalization
Good
Green tax
Greenwashing
Lawsuit
Manifest destiny
Market-based economy
Neoclassical economics
Nongovernmental organizations
Nonmarket value
Option value
Policy
Public policy
Public trust doctrine
Regulatory law
Scientific value
Second wave
Service
Statutory or legislative law
Subsidy
Sustainable development
Third wave
Tragedy of the commons
United nations
Use value
World bank
World trade organization

Chapter 9

HUMAN POPULATION

Chapter Topics

In the 1970s, many people were concerned about population growth. One country in particular was worried that it might not be able to support its growing population. In 1962, China had experienced a famine that killed about 30 million people. By 1970, it was still facing food shortages, imperiling its ability to feed its 818 million residents. The country began to adopt a series of policies designed to reduce its population growth rate, culminating what became known as the "One Child Policy" in 1979.

Families in most regions of China were restricted to having only one child. Incentives were put in place to encourage family planning. Awards, job promotions, and even better apartments were offered to those who voluntarily received a vasectomy or tubal ligation. A massive propaganda program was launched promoting the idea of one child being best and parents can do a better job raising one child instead of two or more. However, some provinces took a more draconian approach. Fines, some more than an average worker's yearly salary were imposed for having a second child. Allegations of forced sterilization began to surface in parts of the country.

In terms of its stated goals, the One Child Policy was successful in reducing China's growth rate. Its current populations of 1.4 billion would have been much higher without the program. The country's fertility rate dropped from 5.8 children per woman to 1.6 and its growth rate fell from 2.8% to 0.5%. However, Once Child also had many unintended consequences for the country, many of which stem from a cultural preference for sons over daughters. This preference is common in many preindustrial societies, especially those that lack a social safety net for the elderly. In agricultural-based societies, farmers expect to continue working the land for as long as they are able. If they are fortunate enough to live to old age, the eldest son is expected to take care of them when they're too old to work on the farm any more.

The expectation is that the eldest son will care for elderly parents and then inherit the farm upon their death. Daughters, on the other hand, would move into their husband's' home and probably help take care of their elderly in-laws. Like many cultural mores, this preference carried over into industrial times, even after fewer and fewer families were farmers.

So, the cultural preference for sons combined with the One Child Policy resulted in parents using ultrasounds to determine the gender of the baby and performing sex-selective abortions. Infant daughters would often be abandoned with the fortunate ones ending up in state-run orphanages. Foreign adoption became a big business in China, where infertile couples from western nations would come to China to adopt baby girls, as girls were all that was available in the orphanages. As a result of abortion, infanticide, or removal from the country through adoption, China began to experience a gender gap. China soon had 116 male births reported for every 100 female births.

In the 40 years since Once Child was adopted, this gender gap has resulted in many young men in China being unable to find a wife. Some have resorted to traveling to other countries in search of wives. It has also resulted in a black market where young women in China and neighboring countries are kidnapped and trafficked into forced marriages.

China is also experiencing an aging population. In 1964, about 6% of China's population was aged 60 and above. By 2017, it had risen to 17% and is projected to reach 36% by 2050. This shift toward an older population is expected to create a worker shortage and put a strain on China's social services as the need for elder care increases.

To combat these problems, China began to reverse many of its earlier policies. It ended foreign adoption and in 2015, officially changed One Child to a Two Child Policy. The government is now engaging in a propaganda campaign that two children are best now. The hope is that this will prevent the impending worker shortage and correct the gender gap. However, so far, it appears that four decades of One Child propaganda has left a lasting impression on the Chinese people and few couples today are opting to have a second child.

Arthimedes/Shutterstock.com

Crystal Eye Studio/Shutterstock.com

China's one child policy has been replaced with a two child policy.

9.1 The Current State of the Human Population

As of 2019, the global population stands at 7.7 billion. China is currently the world's most populous country at 1.4 billion people. India is second with 1.3 billion. This means that close one out of every three people on Earth lives in either India or China. India has not adopted a stringent program like China's One Child Policy and is poised to overtake China as the world's most populous nation in 2020. The United States is in a distant third place with about 330 million people.

The current population growth rate is at about 82 million people per year or about 1.08%. This comes out to about 2.5 people added to the global population each second. For most of the human history, the global population was well below this number. Two events have caused the world population to surge. The first was the Agricultural Revolution approximately 10,000 years ago. The second surge began in the later eighteenth/early nineteenth century as a result of the Industrial Revolution. Both revolutions caused a shift which resulted in an increase in the global carrying capacity. The Agricultural Revolution produced a higher level of food production, making it possible to feed a growing population. With the Industrial Revolution, increases in food production combined with improvements in medical care and public sanitation caused a drop in the death rate and spurred a higher population growth rate.

Two thousand years ago, the global population was estimated at about 170 million. One thousand years ago, it reached 265 million. It took until 1800 to reach a global population of 1 billion, but it took only 103 years to double it to 2 billion by 1930. Thirty years later, we hit 3 billion. We doubled that again to 6 billion in 1999 and reached 7 billion by 2011.

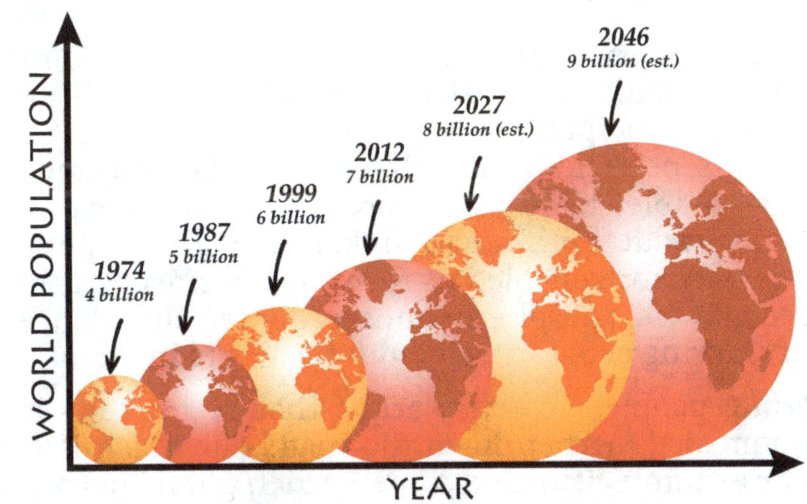

desdemona72/Shutterstock.com

Overall, the rate of population increase peaked in the 1960s and has been declining for the past 50 years or so. Note that the world population is still rising, it's just rising a slower rate than it was before. The population growth from about 2,000 years ago until about 1964 demonstrated a clear exponential growth rate or a J-curve (Chapter 3). An exponential growth rate is not sustainable in the long term. However, there remains considerable debate as to what the ultimate limit of the human population can be.

First, we need to note that the rate of population growth is not evenly distributed. Many industrialized nations have very slow growth rates. Some, such as Canada, Great Britain, and France have relatively stable populations where the growth rate is close to zero. Some, including Japan, Germany, Russia, and many other countries in Eastern Europe have negative growth rates, which means their populations are currently shrinking. As a rule, the poorest countries, especially those in Africa and Central Asia, have the highest growth rates. This means that the vast majority of the next 1 million people to be born will be in countries that are least equipped to care for them. This may seem counterintuitive as one might assume that since wealthy countries can afford to care for more children, they would have more children. But, as we will soon see, the actual trend is that as societies become wealthier, parents focus more on providing their children with a higher quality of life instead of having a larger quantity of children.

One of the first people to raise concerns about the global population growth was Thomas Malthus. In his 1798 work, *An Essay on the Principles of Population*, Malthus projected the current estimates on population growth. He reasoned that while populations can grow exponentially, food production could only increase by cultivating additional land. After crunching some numbers and the population growth rate of his time, Malthus predicted that by the 1950s, the global population would reach a point that the only way to feed it would be to convert all of the land surface area in the world into farmland. Since that would leave no land left for people to live on, the human population would likely crash long before that point. In his opinion, humanity would be doomed to suffer war, disease, and starvation as these were the only factors that limited human population growth. The more the population grew, the bigger the ultimate crash would be.

If these concerns sound familiar to you, then perhaps you have seen the recent Avengers movies in which the main villain, Thanos, raised similar concerns on the universe as a whole.

Of course, the 1950s have come and went and humanity has managed to avoid the global population crash predicted by Malthus. The reason for this is that Malthus did not anticipate the advances of the Industrial Revolution. Mechanization, fertilizers, pesticides, and other modern farming techniques of the Green Revolution have enabled us to grow more food per acre than Malthus could have anticipated. Food production has actually kept ahead of the population growth rate. Technology has therefore enabled us to avoid the crash Malthus predicted. We'll get into the reasons why, despite an unprecedented abundance of food, there are still millions of starving people throughout the world in Chapter 11.

Some people, however, argue that technology has only pushed the inevitable doomsday further down the road. The crash is still inevitable, just forestalled. For example, Paul and Anne Ehrlich published a book entitled *A Population Bomb* in 1968. Their "neo-Malthusian" predictions mirrored Malthus 170 years earlier. Like the Ehrlichs, many warn that we cannot depend on technology to keep ahead of the population growth curve. Fortunately, none of them are advocating the solution Thanos tried in the Avengers movies, but they do argue that humanity should reduce the global birth rate in order to avoid a catastrophe. They point to the fact that our current consumption of natural resources exceeds the **biocapacity** of the Earth by about two-thirds and state that only a significant reduction in the population growth can bring us back to a path to sustainability. Others point out that even if technology can continue to increase food production at pace with the population growth, the average quality of life for people will decline as there will be less resources, less space, and less wealth available for each person. Overcrowding will still be a problem even if everyone gets enough to eat.

Arthimedes/Shutterstock.com

The population "bomb" as envisioned by the neo-Malthusians.

So, how big will the global population reach in this century? If the current growth rate continues, the world population will reach about 12 billion by 2050 and 20 billion by the year 2100. However, as we noted, the growth rate is slowing down, and if that continues, the global population will hit 10 billion by 2050 and then level off at about 11 billion by the end of the century. That represents the most likely scenario. On the other hand, if people adopt a more aggressive effect in reducing the birth rate as advocated by the neo-Malthusians, the population will peak at 9 billion by 2050 and actually start to drop back down to about 7 billion by 210.

9.2 The Effect of Populations on the Environment

More people mean more demand for food and other resources. This means more overall consumption of resources. However, the size of the population is not the only factor that influences its effect on the environment. The **IPAT Model** illustrates the total impact of a population as the product of three factors:

$$I = P * A * T,$$

where I is the impact of the population on the environment, P is the size of the population, A is the affluence or the per capita resource consumption of the population, and T is the use of technology. Note that technology could either increase the impact on the environment through an increase in the demand for energy or decrease the impact as new technologies enable a society to reduce the amount of pollution generated.

For example, a resident of a wealthy country like the United States has the same impact on the environment as 3.4 people in China, 8 people in India, or 14 people in Afghanistan. Note that despite having high poverty rates, China and India both have a greater per capita impact on the environment than Afghanistan. In addition to being the two most populous countries in the world, they are also the two most rapidly industrializing countries in the world. This has resulted in both countries have a growing middle class and with it, a growing per capita resource consumption. This combined large population and affluence may result in India and China surpassing the United States in its impact on the environment.

9.3 The Demographic Transition

Demography, the study of human populations, often looks at many of the same factors biologists examine with populations in ecosystems:

- Size
- Distribution
- Density
- Age structure
- Sex ratio
- Rates of birth, death, emigration, and immigration

As noted above, human populations vary in size and distribution, with some countries being much more populated than others. Density also varies greatly from one country to another. The table below compares some of the more populous countries in the world.

Country	Population Size	Area (km^2)	Population Density
Japan	126 million	377,973	334/km^2
United States of America	325 million	9,833,520	33.6/km^2
Russian Federation	146 million	17,098,246	8.4/km^2
Indonesia	268 million	1,904,569	138/km^2
China	1.4 billion	9,596,961	145/km^2
India	1.3 billion	3,287,263	403.7/km^2
Brazil	210 million	8,515,767	25/km^2
Pakistan	213 million	881,913	244.4/km^2
Mexico	127 million	1,972,550	61/km^2
Bangladesh	161 million	147,570	1,106/km^2

Note that while Japan has less than half the population of the United States, it has 10 times the population density due to its much smaller land area. On the other hand, Russian, which has large areas in Siberia with almost no population in it, has one of the lowest population densities.

In human populations, age structure is usually illustrated in a **population pyramid or age structure diagram**, which break down the relative number of people into three main age categories:

- Prereproductive age: 0 to 15 years old
- Reproductive age: 16 to 45 years old
- Postreproductive age: >45 years old

A population period with a wide base has a relatively large population in the prereproductive age. This means the average family is having a lot of children and is experiencing a rapid growth and the pyramid is **expansive**. A population pyramid that shows an even distribution among the three age categories is a relatively stable population, in which the number of births is roughly equal to the number of deaths in the population. Such a pyramid is called a **stationary** pyramid. Finally, a population pyramid with a narrow base and wide top is a population experiencing decline. The number of births is not keeping pace with the aging population and the pyramid is considered **constrictive**.

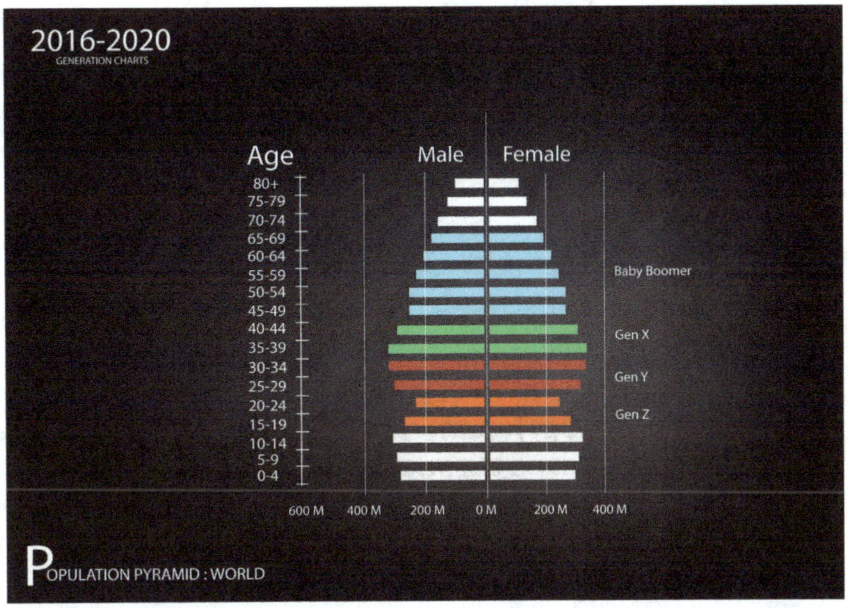

lamnee/Shutterstock.com

World population pyramid 2016-2020.

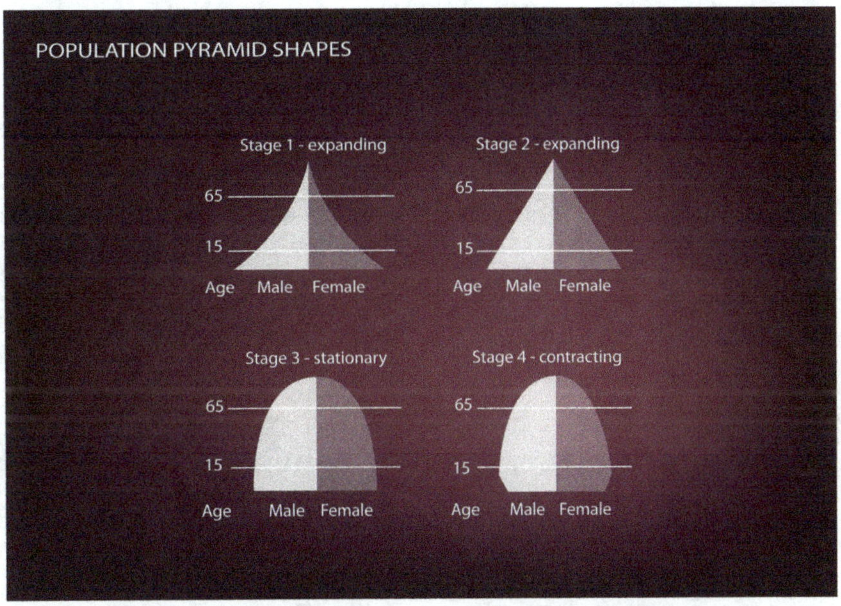

lamnee/Shutterstock.com

A rapidly growing population has a strong workforce, but it may be growing too fast to support itself. So, it may experience a high poverty rate and a high crime rate. On the other hand, more people could also mean more ideas, more innovation, and more solutions to problems being developed. Conversely, a declining population will likely have a low unemployment rate. It may even be experiencing a worker shortage and may need to encourage immigration to replace retirees. This is the current situation in some European countries today. An aging population may also enjoy a lower crime rate, as violent crime is mostly committed by the young (how many 80-year-old muggers have you seen?). Social services, however, may be stressed. Older people tend to have more health problems and therefore require more money to be spent on their healthcare. A shrinking workforce may not have enough people paying into the system to maintain these services.

Overall, the global population is getting older. Today, the median age worldwide is 28. By 2050, it is expected to be 35.

In terms of sex ratios, there is a slight skewing of births in favor of males with 106 male births reported for every 100 female births. Some suggest this is an evolutionary adaptation as males are more prone to death in all age categories and have a slightly shorter average lifespan than females do. This skewing then tends to even the numbers out in adulthood.

Compared to developing countries, developed countries tend to have a higher **life expectancy**, the average number of years a person is expected to live, and a lower infant mortality rate. Industrialized or developed nations include the United States of America, Canada, Japan, Australia, New Zealand, and most of Europe. All of these countries have passed through a series of four stages of population growth that make up the **Demographic Transition Theory**. These stages are the preindustrial stage, the transitional stage, the industrial stage, and the postindustrial stage.

1. The **preindustrial stage** begins when nations are still mostly agricultural-based. This stage is characterized by rampant disease, poor sanitation, little healthcare, and little food security. Birth and death rates are both high during the preindustrial stage, maintaining a relatively stable population.
2. During the **transitional stage**, healthcare and sanitation improve and food production becomes more reliable. As a result, death rates begin to drop. However, there is a lag in as people maintain previous habits, so birth rates remain high. The transitional stage is the period of most rapid population growth.
3. At the **industrial stage**, people begin to adjust the changing circumstances and birth rates begin to decline. The population growth rate slows down.
4. Finally, in the **postindustrial stage**, the society reaches a new equilibrium where birth and death rates are both much lower. The population as whole enjoys a higher standard of living compared to the preindustrial stage.

All of the wealthy, industrialized nations have reached the postindustrial stage, including the United States. However, the United States does have a slightly higher birth rate compared to many other industrialized nations due to immigration. In contrast, the less wealthy industrializing nations have yet to move through the four stages of the demographic transition. In fact, some have populations that are so high that they are experiencing what is known as **demographic fatigue** and many expects doubt that these nations will be able to transition. Their populations are so large that the resources needed to feed their people are preventing them from transitioning. This has resulted in an ongoing debate as to whether these nations need economic aid to help them transition or most access to family planning services in order to reduce their birth rates. The current global consensus is that many of these nations, especially the poorest ones in sub-Saharan Africa and Central Asia will need both.

9.4 Factors That Influence Population Growth

Just as in ecosystems, the four factors that influence population growth are birth, death, immigration, and emigration rates. Births and immigration add individuals to populations while deaths and emigration subtract individuals. Historically, infant mortality rates have been very high and as a result, families tended to have high birth rates as a hedge against the high number of children dying young. As infant mortality rates dropped, populations tended to rise. Today, there is a high correlation between industrialization and lower infant mortality rates. As a country becomes more industrialized, infant mortality rates drop. Overall, global infant mortality has been dropping for several decades, even in the poorest countries as improvements in sanitation and medicine have spread across the globe.

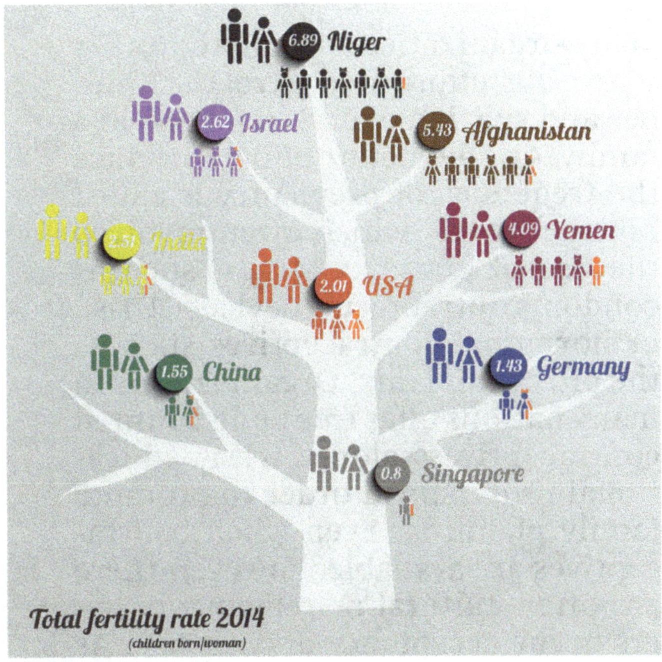

SyzSV/Shutterstock.com

A nation's **total fertility rate (TFR)** is the average number of children born per woman in their lifetime. To keep a population stable, a country needs to have a TFR of at least the **replacement fertility**, which for humans is 2.1, reflecting the fact that infant mortality is not zero. If it was, then replacement fertility would be exactly 2, just enough to replace the mother and father. Factors such as industrialization, women's rights, and improved healthcare have driven TFRs down in the developed nations. Currently, nearly every European nation has a TFR below replacement. According to the CIA's World Fact Book, Niger currently has the largest TFR at 6.49, while Singapore has the lowest at 0.83. Meanwhile, the United States has a TFR of 1.83, which is below replacement. The United States would have a shrinking population today if not for immigration.

The table below gives the TFR of regions throughout the world from 2010 to 2015 as compiled by the Pew Research Center. Currently, sub-Saharan Africa has the highest TFR in the world, while Europe has the lowest.

Total Fertility Rates by Region and Religion, 2010-2015

Number of children an average woman is expected to have in her lifetime

	All Religions	Muslims	Christians	Unaffiliated
Sub-Saharan Africa	4.8	5.6	4.5	4.3
Middle East and North Africa	3.0	3.0	2.5	*
Latin America and the Caribbean	2.2	*	2.2	2.3
Asia-Pacific	2.1	2.6	2.3	1.6
North America	2.0	2.7	2.1	1.6
Europe	1.6	2.1	1.6	1.4
World	2.5	3.1	2.7	1.7

* Insufficient data for reliable estimates.
Regional differences in fertility rates for other religious groups are discussed in Chapter 2.
Source: The Future of World Religions: Population Growth Projections, 2010-2050
PEW RESEARCH CENTER

"The Future of World Religions: Population Growth Projections, 2010–2050." Pew Research Center, Washington, D.C. (April 2, 2015) https://www.pewforum.org/2015/04/02/main-factors-driving-population-growth/.

There are a several factors, both economic and cultural, that affect a nation's TFR. These include

- Access to contraceptives
- Acceptance of contraceptive use
- Level of women's rights
- Cultural influences, such as television programs
- Level of affluence
- Importance of child labor
- Availability of governmental support for retirees

Pavel Vinnik/Shutterstock.com

Many organizations promote **family planning**, efforts to control the number and spacing of children born in a family. Any deliberate effort to reduce the frequency of pregnancy is called **birth control**, which often involved the use of **contraceptives**, such as condoms, intrauterine devices (IUDs), or hormonal contraceptives such as the pill or implants. Organizations in many nations offer free or discounted contraceptives along with educational programs in order to promote family planning. Even when contraceptives are available, however, there are often cultural or religious barriers to the acceptance of their use. As a result, contraceptive usage rates can range from 84% in China to less than 10% sub-Saharan African nations.

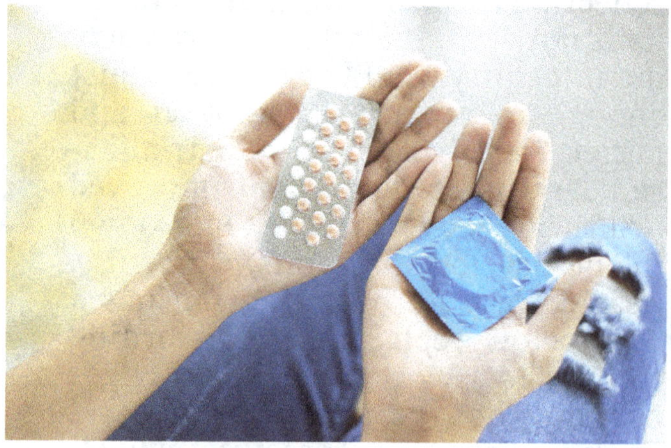
Poring Studio/Shutterstock.com

Access to family planning services gives women more control over their **reproductive window**, the period of time in their lives when they are able to conceive. A woman's reproductive window generally ranges from the beginning of menstruation until menopause. In societies where women have fewer rights, including preindustrial societies, women tend to get married at younger ages and have their first pregnancies at a younger age as well. As women gain more rights and more independence, they can choose to delay their first

PandaStudio/Shutterstock.com

pregnancy until a time of their choosing and choose how many children they will have in their lifetime.

Fertility rates, therefore, tend to drop as women gain more rights and have more opportunities to earn incomes outside of the family. The equality in decision-making and the more educational opportunities women have, the lower a nation's TFR becomes. There is a strong correlation between women having access to education and lower TFR. Developing nations that have higher rates of female secondary school enrollments tend to have lower TFRs.

Cultural influences, such as television can impact TFRs. For example, surveys have found that Internet searches for terms like "contraceptives" increase the day after programs like MTV's "16 and Pregnant" air, indicating that seeing other teens dealing with pregnancies may motivate teens to research ways to avoid pregnancy. In other countries such as Brazil, programs known as Telenovelas (sort like soap operas in the United States) that feature affluent, independent women can influence women to take more control over their fertility.

The degree of acceptance of child labor also influences in TFR. In preindustrial, agricultural-based societies, more children mean more farm labor. A large family can cultivate more land, harvest more crops, and raise more animals. Children therefore contribute to the family's prosperity. However, mechanization reduces the need for farm hands, while industrialization creates a demand for factory workers. As populations move from the farms to the crowded cities looking for work, a child becomes less an economic benefit for the family and more of cost. Families tend to think in terms of "can we afford another child?" instead of thinking of large families as the path to prosperity. This is especially true if laws against child labor are passed. Nations where children are sent to work in factories and thus contribute to the family's income, still tend to have high TFRs, even as they industrialize.

StanislavBeloglazov/Shutterstock.com

Child labor in Nepal. Many developing countries still rely on child labor, even as they industrialize.

iQoncept/Shutterstock.com

Social safety nets relieve parents' expectations that their adult children will take care of them in their old age.

As noted earlier, in preindustrial societies, parents depend on their adult children to care for them in their old age. However, as programs such as Social Security and Medicare that provide support for the elderly are adopted, there is less pressure to have children as an insurance for old age. As a result, TFRs go down.

There is also a strong correlation between poverty and high TFRs. Poverty promotes a higher population growth and a high population growth exacerbates poverty. Higher populations put pressure to cultivate more land and therefore impacts on the environment increase. It is important to remember, however, that the size of a population is only one factor in the IPAT model. Affluence and the use of technology can increase a society's per capita impact on the environment more than just the number of individuals.

Nestor Rizhniak/Shutterstock.com

More educational opportunities for women tend to reduce TFR.

9.5 Discussion Questions

1. What were the unintended consequences of China's One Child Policy?
2. What factors did Malthus not anticipate in his predications about population growth?
3. What challenges and advantages might a growing population experience? What about for a shrinking population?
4. Describe each stage in the demographic transition theory.
5. Why might many developing countries not go through the stages of the demographic transition?
6. What factors influence a country's TFR?
7. What factors have enabled women to gain greater control of their reproductive window?

9.6 Key Terms

Age structure diagram
Biocapacity
Birth control
Constrictive population
Contraceptives
Demographic fatigue
Demographic Transition Theory
Expansive population
Family planning
Industrial stage
IPAT Model
Life expectancy
Population pyramid
Postindustrial stage
Preindustrial stage
Replacement fertility
Reproductive window
Stationary population
Total fertility rate
Transitional stage

Chapter 10

SOIL AND SOIL MANAGEMENT

Chapter Topics

From the mid-nineteenth through the early twentieth centuries, American farmers moved into the midwestern plains states, where they converted the native grasslands into some of the most productive farmland in the world. However, repeated plowing loosened the soil and made it more prone to wind erosion. Then, the 1930s, the region was hit with an extended drought. Wind storms blew across the plains picking up the soil and transporting as far as upstate New York, creating what became known as the **Dust Bowl**. In just a few years, farms lost as much as four inches of topsoil. Crops dried up and many farmers went bankrupt and lost their land. This could not have come at a worse time for the United States. Already, the country was facing record unemployment at the height of the Great Depression and now America had to contend with the collapse of the midwestern farming industry.

To combat the Dust Bowl, Congress passed the Soil Conservation Act of 1935 and created the Soil Conservation Service to work with farmers to reduce erosion.

The lesson of the Dust Bowl shows us the importance of soil in our food security and how its loss can devastate communities.

Everett Historical/Shutterstock.com

The Dust Bowl in Kansas, 1935.

10.1 Soil Characteristics

Soil is defined as a mixture of disintegrated minerals, partially decomposed organic material, microorganisms, water, nutrients, and air. Breaking down its components, soil is made up of 55% minerals, 45% pore space, and 5% organics. Soil makes up a critical part of all terrestrial ecosystems and is home to a variety of organisms. These include detrivores and decomposers such as bacteria and fungi that help recycle nutrients like nitrogen and phosphorus. Earthworms, insects and other arthropods, and other borrowing organisms make their home in the soil and help give it a loose, clumpy structure that promotes root infiltration and nutrient absorption by plants.

Soil begins with the **parent material** or solid rock. Parent material may come from lava or volcanic ash that has hardened, sediments that have accumulated and solidified, or **bedrock**, the solid layer of rock that forms the basis of the Earth's crust. Soil formation begins during primary succession (Chapter 4) as wind, water, and organisms such as lichens began a process called **weathering**, in which the rock is worn away into tiny particles. Physical weathering is caused by friction from wind or rain, or by cycles of freezing and thawing. Chemical weathering can be caused by water or gases altering the rock and separating it into smaller particles. Finally, biological weathering can result from acids secreted by lichens or by plant roots rubbing against the rocks.

As weathering breaks the parent material into smaller particles, these particles mix with partially decomposed organic material or **humus**. Humus is a rich source of nutrients and helps the soil retain moisture, making it vital for plant growth and productivity.

MH Art/Shutterstock.com

Humus, a good source of nutrients for soil.

Climate plays a big role in weathering as warm, moist climates promote more rapid weathering. Organisms like plants and decomposers influence the organic material that mixes with the soil. Topography can also influence exposure to wind, sunlight, water, and other factors that can affect weathering and soil transport. In addition, the composition of the parent material can influence the rate of weathering and how the soil forms. Soil formation can take centuries or even thousands of years, making it a renewable resource, but one that takes a long time to renew itself.

One of the most important characteristics of soil is the **soil texture** or the relative proportions of different sizes of particles. Soil scientists divide soil particles into three categories: sand, silt, and clay. **Sand** particles are the largest, ranging from 2.0 to 0.063 mm in diameter. **Silt** makes up the medium-sized particles ranging from 0.063 to 0.004 mm in diameter. **Clay** particles are then any particles less than 0.004 mm in diameter.

Different soil particle sizes have different properties that can affect how productive the soil is. For example, larger particles have more spaces in between them. This allows for more air and water to infiltrate those spaces. If you've ever poured water on the sand at the beach, you can see how rapidly the water seeps through the sand particles. Conversely, water has a harder time penetrating through layers of clay. On the other hand, smaller particles have more surface area relative to their volume. Nutrient ions such as nitrates and nitrites will cling to the surfaces of particles, as does water. As a result, soils that are higher in clay content have a higher **nutrient holding capacity** and a higher **water holding capacity**. High clay content soils retain nutrients better than sandy soils do.

On the other hand, high clay content soils are not as workable as more sandy soils. **Workability** is a measure of how easy a soil can be cultivated. Think about what happens when clay dries out. It becomes hard and brick-like. This makes clay a much less workable soil compared to sand. Therefore, neither soils with high sand content nor high clay content are desirable for farming. Instead, a more balanced mixture is desired. For example, one soil type called loam consists of 40% sand, 40% silt, and 20% clay. If loam has a high organic content, it can make an ideal soil for agriculture because the mix of particle sizes enables it to retain nutrients and water, be workable, and allow enough water and air to infiltrate the soil spaces to support crops.

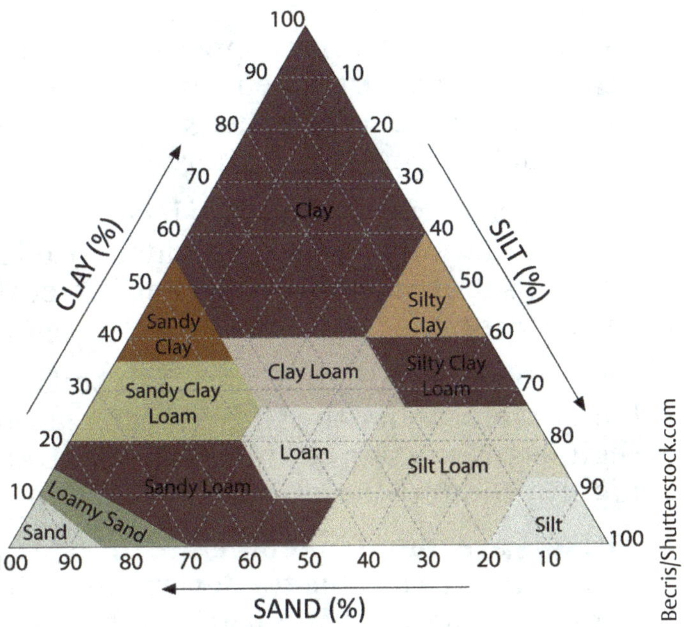

Different soil particles have different properties that affect productivity.

The following table summarizes the properties of the different soil particle sizes.

Particle Size	Water Infiltration	Aeration Capability	Water-Holding Capacity	Nutrient-Holding Capacity	Workability
Sand	Good	Good	Poor	Poor	Good
Silt	Medium	Medium	Medium	Medium	Medium
Clay	Poor	Poor	Good	Good	Poor

Roots need to breathe. If they don't get enough air, they cannot carry out their metabolic processes. This makes the **aeration capability** of soil important. If the soil becomes too compacted, pressing the soil particles too close together so that air cannot infiltrate the soil can kill the plants. If you've ever overwatered a plant, you've seen what happens when roots don't get enough air. Too much water fills up all the spaces between the soil particles, suffocating them.

Soil pH is another important limiting factor. Remember that the pH scale ranges from 0 to 14, where a pH of 7 is neutral, below 7 is acidic, and above 7 is basic or alkaline. Most plants grow best in soils that have a neutral pH. Although there are plant species that prefer acidic or alkaline soils. For example, blueberries love acidic soils. If the soil pH is not suitable for the crops, additives can be used to correct it. For example, if the soil is too acidic, lime can be added to raise the pH to a more suitable level.

Salinity or salt content can also affect plant growth. Ever put salt on a slug? If you have, you've seen what happens: They bubble, shrivel up, and die. Slugs have a very thin skin and need to keep it moist. That's why they're covered in slime. Salt draws the moisture out of the slug's cells, killing it. The same thing can happen to roots. Too much salt pulls the water out of the roots, causing them to wither and die.

Soil characteristics can vary by climate and different soil types can be found in different biomes. Soils are divided into four major types of orders. Each order has its own set of properties that can affect their productivity and their usefulness in agriculture.

Mollisols are dark, nutrient-rich soils that tend to be found in temperate grasslands in places like the Midwestern region of the United States as well as temperate regions in Eurasian countries such as Ukraine, Russia, and Mongolia. Mollisols are highly productive, rich in humus and minerals, and can be farmed repeatedly.

Alfisols are found in temperate forest regions. They tend to be moister than mollisols and are suitable for agriculture. However, they will require more fertilizers or organic material applications than mollisols.

Oxisols are the soils found in tropical and subtropical regions. The warm, moist conditions promote rapid decomposition and leaching of nutrients. As a result, oxisols are poor in nutrients compared to other soil types. They also tend to be acidic. Slash-and-burn techniques, where the vegetation is cut down and burned to provide a quick fertilizer can enable these soils to be used for agriculture for a few years. However, the heavy rains will eventually leach the minerals and nutrients away, leaving a hardpan that is no longer productive.

Aridisols are found in deserts and other drylands. They tend to be loose with little vertical structure. They are often poor in organic material, but some may support enough vegetation to be used as rangeland or pasture for feeding animals. They can be used for growing crops, but require irrigation. However, the dry, warm conditions promote rapid evaporation resulting in salinization of the soils, eventually rendering them unproductive.

Finally, there are **hydric soils** that are found in wetlands and other aquatic ecosystems. These are usually not suitable for agriculture and don't form a soil type all their own. They are often used on maps to mark the boundary between wetland and dryland systems.

A **soil profile** is a cross-section of soil layers. In a typical soil profile, each layer or **soil horizon** has its own properties. Analysis of the soil profile and the properties of each horizon can tell a lot about the soil's productivity.

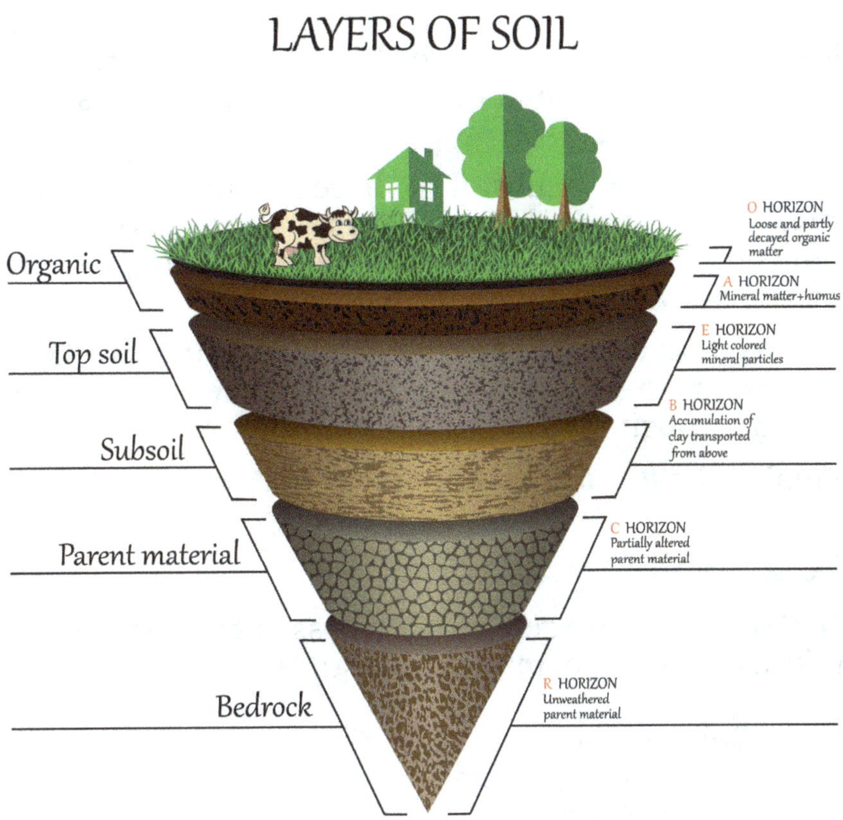

Ellen Bronstayn/Shutterstock.com

A soil profile.

The topmost layer is the **O horizon** or litter layer. This is where dead organic material or detritus accumulates and decomposes into humus. Due to rapid decomposition and leaching, oxisols tend to have very little O horizons while mollisols have the deepest.

Antonov Roman/Shutterstock.com

Soil layers.

Below the O horizon is the **A horizon** or topsoil. Plant roots infiltrate this layer, so this is where plants absorb their nutrients and water. The A horizon tends to be rich in humus and is dark in color. The thickness of the A horizon varies with soil type and biome. For example, mollisols tend to a deep, rich A horizon, while oxisols and aridisols have very thin or even nonexistent A horizons. Alfisols also tend to have well-developed A horizons.

The **E horizon** or zone of leaching is the layer that has lost a lot of minerals due to leaching. The letter E stands for eluviation or the process of leaching. E horizons tend to be pale in color due to the loss of minerals. Alfisols tend to have well-developed E horizons.

The **B horizon** or subsoil is where the minerals such as iron and aluminum oxides that leached from down from the A and E horizons accumulate. The B horizon tends to be reddish or yellow in color due to the presence of these oxides. Oxisols often have very pronounced B horizons.

C horizons consist of the weather parent material. Usually, the parent material represents the original foundation of the landscape, weather from volcanic ash or glacial deposits.

Finally, the **R horizon** consists of the solid bedrock of the Earth's crust.

10.2 Soil Degradation

Soil degradation refers to any loss of the quality or productivity of soil. This can take a variety of forms, including:

- Soil erosion
- Nutrient depletion
- Water scarcity
- Salinization
- Waterlogging
- Chemical pollution
- Changes in soil structure and pH
- Loss of organic matter from soil

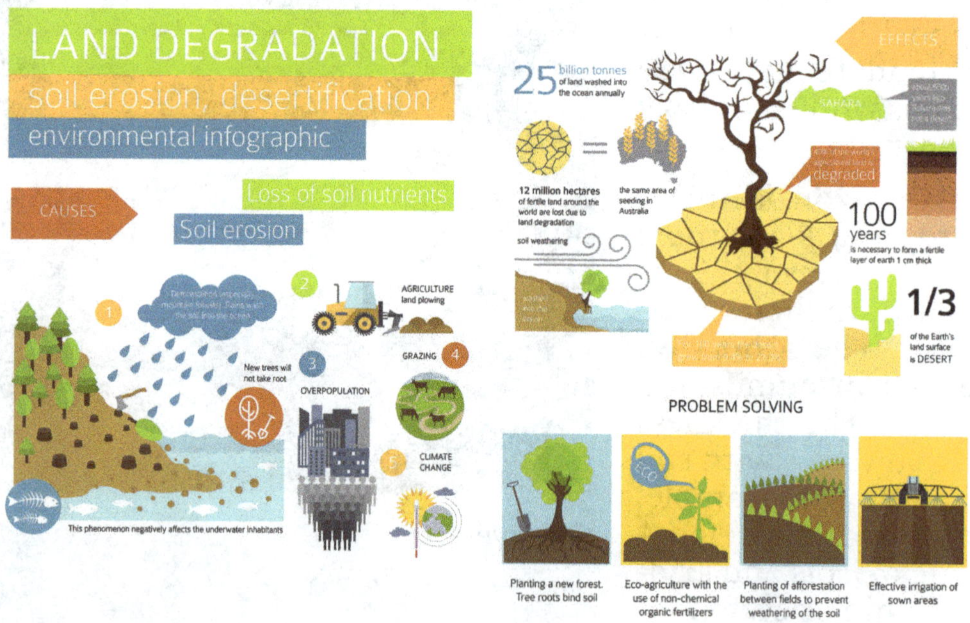

Olha1981/Shutterstock.com

Erosion is the removal and transport of a material by wind and water action. The other end of the process of erosion, **deposition**, occurs when the material arrives at a new location. Erosion and deposition are natural processes and, indeed, are the main mechanism through which fresh soil accumulates and builds up on the land. However, human activity can accelerate the process, resulting in the loss of productive topsoil. Overcultivation of farmland, overgrazing of rangeland, and clear-cutting of forests are examples of activities that can worsen erosion. Basically, any activity that removes the vegetation cover from the land or loosens or exposes the soil to more wind and water can accelerate erosion. Windy or rainy climates can increase erosion as can areas with slopes be more vulnerable to erosion.

vivatchai/Shutterstock.com

Kaentian Street/Shutterstock.com

Supertrooper/Shutterstock.com

Examples of soil erosions.

The other end of erosion, deposition, can cause problems as well. Too much sediments can fill in ponds, clog streams, and increase flooding. In addition, suspended sediments can block sunlight in aquatic ecosystems and inhibit photosynthesis by the aquatic plants. Finally, sediments can choke the gills of fish and clog the feeding and breathing passages of invertebrates.

Growing crops inevitably removes nutrients from the soil. This loss of nutrients can be offset by using inorganic or organic soil. However, as the humus component in soil decomposes and organics are depleted, the soil starts to lose its structure. The soil will also lose nutrient-holding capacity. This process, **mineralization**, results in a soil that only contains the inorganic mineral particles of silt, sand, and clay. As the soil's nutrient-holding capacity declines, inorganic fertilizers become more prone to runoff, resulting in eutrophication and hypoxia in coastal communities

(Chapter 2). While inorganic fertilizers have value in their ability to provide plants with large quantities of nutrients, organic fertilizers add to the humus in the soil, improve the soil structure, and reduces loss of nutrient-holding capacity. A good strategy combines the use of both inorganic and organic fertilizers in controlled amounts, thus gaining the benefits of both while reducing some of the negative consequences.

Maksimilian/Shutterstock.com

Overgrazing in India.

Overcultivation can increase erosion through excessive plowing. Traditionally, plowing has been seen as a vital part of agricultural. Primarily, it is used to remove weeds before planting. Plowing also improves soil aeration and **water infiltration**. Plowing loosens the soil, making it more vulnerable to wind and water erosion. Cultivation also results in long periods of time, both after planting and after harvesting, where the soil is bare and vulnerable to erosion. Reduced tillage programs, such as **low-till** or no-till, have been adopted in many areas to reduce erosion from overcultivation.

Overgrazing can also lead to erosion and other forms of soil degradation. Too many animals will strip the grass from the land, exposing the soil to wind and water. In addition, as the animals trample around the pasture, they can compact the soil, inhibiting regrow of the grass. This often leads to weeds and other less desirable plants moving in and displacing the grass.

Trees also reduce erosion. They absorb a lot of water and their roots help hold the soil in replace. Clear-cutting, therefore, can also contribute to increased erosion, especially in rainy areas with steep slopes such as the Pacific Northwest. Other human activities that remove the native vegetation such as strip mining or building construction, can also increase erosion. Covering the ground with impervious materials like concrete or blacktop can also increase erosion. These materials increase runoff and concentration of the flow of water into exposed areas.

Water scarcity can be a problem in drylands or places where the climate promotes rapid evaporation. Soils in drylands are particularly vulnerable to differential erosion by wind. Wind erosion tends to pick up the lighter soil particles like silt and clay before it picks up the sand particles. Loss of the smaller particles can lead to **desertification**, where the soil becomes mostly sand. Since sand has a poor nutrient-holding

bqmeng/Shutterstock.com

Desertification.

and water-holding capacities, soil productivity declines. While desertification doesn't actually turn the land into deserts, a similar process does occur in deserts due to sparse vegetation cover. This is why soils in deserts are mostly sand. Continued wind erosion can lead to **desert pavement**, where the land becomes a hard, gravely pack that is difficult to cultivate. Over time, organisms called cryptogams (algae, lichens, and mosses) can grow on the desert pavement, creating a **cryptogamic crust** that covers the soil. While cryptogamic crust can reduce erosion, it can also inhibit water infiltration and seed germination. Cryptogamic crust can be broken up by the trampling or livestock or direct human action. However, if efforts aren't made to protect the soil below, it will become vulnerable to wind and water erosion again.

While irrigation can make soils in dry climates more productive, excessive irrigation can lead to problems like **waterlogging** and **salinization**. Irrigation makes up about 70% of our water consumption and much of it is unsustainable. The traditional method of flood-and-furrow irrigation, where water is allowed to flow into furrows, can waste over half the water to runoff or evaporation. Too much water flooding the fields can waterlog the soils, filling in all the spaces between the soil particles and suffocating the roots.

nd3000/Shutterstock.com

Andrii Yalanskyj/Shutterstock.com

Tradition and drip irrigation systems.

Hussain Warraich/Shutterstock.com

Soil that has experienced salinization.

In addition, all water sources contain some salt. The difference between what we call freshwater versus saltwater is that in saltwater, there is enough salt that we can taste it. As irrigation water evaporates, it leaves whatever salts it contained behind in the soil. Excessive irrigation, especially in warm climates that promote rapid evaporation, can lead to the buildup of salt, or salinization, in the soil. As noted in Section 10.1, too much salt in the soil draws the water out of the roots, killing them.

Acid rain can lower the pH of soil. This can have a variety of effects. First, the lower pH itself can harm many crops directly. Second, it can cause key nutrients like calcium, magnesium, and potassium to separate from the soil, making them more vulnerable to leaching. The same can happen to the organic materials in the humus, resulting in mineralization. Finally, the lower pH can free up toxic chemicals in the soil like aluminum and mercury.

A relatively new threat to soil quality is nanoparticles. Nanoparticles are very tiny particles in the range of 1 nm in diameter (a nanometer is 1 10E-9 m). Silver nanoparticles, for example, have been used for their antibacterial properties. Research has shown that these nanoparticles can also kill beneficial soil bacteria. The study of nanoparticles is still an emerging science and the effects of even small amounts of these particles are not well known.

Johann Stubhan/Shutterstock.com

Crop Rotation

10.3 Soil Conservation and Restoration

Since the Dust Bowl, efforts have been promoted to protect the soil and restore soil quality in areas where it has become degraded.

Crop rotation is the practice of planting alternating crops in a field from one year to another. For example, a farmer may plant corn in a field one year and then switch to hay or soybeans in another. Crop rotation can make farms less vulnerable to disease and pest outbreaks since different crops are more vulnerable to different pests. In addition, legumes such as soybeans contain nitrogen-fixing bacteria and will increase the amount of biologically available nitrogen in the soil. This is turn can reduce the need for additional fertilizers.

Ozgur Coskun/Shutterstock.com

Contour Farming

Farmers in hilly regions practice **contour farming**, where they plow perpendicular to the slope of the hill. The rows of furrows will then stop the flow of water downhill, thereby reducing water erosion.

Similar to contour farm is a practice adopted in mountainous regions like the Andes or the Himalayas call **terracing**. In terracing, flat platforms are cut into the sides of the mountain, creating what looks like a staircase used by giants. Each platform serves to capture water and prevent it from flowing downhill.

Intercropping is similar to crop rotation, except instead of alternating crops by year, the different crops are planted in alternating rows. This added variety in plant life reduces pest and disease outbreaks. In addition, rows of legumes can be used to restore nitrogen to the soil, while taller crops reduce wind erosion.

Vietnam Stock Images/Shutterstock.com

Terracing

In areas wind erosion is of high concern, rows of trees or hedges can be plants as **shelterbelts**. These serve as windbreaks to reduce the force of the wind. The trees can also provide habitats for beneficial species or be used to grow fruit, making them an additional source of revenue.

Alchemist from India/Shutterstock.com

Intercropping

Finally, **conservation tillage** can reduce erosion by cutting down on excessive plowing. **No-till** eliminates plowing altogether. In a traditional plowing operation, a farmer will plow deep furrows in the soil, then come back around with the seed hopper to deposit the seeds, and then make a third pass to close the furrows. Some areas with persistent weeds may require plowing three or more times to remove all of the weeds. In a no-till operation, first, the field is sprayed with weed killer and then a single apparatus performs the entire process in a single pass. First a blade cuts a small hole in the ground, then a seed is dropped into the hole. Fertilizer may also be sprayed into the hole before the apparatus closes it up. The previous year's crop residue is also left in place. This provides a cover for the soil during the off-season as well as providing a source of nutrients as it decomposes. This cover means the soil is never exposed to wind and water erosion.

Thomas Kloc/Shutterstock.com

Shelterbelts

Proponents of no-till point to the reduced erosion and fertilizer use compared to traditional plowing. Often, no-till can produce higher yields as well. Critics of no-till, however, point to the increased use of herbicides (weed-killers) as a substitute for plowing. This often requires the use of genetically modified crops (Chapter 11) that are herbicide resistant. Proponents, however, argue that the benefits of no-till outweigh the overall risk of the herbicide use.

A variation of no-till, low-till, is being practiced in Asia. In low-till, only a single pass is made to plow the field as well as plant the seeds and spread the fertilizer, instead of the usual 6 to 12 times as done in the past. Like no-till, low-till often requires the use of herbicides to control weeds.

Kraiwut_K/Shutterstock.com

In no-till, the previous year's crop residue is left in place as a cover to protect the soil from erosion.

In most developed nations, including the United States, agriculture is subsidized by the government. In the United States, about one-fifth of a typical farmer's income comes from taxpayer subsidies. While subsidies can sometimes promote sustainable practices, they can also be used to reduce erosion. For example, the 1985 Farm Bill created the **Conservation Reserve Program**, which pays farmers to plant cover vegetation on erosion-prone lands and put that land into conservation reserves.

chinahbzyg/Shutterstock.com

A no-till process.

The Natural Resources Conservation Service (NRCS), formerly known as the Soil Conservation Service, was created in response to the Dust Bowl. Today, the NRCS provides funding and education programs to work with farmers in developing soil conservation plans and other efforts to reduce soil degradation. Studies conducted by the NRCS indicate that the loss of soil due to erosion has dropped from 3.36 billion tons in 1982 to 1.72 billion tons in 2010, indicating that improved soil management practices have been successful.

The NRCS also provides information to ranchers to remove unwanted plants from their pasture land and replant beneficial grasses. They also promote moving herds around to different locations in order to reduce overgrazing.

Soil that has been degraded can be restored. One of the first steps involves returning organic material. Compost and manure can be used to build up fresh humus and earthworms can be used to improve soil structure. Vegetation can be replanted to establish root systems. If the soil has been contaminated with pollutants, organisms can be used to treat the soil through **bioremediation**. For example, bacteria can be used to break down petroleum products or plants can be used to absorb the pollutants from the soil. Once the pollutants have been taken up, the plants can be harvested and disposed of through incineration.

10.4 Discussion Questions

1. Describe the events that led up to the Dust Bowl and its impacts on Midwestern farmers.
2. What are the main components of soil and where do each of them come from?
3. Describe the five properties of soil and how different particle sizes affect them.
4. How do mollisols differ from oxisols?
5. What are the main causes of erosion?
6. Describe the processes that can lead to desertification and desert pavement.
7. What is salinization and how can it affect soil productivity?
8. What are the pros and cons of no-till agriculture?
9. How has the Natural Resources Conservation Service reduced soil degradation in the United States?
10. How does a low soil pH affect plant growth?**Crop Rotation**

10.5 Key Terms

A horizon
Aeration capability
Alfisols
Aridisols
B horizon
Bedrock
Bioremediation
C horizon
Clay
Conservation Reserve Program
Conservation tillage
Contour farming
Crop rotation
Cryptogamic crust
Deposition
Desert pavement
Desertification
Dust Bowl
E horizon
Erosion
Humus
Hydric soils
Intercropping
Low-till
Mineralization
Mollisols
No-till
Nutrient holding capacity
O horizon
Oxisols
Parent material
R horizon
Salinity
Salinization
Sand
Shelterbelt
Silt
Soil
Soil degradation
Soil horizon
Soil pH
Soil profile
Soil texture
Terracing
Water holding capacity
Water infiltration
Waterlogging
Weathering
Workability

Chapter 11

AGRICULTURE AND FEEDING THE GLOBAL POPULATION

Chapter Topics

In the 1990s, the African nation of Somalia was wrecked with a famine that was caused by a combination of civil war and drought. The civil war brought the government to a virtual collapse, turning Somalia into a failed state. The international community sent food aid to the stricken country. However, various warlords, who controlled much of the territory in Somalia, would seize the food, sometimes taking it right off the ships as they arrived in the port of Mogadishu. In 1992, the United States and other nations sent military personal to ensure the food got to the starving people and the attempt to stabilize the country. In 2019, history began to repeat itself. After two years of drought, Somalia is again facing food shortages. Now, as in the 1990s, the food aid is not getting to the people the most. This time, a radical organization called al-Shabaab, an affiliate of al-Qaeda, controls large territories in the control and is again preventing food from being distributed.

Since the time of Thomas Malthus, concerns have been raised about whether or not we can produce enough food to feed an ever-growing population. The good news is that technology has increased food productivity in ways Malthus could not have anticipated. Despite local droughts and crops failures in countries like Somalia, overall, the world produces more food than ever. Indeed, globally, there is no food shortage. The wealthier nations produce more food than they can consume even as poorer nations have had food shortages. Throughout the twentieth century, most famines were caused by social factors such as war and poverty.

With a global population of 7.7 billion and growing, the need to increase food production continues even as the amount of cultivated land has decreased. Many people argue that technology, including genetic engineering, will continue to keep pace with the population growth as we approach 9 billion by the year 2050 and 11 billion by the end of the century. The specter of global warming, however, raises the possibility that we may not be able to produce enough food to feed a growing population. Droughts in many regions, including sub-Saharan Africa, are expected to become more frequent and last longer. This would lead to more crop failures and more political instability in these countries. This may result in millions of people becoming climate refugees as they abandon their homes and immigrate to other countries. Others may turn to radical terrorist organizations like al-Qaeda and al-Shabaab. For these reasons, the US Department of Defense considers global warming to be an issue of national security. Finding ways to ensure food security to people in the poorest nations will be a major challenge in the twenty-first century.

11.1 Subsistence versus Industrial Agriculture

Steve Allen/Shutterstock.com

Humans first discovered agriculture about 10,000 years ago and for most of that time, they practice **subsistence agriculture**. Before this, all humans were living a hunter-gatherer lifestyle in which they would search for food in the areas around them. Hunter-gatherers were mobile. If food and other resources became scarce in one location, they would pick up and move to another. Often, they would follow herds of game animals as they migrated across the land. With the discovery of agriculture, people became more tied to one location. Growing crops requires people to live near them to tend the crops and protect them from pests until it was time to harvest them.

Subsistence agricultural involves farming with human and animal muscle as the primary source of power. It uses simple machines and hand tools to plant, tend, and harvest crops. Most often, subsistence agriculture is practiced on small family farms that plant **polycultures** (multiple crop species planted on the same plot) in small plots. Much of the food grown is for the family's personal consumption with a small surplus for trade. Subsistence agriculture is still practiced today in many developing nations and in small communities like the Amish in North America.

In much of the developed world, traditional agriculture has given way to **industrial agriculture**, which replaces human and animal muscle with powered tractors and other mechanization. Synthetic fertilizers, pesticides, and irrigation are used in large quantities in order to boost production. Most modern agriculture operations focus on planting **monocultures**, where large tracts of land are devoted to growing a single plant species. While monocultures can produce large quantities of a crop efficiently, they are also more vulnerable to pest and disease outbreaks compared to polycultures. In addition, the industrialization of agriculture has involved greater consumption of fossil fuels, pollution from runoff, and more resource consumption compared to traditional agriculture.

The focus on monocultures can also result in the loss of crop diversity. With more and more farmers devoted to growing narrow strains of crops, many traits may be lost. Many crop strains have gone extinct. Even some breeds of domesticated animals have become extinct as their traits were no longer desired. However, some of these strains could hold the key to resistance to certain

Suwin/Shutterstock.com

Valentin Valkov/Shutterstock.com

Theerasak Namkampa/Shutterstock.com

Global Seed Vault Svalbard.

diseases like the blight that struck potato crops, causing the Irish Potato Famine of the nineteenth century. As a hedge against the next major crop disease outbreak, **seed banks** have been established throughout the world. These facilities store seeds of rare crop strains and periodically replant them to keep the strains viable. In addition, in some countries like India, farmers have formed seed-sharing cooperatives to preserve strains of crops like rice.

We depend on agriculture to provide the food we eat as well as fibers like cotton and wool for our clothing. **Agriculture**, the practice of raising domesticated plants and animals for human use and consumption, is vital to the human civilization. Agriculture requires a lot of land, including **cropland** for raising plants and **pasture or rangeland** for the grazing of animals. Today more than out of every three acres of land on Earth are now devoted to raising crops or animals. In addition to land, agriculture requires inputs from the sun, soil, water, nutrients, and pollinators in order to produce the food and fiber we need.

11.2 Food Security and Famine

Food security is the guarantee of a supply of safe, nutritious food. Despite the increase in food production brought on by modern agricultural practices, hundreds of millions of people throughout the world still do not get enough calories needed to maintain good health. **Undernutrition**, or the lack of sufficient food energy, is the problem that plagues many developing countries. While droughts and crop failures can contribute to global **hunger**, **famines**, instances of widespread hunger, are mostly the result of social problems such as poverty and war. Climate change, however, is expected to cause increases in droughts and crop failures in many areas of the world where food production is already close to the edge. While most people suffering from undernutrition live in developing countries, nearly 50 million people in developed countries are considered "food insecure."

Geerte Verduijn/Shutterstock.com

Dead cattle in Kenya as a result of drought and famine.

Another problem facing the food insecure is **malnutrition**, or the lack of certain nutrients, vitamins, or minerals in one's diet. For example, kwashiorkor is the result of a protein-deficient diet. Symptoms include a wasting of the limbs, loss of hair, bloated stomach, and mental disabilities that could be permanent.

Marasmus is another form of protein deficiency that is combined under nutrition. A deficiency of iron can cause anemia, while a lack of vitamin A in the diet can lead to childhood blindness.

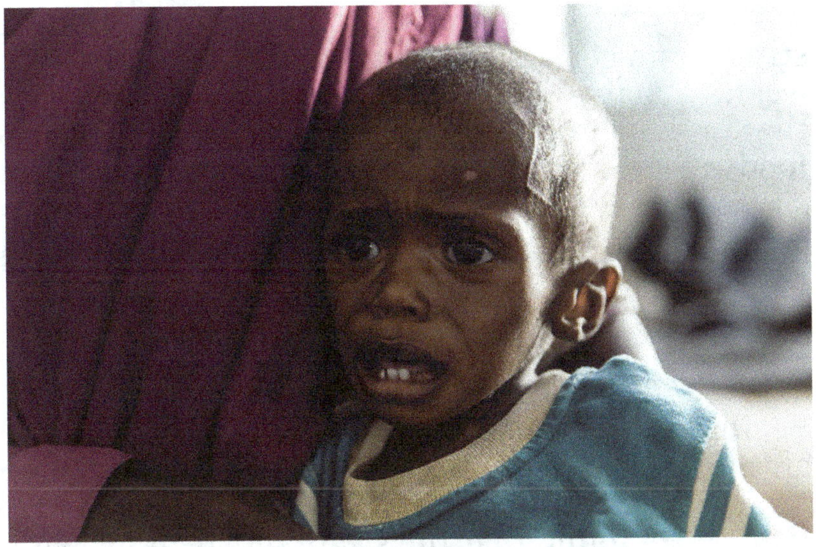

mustafa olgun/Shutterstock.com

One form of malnutrition that is somewhat uniquely human is scurvy. Most mammals have the ability to manufacture vitamin C. However, primates, including humans cannot. The genes for making vitamin C are there, but they were switched off millions of years ago. Most primates live in tropical climates where there are abundant fruits and other sources rich in vitamin C. So, these animals get all the vitamin C from their diet. Humans who live in areas where fruit is scarce, however, can be vulnerable to scurvy, which is caused by a lack of vitamin C. Your cat will never get scurvy even though he never eats oranges, but you could. In the days before refrigeration, scurvy became associated with sailors. During sea voyages that lasted several months, where they did not have access to fresh fruits, sailors became vulnerable to scurvy. The British Navy found they could ward off scurvy by keeping barrels of lime juice on the ships and requiring their sailors to drink it.

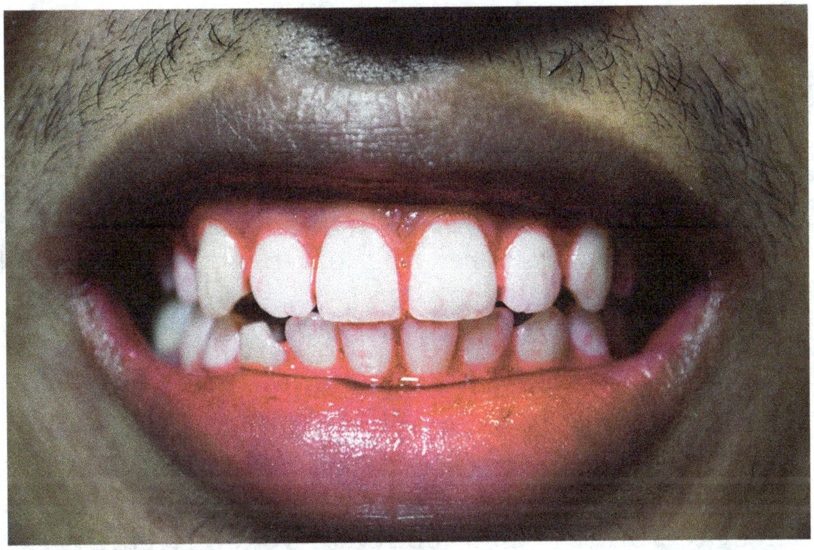

Pantakan Sakda/Shutterstock.com

Sore gums is a sign of scurvy

Poznyakov/Shutterstock.com

In developed countries like the United States, another dietary problem is reaching epidemic proportions: **overnutrition**, or the consumption of too many calories. Overnutrition can lead to obesity, which increases the risk of ailments such as diabetes, heart attack, stroke, and certain forms of cancer. Overnutrition is not just a problem for the wealthy. Many low-income communities are classified as **food deserts**, where there are little or no places to purchase fresh produce. Often, the main sources of food are junk foods that are high in calories but low in nutrients. As a result, obesity is a growing health problem in some of the poor communities in developed countries. In addition, scurvy appears to be making a comeback in some of these food deserts.

11.3 The Green Revolution

Beginning in 1940, teams of agricultural scientists like Dr. Norman Borlaug began an effort to tackle world hunger in what became known as the **Green Revolution**. Borlaug began his efforts in Mexico, where he studied many cases of wheat crop failure. He found that the strain of wheat grown in Mexico was very tall, which made it vulnerable to being knocked over in heavy windstorms that were common in some parts of Mexico. Through crossbreeding, Borlaug was able to produce a hybrid dwarf strain of wheat that had a higher yield and was left vulnerable to wind damage. Within a few years, Mexico went from being a net importer of wheat to a net exporter.

Borlaug and his team then introduced higher-yielding strains of wheat and rice crops into other countries such as India and Pakistan. Norman Borlaug's research was so effective in increasing food production that he was awarded the Nobel Peace Prize in 1970.

While the Green Revolution did reduce global hunger, the techniques Borlaug's team introduced to the developing nations required more use of mechanism, pesticides, fertilizers, and irrigation. This resulted in an increase in the environmental impacts in these countries, including pollution of water supplies and soil degradation. It also required more inputs of fossil fuels, increasing air pollution. These techniques also tended to benefit wealthy land owners less than the poorer subsistence farmers in these countries. Finally, these techniques for increasing crop yields have been just about maxed out in their potential. For example, one cannot keep increasing crop yields just by dumping more and more fertilizers on the land. Eventually, they reach a point of diminishing returns, where each ton of fertilizers added results in a smaller and smaller increase in production. As a result, scientists are now focused on genetic engineering (Section 11.7) as a means of increasing food production.

11.4 Irrigation and Fertilizer Use

Irrigation, the artificial supplying of water to grow crops, has turned many lands that would have been unproductive into productive farmlands. It is necessary to crop water-intensive crops like rice in arid regions. However, irrigation has a number of environmental impacts, including depletion of water supplies. Irrigation is the single largest human use of water and depletes many sources. For example, so much water is withdrawn from the Colorado River by states in the Western United States that it most over runs dry before it reaches the Gulf of California.

uparmar34617/Shutterstock.com

Excessive irrigation can also lead to waterlogging of crops and salinization, or the build up of salts in the soils. Too much salt in the soil will draw the water out of the roots (Section 10.2).

Juri Bondar/Shutterstock.com

One way to combat excessive irrigation is by matching the crop to the climate. For example, not growing rice in an arid region. Another is to avoid the traditional **flood-and-furrow method**, in which more than half of the water can be lost to runoff of evaporation. Flood-and-furrow, as the name implies, involves digging furrows in the field and flooding them with water. An alternative to flood-and-furrow is **drip irrigation**, which involves running a series of hoses or pipes through the field in order to supply each plant with just enough water that it needs. This reduces both runoff and evaporation. It also reduces the risk of waterlogging and salinization.

Fertilizers, which are substances that can supply nutrients to promote plant growth, have also been used to boost crop production. **Inorganic fertilizers** are manufactured from mined minerals or produced through industrial fixation of nitrogen. Inorganic fertilizers are designed to dissolve quickly and release lots of nutrients to crops in a short period of time, giving them a rapid boost in growth. However, the fact that they dissolve readily also makes them more prone to runoff, which carries them into rivers and streams. As these nutrients travel through the waterways, they eventually promote algae growth, which leads to eutrophication and the formation of dead zones along the coast (Chapter 2).

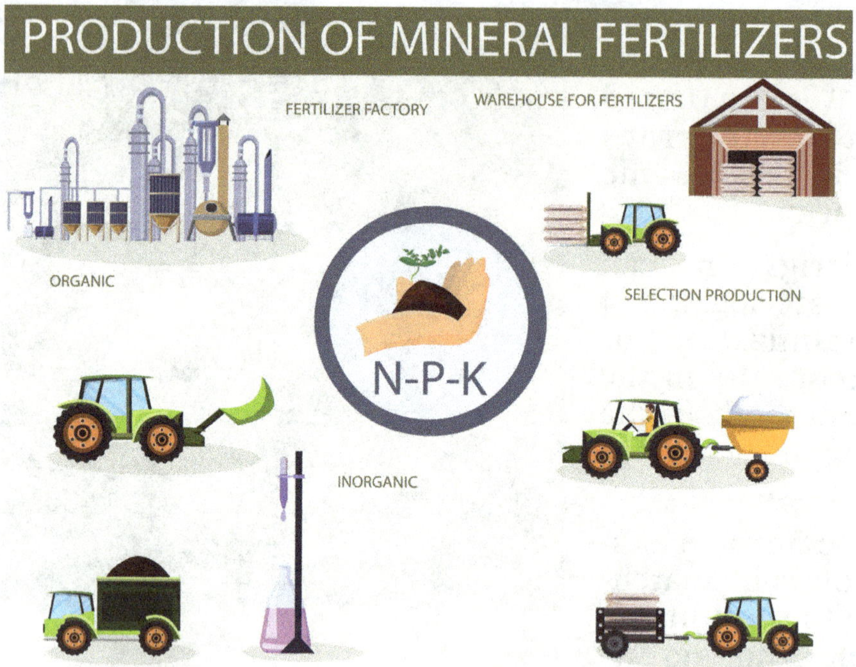

Unitone Vector/Shutterstock.com

Inorganic fertilizers.

Hemerocallis/Shutterstock.com

Organic fertilizers are produced from the remains and wastes of living organisms. They can include animal manure, crop residue (sometimes called "green manure"), fresh vegetation, and charcoal. **Compost** can also be used as a fertilizer and is produced by decomposing food scraps or other organic material that is broken down by decomposers to produce a dark, humus-like material. Food processing waste, pulp mill wastes, and even treated sewage sludge or biosolids can also be used as fertilizer. While organic fertilizers release their nutrients slowly over time and do not promote the rapid growth that inorganic fertilizers do, they have other benefits. For example, they improve soil structure, water-holding capacity, and nutrient-holding capacity. They are also less vulnerable to runoff compared to inorganic fertilizers. As a result, it is often recommended that farmers add a mixture of both organic and inorganic fertilizers to gain the benefits of both while minimizing the negative consequences.

Because the soils in tropical rainforests are often nutrient poor, farmers in those biomes often practice **swidden or slash-and-burn agriculture**. In swidden agriculture, the native vegetation is cut down and burned. The ash is then mixed with the soil to produce a quick fertilizer. However, the heavy rains and warm temperatures still cause the soil to lose nutrients quickly, so a given plot can only be farmed for a few years before they have to move on. When practiced on a small scale, slash-and-burn can be sustainable and the forest can reclaim the cleared land over time. However, if too much land is cleared at once, the forest will not be able to reclaim the land. Over time, the edge of the forest can be pushed further and further back, converting the cleared land into less productive scrubland.

Farmers can avoid excessive fertilizer use by practicing **precision agriculture**, in which they monitor the nutrient levels in their soil and only add when it is needed. Many universities and state departments of agriculture will provide farmers with agronomy guides that will help them determine how much nutrients each crop needs and how much fertilizer should be added for optimum growth. In addition, practices like crop rotation and intercropping (Section 10.3) can incorporate legumes into the farming practice to increase the amount of available nitrogen in the soil.

Fertilizers can also be added to the water in drip irrigation, thereby giving each plant just enough of the nutrients they need.

Hennadii H/Shutterstock.com

Florian Giesler/Shutterstock.com

Slash-and-burn in Laos.

11.5 Pest Management

A **pest** is any organism that damages crops or livestock or spreads disease. Animal pests can include a variety of insects like grasshoppers, rodents like rats or mice, or even birds like crows. Weeds are any plants that act as pests by competing with crops for water, sunlight, nutrients, or space. Since the monocultures favored by industrial agriculture are particularly vulnerable to pest outbreaks, farmers rely on chemical pesticides to control them.

Malachi Jacobs/Shutterstock.com
Raul Baena/Shutterstock.com
kaysona Global/Shutterstock.com
Aleksandar Cholanchevski/Shutterstock.com

Examples of pests: locusts, rats, crows, and weeds.

A **pesticide** is any chemical manufactured for the purpose of killing pest populations. The suffix "-cide" means to "to kill," as the word "homicide" means to kill a person. The chemical industry produces many pesticides to target specific pests. For example, insecticides are used to kill insects, herbicides kill weeds, fungicides kill fungi, and rodenticides kill rodents. Unfortunately, pesticides can also kill nontarget species. For example, DDT, once widely used to control insect pests, have harmed populations of birds of prey like eagles and owls. Many insecticides also kill beneficial species like pollinators and are suspected to be responsible for colony collapse disorder. Often, pesticides also kill the parasites and predators of pests.

Jinning Li/Shutterstock.com

Aerial drone spraying pesticide on a field.

The first generation of pesticides used in agriculture were compounds containing heavy metals like lead and mercury. These toxic heavy metals were effective in controlling pest populations until they weren't. After a while, pest populations become resistant to the pesticide through natural selection. For example, the first time a pesticide is applied, it may kill over 90% of the pest population. However, those that survived had traits that enabled them to metabolize or excrete the pesticides before they could be killed. Then, they become the breeding population for the next generation of pests. With each pesticide application, the population becomes more and more resistant, making the pesticide less effective in killing the majority of pests. To combat **pesticide resistance**, applicators initially increased the dosage applied. However, eventually, they were applying dosages that were hazardous to human health, so they had to switch to second-generation pesticides that were made form synthetic organic compounds.

Today, we are using third- and fourth-generation pesticides, which over time, will lose their effectiveness. Many weed and insect species are now resistant to multiple pesticides. This has put us on a **pesticide treadmill**, in which scientists have to keep researching new and often more toxic pesticides to replace ones that have become less effective.

One alternative to chemical pesticides is **biological control** (biocontrol) in which the pest's natural enemies control it. It could be a disease, such as rabbit hemorrhagic disease, which is used to control rabbits that are damaging the grasslands in Australia. Or it could be a parasite or predator, such as the parasitoid wasps that lay their eggs on the backs of caterpillars. When the eggs hatch, the wasp larvae emerge and eat the caterpillar, making them effective in controlling them.

However, sometimes introducing a species to control a pest can backfire. Sometimes, the predator may decide it is easier to go after native species that were not the original target. As we saw in Chapter 5, when the mongoose was introduced into Hawaii to control rats, it instead fed on bird eggs, putting native Hawaiian bird populations at risk. Sometimes, they can jump locations, going from controlling a pest in one location to becoming pests in another. For example, the cactus moth was introduced into the Caribbean Islands to control the invasive prickly pear cactus. However, the cactus moth soon made its way to America, where it is now attacking native cacti. For these reasons, any **biocontrol** measures should be implemented with careful planning and research to try to identify any potential downsides before introducing a new species.

Jennifer Bosvert/Shutterstock.com

Ladybugs are effective in some biocontrol measures.

We can also use the pest's own biochemistry against them. For example, a juvenile hormone can be applied to insects to prevent molting or moving into the pupil stage. These insects would then just keep gaining weight until they collapse. Sex hormones can be used to lure insects into traps or sprayed around an area to confuse males. The male can't find a female to mate with if everything around them smells like a female. The advantage of using hormones for biocontrol is that they are species-specific and do not harm target species.

Males can also be sterilized by radiation and then released to mate with females in the wild. This can disrupt the entire breeding cycle.

Another tool in biocontrol is the bacterium **Bacillus thuringiensis (Bt)**. This is a naturally occurring soil bacterium that produces a protein that kills many species of caterpillars, as well as some fly and beetle larvae. The spores of Bt can be sprayed on crops to kill insects without harming people or beneficial species. The genes that produce the Bt toxin can also be inserted into the genome of plants through genetic engineering so that they can produce the toxin themselves.

Integrated pest management (IPM) is a holistic approach to controlling pests that involves studying the pest's life cycle, looking for factors like where is the pest most vulnerable and what changes can be made into the environment to make it least attractive to the pest? Instead of relying on one tool to control the pest, IPM adopts a variety of techniques in order to maximize effectiveness. These tools can include

- Chemical pesticides in a targeted, limited manner
- Biocontrol
- Habitat alteration
- Crop rotation
- Transgenic crops
- No-till farming
- Mechanical pest removal through traps or other methods
- Monitoring pest populations and targeting where the pests breed
- Altering the environment to attract predators or parasites that attack the pest
- Nets or other covers making it difficult for the pests to reach the crops

Catherine Eckert/Shutterstock.com

Crop scouting as part of an IPM program.

Thanakrit Homsiri/Shutterstock.com

Covering crops for IPM.

Thanakrit Homsiri/Shutterstock.com

Flowers used to repel certain pests.

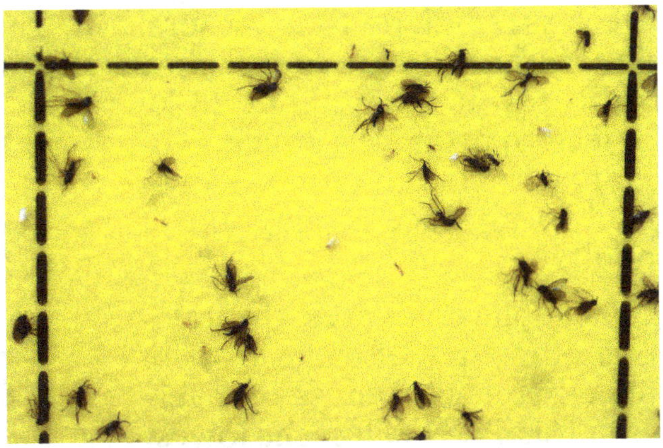

Catherine Eckert/Shutterstock.com

Sticky traps.

IPM can also involve rotating different pesticides from year to year in order to reduce the chances of the pest population developing resistance to any one pesticide. While IPM programs require more planning and preparation than conventional pesticide use, studies have found that overall, farmers save money through reduced pesticide use. Indonesia has managed to increase rice production through an IPM program that reduced overall pesticide use. In the United States, studies have found that farmers using IPM produce crop yields that are comparable to conventional pesticide use. This has resulted in a net increase in profitability.

11.6 Ranching, CAFOs, and Aquaculture

Bob Pool/Shutterstock.com

In preindustrial societies, meat makes up a much smaller portion of the human diet. Meat is expensive to produce and animals are often more valuable for their milk, eggs, or labor. However, as societies industrialize and become more affluent, the demand for both meat and other animal products increases. Worldwide, global meat production has increased by a factor of four or five since 1961.

Because consuming meat involves eating one step higher on the food chain than consuming plants, eating meat is a less efficient means of getting our calories. The rule of 10% (Chapter 3) states that only about 10% of the available energy moves from one level in the food chain to another. So, it can take 10 times as much energy to produce the same number of calories of meat as it does calories of plants. Growing grain and then feeding it to a cow results in most of the energy in the grain being consumed by the cow's metabolism, making it unavailable for human use.

Consumption of meat also comes with increased impacts on the environment because we have to account for both the impact of the animals themselves but also the growing of the plants needed to feed them. Some forms of meat production are less efficient than others. For example, as the chart below indicates, it takes about 25 kg of feed to produce a single kilogram of beef, but only 3.3 kg of feed to produce a kilogram of poultry.

This table compares the relative efficiency of each form of meat in terms of energy conversion. Again, beef has the lowest levels of efficiency, while eggs have the highest. The next table below illustrates the amount of land area required to produce a gram of animal protein and once again, beef (and mutton) consumes the most land.

Finally, this table shows the amount of greenhouse gas emissions for each food type.

Because of the rising demand for meat, eggs, and dairy, the farming industry is turning more and more toward **concentrated animal feeding operations (CAFOs)**, also known as factory farms or feedlots. In CAFOs, animals are packed into high-density pens or cages and feed a diet designed to maximize production. CAFOs have the advantage of being able to produce large quantities of meat, eggs, and dairy efficiently and they reduce the impact on pasture land. To allow the same number of animals to graze on fields would require a much larger amount of land area and would make problems like overgrazing worse.

Critics of CAFOs, however, argue that they are cruel to the animals as they spend their

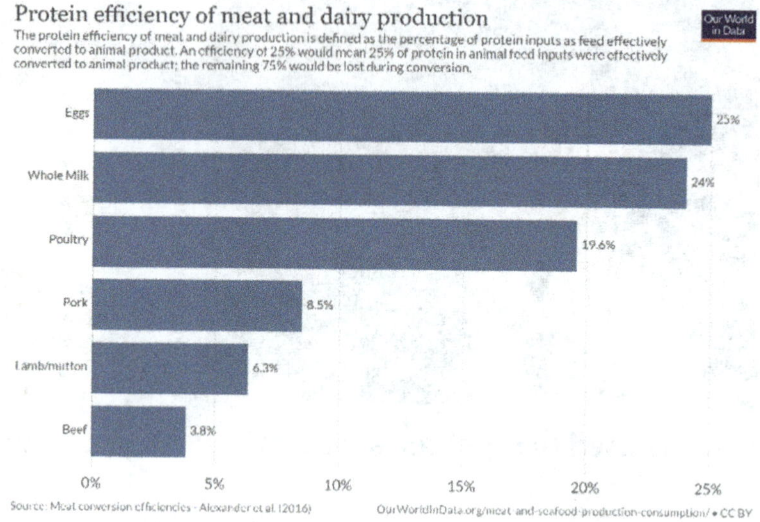

Source: Our World in Data. https://creativecommons.org/licenses/by/4.0/

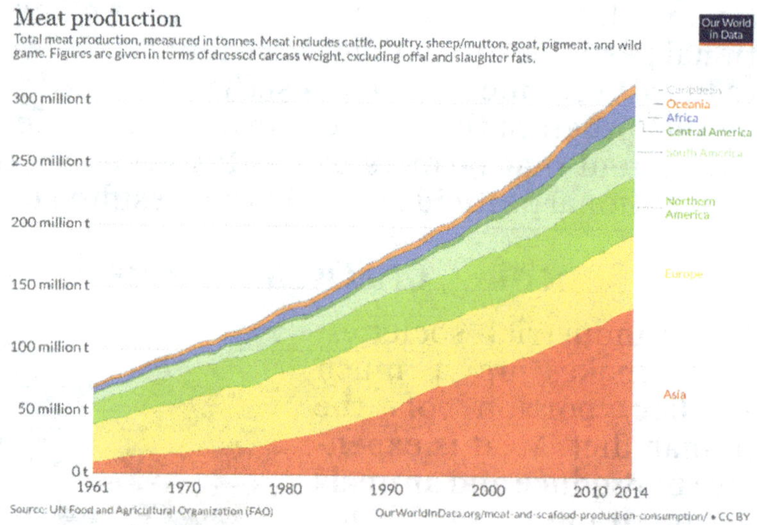

Source: Our World in Data. https://creativecommons.org/licenses/by/4.0/

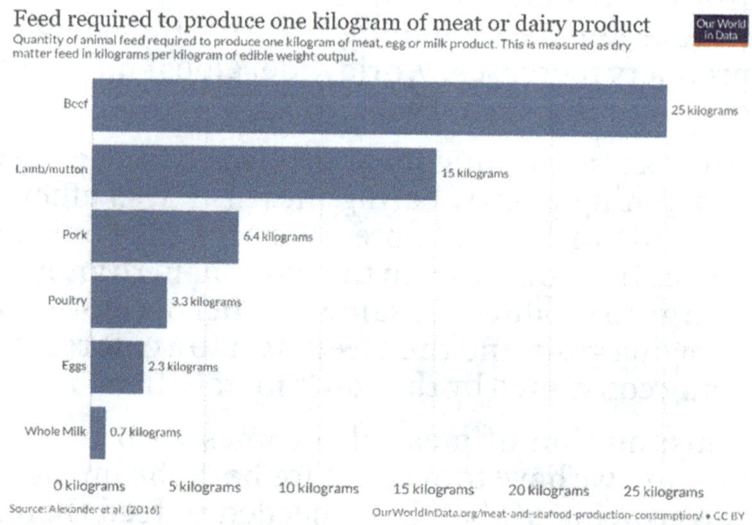

Source: Our World in Data. https://creativecommons.org/licenses/by/4.0/

entire lives in very confined spaces and put too much stress on them. Many poultry CAFOs take the precaution of filing down the beaks of chickens in order to prevent them from attacking each other. CAFOs also generate large quantities of manure that must be disposed of carefully as they could contaminate surface and drinking water supplies and spread waterborne illnesses. Often, the manure is collected in pools called "lagoons." In storm-prone areas, these lagoons are at risk of overflowing. In addition to spreading pathogens, animal manure is rich in nitrogen and phosphorus and can contribute to eutrophication.

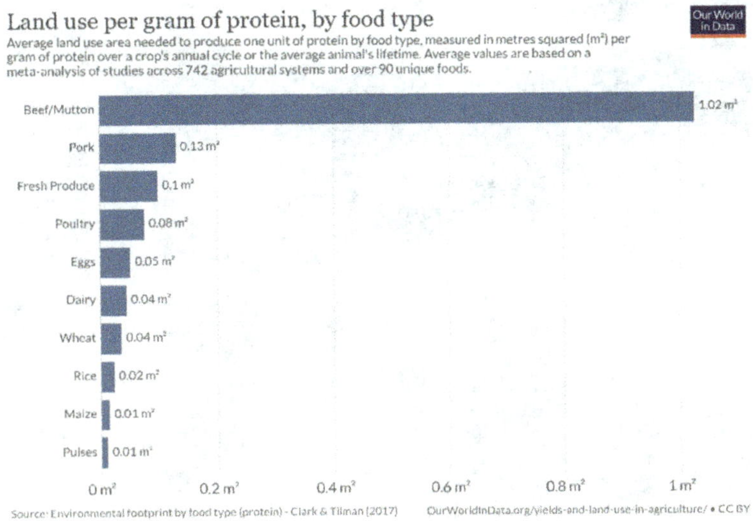

Source: Our World in Data. https://creativecommons.org/licenses/by/4.0/

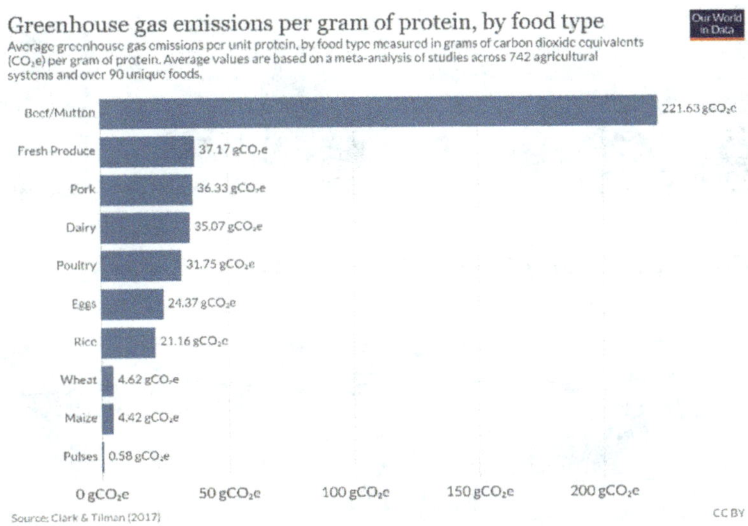

Source: Our World in Data. https://creativecommons.org/licenses/by/4.0/

CAFOs also often employ hormones such as the bovine growth hormone, which is used to increase milk production. Also, because CAFOs involve large numbers of animals in confined areas, disease outbreaks are common and require frequent use of antibiotics. Overuse of antibiotics can result in producing antibiotic-resistant bacteria strains, just as overuse of pesticides can produce pesticide-resistant pests. Many bacterial infections are now resistant to multiple antibiotics making them more dangerous than they were just a few decades ago.

Controlling infections, however, is not the only reason why antibiotics are used in CAFOs. If you've ever had to take an antibiotic treatment for an extended period of time, you may have noticed you put on a little weight afterward. While the reasons this happens are unclear,

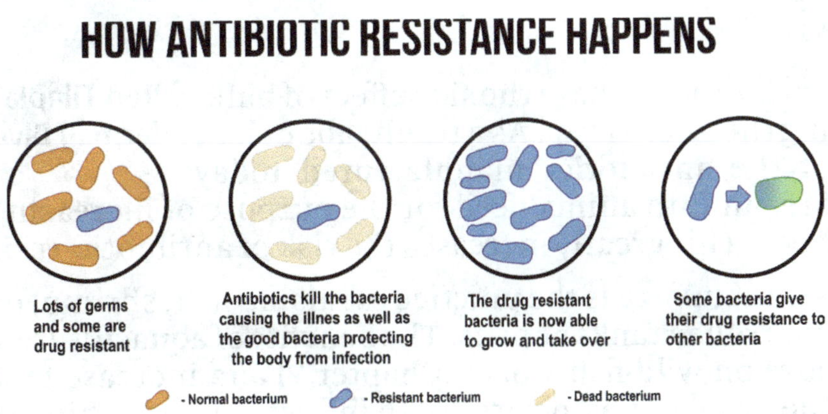

ducu59us/Shutterstock.com

Lost_in_the_Midwest/Shutterstock.com

Joseph Sorrentino/Shutterstock.com

Joseph Sorrentino/Shutterstock.com

Cow and turkey CAFOs. A cow CAFO with a mound of manure.

leo w kowal/Shutterstock.com

Atlantic salmon aquaculture.

tanakornsar/Shutterstock.com

Red Tilapia culture in cage at Aquaculture farm in River from Thailand.

antibiotics do have the side effect of bulking the animals up. As a result, about half of the antibiotics manufactured today are put into animal feed for the purpose of increasing the amount of meat on the hoof. This greatly increases the risk of antibiotic-resistant infections.

Aquaculture is the practice of raising fish, shrimp, or other aquatic organisms in controlled tanks or pens. The benefits of aquaculture are that they reduce the pressure on wild fish stocks (Chapter 7) and increase food security. They also reduce the problems associated with industrialized fishing such as bycatch. On the other

Aerial view of a prawn farm in Thailand.

hand, like other high-density animal operations, they use a lot of antibiotics. They can also degrade coastal ecosystems if they employ pens inside an estuary. They can also carry the risk of spreading diseases among fish populations and introduce invasive species into a new area. Recently, the first genetically modified animal that has been approved for human consumption is a salmon engineered to grow larger and faster. Critics fear that if the salmon escapes from a fish farm, it could outcompete and replace wild salmon.

Despite these concerns, aquaculture is growing rapidly, especially in Asia.

11.7 Genetically Modified Food

Genetic engineering is the practice of actively manipulating an organism's DNA. Organisms that have had their DNA altered are referred to as **genetically modified organisms** (GMOS). Most commonly, GMOs are produced by taking the DNA from one organism and introducing it into the genome of another. Genes that have been moved from one organism to another are **transgenes** and therefore, GMOs are also called **transgenic organisms**. These organisms contain **recombinant DNA** or DNA that has been recombined with that of another organism. The main

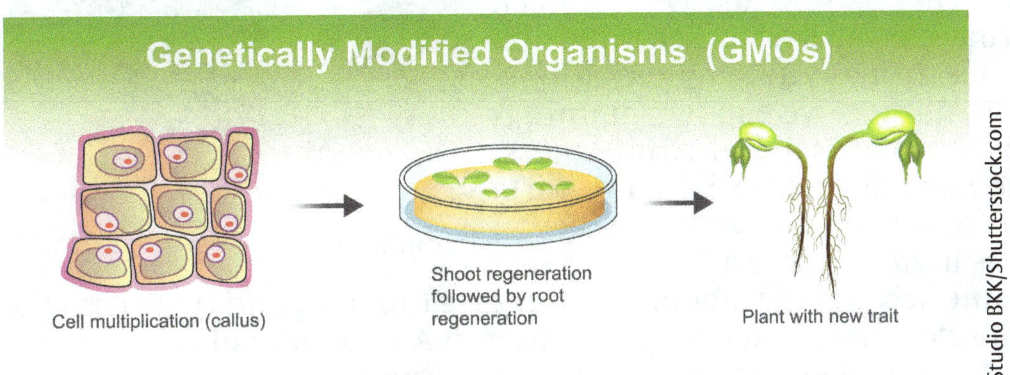

difference between genetic engineering and the practice of selective breeding that humans have employed for thousands of years is that selective breeding can only enhance existing traits in a population. Genetic engineering introduces entirely new traits into an organism. Prior to the era of genetic engineering, the only way new traits could be introduced into a population is through mutations.

Genetic engineering is one example of **biotechnology**, that is, using biological sciences to produce consumer products.

Genetic engineering is not science fiction. In fact, it has become common in modern agriculture. The four most common crops to be genetically modified are corn, cotton, canola, and soybeans. Some of them have even been "stacked," meaning they've received more than one trait introduced through genetic engineering. It's been estimated that 70% of processed foods contain at least one GMO in its ingredients.

One of the most common genetic modifications used in agriculture is introducing the gene that produces the Bt toxin. The Bt gene has been introduced into a variety of crops, including cotton, soy, corn, and sunflowers. The use of the Bt gene has reduced the use of insecticides and saved many farmers money. Critics of the practice argue it increases the risk of insect populations becoming resistant to the Bt toxin, thus reducing its effectiveness. Others, raise the possibility of **horizontal gene transfer**, in which plants like sunflowers could cross breed with their weedy relatives creating "superweeds" that are more difficult to control.

Another common modification is called Round-up Ready, which was developed by the Monsanto Corporation, which also manufactures the popular herbicide Round-up. Round-up Ready makes crops resistant to the active ingredient in Round-up, glyphosate. Round-up Ready has been introduced into a variety of crops, including corn, soy, sugar beets, and cotton. Proponents point to the fact that herbicide resistance is necessary for no-till farming operations. Critics of the practice worry that it will result in the overuse of glyphosate that will harm nontarget species and produce strains of weeds that are resistant to the herbicide. Indeed, there has been an exponential increase in reports of glyphosate-resistant weeds between 2002 and 2015.

Golden rice.

JIANG HONGYAN/Shutterstock.com

Other examples of GMOs include the salmon designed to grow bigger, potatoes resistant to the blight, and papayas resistant to the ring spot virus. One biotech product that has received much attention is golden rice. In many developing nations, children do not receive enough vitamin A in their diets, which can lead to blindness. Since rice is a staple crop in many parts of the world, it makes sense to engineer a crop that is high in vitamin A. Golden rice gets its name because the modification causes the rice grains to turn yellow. So far, a strain of rice that is rich enough in vitamin A to make a difference in children's diet has yet to be developed, but research continues.

Worldwide, the United States is the largest producer of GMOs, followed by Brazil, Argentina, India, and Canada.

Proponents of GMOs argue that their products can reduce pesticide use, reduce poverty and hunger, increase the nutritional value of crops, reduce water use by producing crops that need less water, alleviate pressure to clear more land for agriculture, and increase crop yields. Some of these goals have been partially achieved. For example, Bt crops have resulted in a decrease in insecticide use, but Round-up Ready crops have actually increasing herbicide use.

Many people object to the use of GMOs, with public resistance in Europe being much higher than it is in North America. Some question whether these products are safe and want more stringent testing before GMOs are approved for human consumption. While research has not found evidence that GMOs on the market have any more health impacts than their nonmodified counterparts, many still want a strong precautionary principle applied. They want regulators to require that producers prove that their products are safe before approval, rather than point to an absence of proof of harm. They complain that because of the modifications in many commonly consumed crops like corn and soy, people are being made into guinea pigs on the long-term effects of these GMOs without their consent.

Other critics of GMOs simply do not like the idea of scientist tinkering with the food supply, claiming that it is unnatural. Others fear that we are giving control of our food supply over to large biotech companies such as Monsanto, Bayer, BASF, and DuPont. These companies spend millions of dollars developing their GMO products and expect to make a substantial profit from them. So, they take out patents on these genes and have been willing to take farmers to court who knowingly replant or harvest modified crops without paying them licensing fees. They also often require farmers to sign legal contracts barring them from replanting the seeds from the previous year's harvest. This requires them to purchase new seeds from the company every year.

11.8 Sustainable Agriculture and Organic Agriculture

The goal of **sustainable agriculture** is to reduce the impacts of agriculture while still maintaining food security. Low-input agriculture involves finding ways to reduce the amount of fossil fuels, inorganic fertilizers, and chemical pesticides used in growing food. On the other hand, **organic agriculture** involves eliminating the use of inorganic fertilizers, chemical pesticides, hormones, and antibiotics altogether. The federal Organic Food Production Act of 1990 directed the US Department of Agriculture to develop standards for which farmers can sell their products as certified organic.

Surveys show that most Americans buy some organic foods at least some of the time, so there is a market for them. Research has indicated that organic farming results in improvements in soil structure and microorganism activity. Organic farming, however, is less efficient than traditional methods and governments offer subsidies to help farmers make the transition. Less-efficient farming methods means organic food is more expensive for the consumer and organic produce is often less attractive than produce grown using pesticides. This is because many pests do not make the produce inedible but can degrade its appearance, so much pesticide use is for "cosmetic" purposes. For example, organically grown apples may be a duller red color and have more brown spots. Despite this, there is no real difference in the taste.

ANEK SANGKAMANEE/Shutterstock.com

Organic farming in Thailand.

No study, however, has found that organic foods are safer or more nutritious than traditionally grown crops.

Another way to reduce the environmental impact of agriculture is to buy local. The average food product in America travels about 1,000 miles from farm to plate. Some people call themselves "locavores" because they only eat food grown locally (no bananas for them unless they live in a tropical region!). One way to buy local is shopping at **farmers markets**, where customers can often have the opportunity to interact with the farmer and learn about some of their practices. Another way to buy locally grown food is through **community-supported agriculture**, where customers pay the farmer in advance for a share of their produce. The advantage for the farmer is a guaranteed source of income while the consumer gets fresh produce that they know where it came from. Again, there is an opportunity for Americans to learn how their food is grown and what kinds of fertilizers and pest management techniques are employed.

AYA images/Shutterstock.com

Shopping at farmers market gives people an opportunity to learn where their food comes from.

wow.subtropica/Shutterstock.com

Overall, the goals of sustainable agriculture are

1. Food security
2. Profitability for farmers
3. Conserve resources for future generations

11.9 Discussion Questions

1. What is the difference between subsistence and industrial agriculture?
2. What is the purpose of a seed bank?

3. What are the main causes of hunger throughout the world?
4. What is a food desert and how does it contribute to overnutrition?
5. What is Green Revolution and how did it reduce global hunger in the twentieth century?
6. Describe the problems associated with excessive irrigation and how can they be minimized?
7. What are the benefits and drawbacks of inorganic fertilizers?
8. What is swidden agriculture and why is it unsustainable when practiced on a large scale?
9. How do pests become resistant to certain pesticides?
10. Describe some of the steps involved in an integrated pest management plan.
11. Why do CAFOs use large amounts of antibiotics and what problems can this cause?
12. What are some of the reasons why people object to the use of genetically modified foods?
13. What are the requirements in order for becoming certified organic agriculture?
14. What is community-supported agriculture and what are the benefits to the consumer and the farmer?

11.10 Key Terms

Agriculture
Aquaculture
Bacillus thuringiensis rangeland
Biocontrol
Biological control
Biotechnology
Certified organic agriculture
Community-supported agriculture
Compost
Concentrated animal feeding operations
Cropland
Drip irrigation
Famine
Farmers market
Fertilizer
Flood-and-furrow irrigation
Food desert
Food security
Genetic engineering
Genetically modified organisms
Green Revolution
Horizontal gene transfer
Hunger
Industrial agriculture
Inorganic fertilizer
Integrated pest management
Irrigation
Malnutrition
Monoculture
Organic fertilizer
Overnutrition
Pasture
Pest
Pesticide
Pesticide resistance
Pesticide treadmill
Polyculture
Precision agriculture
Rangeland
Recombinant DNA
Seed bank
Slash-and-burn agriculture
Subsistence agriculture
Sustainable agriculture
Swidden agriculture
Transgenes
Transgenic organisms
Undernutrition

Chapter 12

ENVIRONMENTAL HEALTH AND RISK MANAGEMENT

Chapter Topics

Bisphenol A (BPA) is a chemical that is used in a lot of products, including cans, bottles, toys, dental sealants, and electronics. Studies have found that 90% of Americans have BPA in their bodies, mostly as it leaches from plastics into food and water. BPA is in a class of chemicals known as endocrine disruptors. These are chemicals that affect endocrine system either by mimicking the effects of hormones or interfering with their normal functions. In particular, BPA acts like estrogen in the body, which can result in a variety of impacts on health. BPA has been linked to a number of health problems including infertility, erectile dysfunction, heart disease, Type II diabetes, fetal brain abnormalities, breast cancer, prostate cancer, and asthma.

Burak Tunc/Shutterstock.com

Because of these health concerns, Canada has banned the use of BPA altogether while the United States has placed restrictions on its use. Many food and beverage companies removed BPA from of their packaging and labeling their packaging as BPA-free. Toy manufacturers and manufacturers of products for infants like baby bottles are also phasing BPA out as well. The actions taken over BPA is the result of society evaluating the risks of chemicals through the practice of environmental health.

12.1 The Practice of Environmental Health and Hazard Identification

Environmental health is the study of factors in the environment that impact public health and how these hazards can be reduced. Every day, people face a variety of hazards. While it is impossible for anyone to risk to zero, being aware of potential hazards can help people avoid unnecessary risks. There are four main categories of environmental health risks.

Physical hazards are those that originate from natural processes such as sunlight or natural disasters. Ultraviolet radiation, earthquakes, floods, tornados, and tsunamis are all examples of physical hazards. There is not much we can do to prevent physical hazards, but we can take precautions to reduce our risks. For example, using sunscreen can reduce out risk from UV radiation. Satellite data and radar can give us warnings of hurricanes or other weather-related hazards, enabling people to evacuate or take other preparations. Buildings can be made resistant to damage from Earthquakes. Some, we can make worse, such as clear-cutting slopes, which increases the risk of landslides or building homes in flood-prone areas.

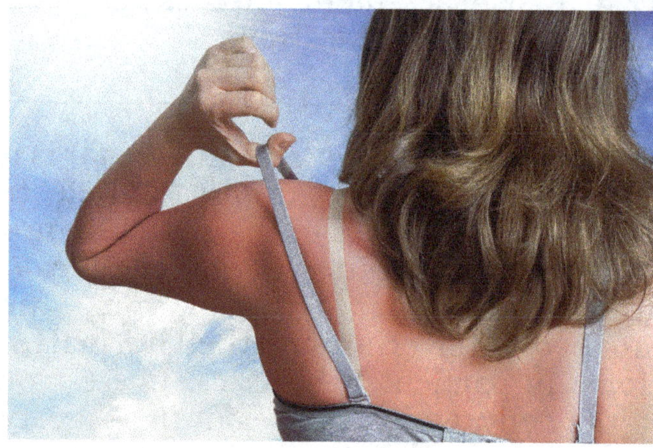

vchal/Shutterstock.com

Chemical hazards include synthetic chemicals such as pesticides, paints, solvents, disinfectants, and pharmaceuticals. It can also include heavy metals like lead, mercury, or cadmium. It can also include natural chemicals like snake venom or poisonous plants.

AStudio-1/Shutterstock.com

Biological hazards are any **infectious diseases** that can be transmitted from person to person. These can be caused by a variety of pathogens, including bacteria, viruses, and protists. Environmental health often focuses on **vectors**, organisms that can transmit a pathogen from one person to another. For example, mosquitoes are vectors for a variety of diseases, including West Nile, malaria, and yellow fever. Ticks are vectors for Lyme disease and rocky mountain spotted fever. Certain fleas that travel on the back of rodents are vectors for the black plaque, which devastated Europe, killing about a third of its population in the Middle Ages.

Finally, **cultural hazards** are hazards associated with individual behavior choices, occupation, or socioeconomic status. These can include a poor diet, smoking, or

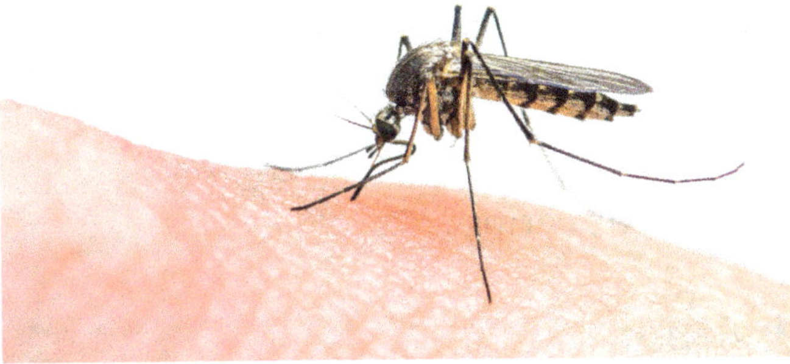

nechaevkon/Shutterstock.com

Cosmin Manci/Shutterstock.com

Mosquitoes and fleas are vectors for several infectious diseases.

drug abuse. In can also include certain workplace hazards such as those faced by firefighters, police, or the military. For example, alcohol abuse can put people at risk for diseases like cirrhosis of the liver. Of particular note is how our societal attitudes toward smoking tobacco have changed. Sixty years ago, cigarettes were advertised on television and were even the primary sponsors of shows like the Flintstones. For decades, there was considerable public debate, much of which was driven by a well-financed effort by the tobacco industry about the risks of smoking. Today, it is universally accepted that not only does direct smoking, but also passive or second-hand smoking carries a significant risk of lung cancer and other diseases. In recent decades, college campuses, shopping malls, and healthcare facilities have banned smoking on their premises. Many states have enacted laws restricting smoking in most places opened to public such as restaurants. Smoking has also been banned from most commercial flights. In short, smoking has gone from something that was ubiquitous in our society to being pushed to the margins as people have come to accept the risks associated even with passive exposure to cigarette smoke.

Olena Yakobchuk/Shutterstock.com

There is a big difference in the primary causes of death in developing and developed nations. In developed nations, the major causes of death are **noninfectious disease** like cancer and cardiovascular disease. Antibiotics, vaccines, and improvement in public sanitation have reduced the prevalence of infectious disease in developed countries. This has resulted in a dramatic drop in the number of deaths by these diseases since the Industrial Revolution. On other hand, the obesity epidemic in many developed countries has put many people at higher risk for noninfectious diseases. Unfortunately, as people have begun to (falsely) question the safety of vaccines, vaccination rates in some developed countries have dropped and infectious diseases like measles have begun to make a comeback.

As better healthcare and sanitation have spread throughout the developing world, deaths from infectious diseases have gone down, but these diseases are still the largest causes of death in these countries. Major infectious diseases in the developing working include respiratory infections like influenza, pneumonia, and tuberculosis, diarrheal and gastrointestinal disease, malaria, meningitis, and (especially in Sub-Saharan Africa) AIDS.

Factors that reduce the spread of infectious disease include vaccines, public sanitation, access to clean water, food security, and education.

12.2 Toxic Substances and the Environment

A toxic substance or **toxicant** is a chemical substance that is injurious to health. **Toxicology** is the study of have toxic substances affect the human body and the environment. **Toxicity** is a measure of how harmful or poisonous a substance is. The old saying in toxicology is "the dose makes the poison" means that the degree of harm a substance a chemical can have is based on the dosage received and the extent of exposure. Some chemicals like mercury are very toxic even at minute dosages. Other substances can have little or no effect at low dosages but can become toxic at high dosages. Some, like Vitamin A, are even beneficial at low dosages but toxic at much higher dosages.

Some of the toxic substances that have been extensively studied in environmental toxicology include cigarette smoke and radon, which are the first and second leading causes of lung cancer, respectively. Cigarette smoke, of course, comes from the burning of tobacco. Radon is a naturally occurring radioactive gas that is produced by radioactive decay of isotopes in the soil. It often seeps into homes and can build up inside. Because it is radioactive, inhaling radon results in exposing the lungs to radiation, increasing the risk of lung cancer.

Asbestos is another respiratory hazard that has been extensively studied. Originally, the friable form of asbestos was widely used as insulation and fireproofing buildings. Asbestos is a fibrous mineral that, when it becomes friable, releases tiny hair-like particles into the air. When inhaled, these particles can irritate the lungs, resulting in a condition called asbestosis. The friable form of asbestos was phased out of the market in the 1970s, but many older buildings still contained it. In the 1980s, many building owners began ripping out asbestos insulation. Unfortunately, this often made the problem worse as the act of tearing the insulation out released more asbestos particles into the air. As a result, the industry shifted to more encapsulation in place to contain the asbestos and prevent it from becoming airborne instead of removing it.

HikoPhotography/Shutterstock.com

Once widely used in insulation, asbestos has been linked respiratory illnesses.

Jamie Hooper/Shutterstock.com

Lead is a heavy metal that is used in making car batteries and other industrial processes. Lead was also used as an additive in gasoline to make engines quieter, as the basis of paints, and in the solder in plumbing. Lead is a neurotoxin that can cause a variety of effects, including permanent learning buildings. Lead was phased out of gasoline starting in the 1970s. Lead solder and lead-based paint were phased out in the 1980s. However, many older buildings may still contain both. Small children and infants are particularly vulnerable to inhaling or ingestion lead from lead paint. And, as the residents of Flint, MI found (Chapter 8), lead solder can still be a health hazard, especially when caustic water is used.

Polybrominated diphenyl ethers (PBDEs) are a class of chemicals used as fire retardants in furniture, carpets, electronics, and plastics. **Phthalates** are also found in many plastic products. Both PBDEs and phthalates, like BPA, are hormone disruptors with PBDEs in particular interfering with thyroid hormones. The European Union has banned PBDEs, although they are still used in the United States.

Marina_D/Shutterstock.com

Richard Peterson/Shutterstock.com

It is important to note that not all synthetic chemicals are hazardous to human health. Many chemicals have a variety of beneficial uses in treating and preventing disease and have contributed to an increase in the standard of living. With every chemical, benefits should be weighed with the risks associated with them in making decisions about what, if any, restrictions could be placed on the use of these chemicals.

Likewise, it is fallacy to assume that all naturally occurring chemicals are harmless. Poison ivy is perfectly natural, but you wouldn't want to rub it all over your skin. That's because poison ivy produces a toxin, a hazardous chemical produced through biological processes. Other naturally occurring hazardous chemicals include petroleum and radon. Ergot is a kind of fungus that grows on rye and produces a chemical similar to Lysergic acid diethylamide (LSD). Some historians believe that the hysteria that led to the Salem Witch trials were the result of the original young girls have hallucinations from eating rye bread contaminated with ergot. In addition, the poison dart frog produces a toxin that many hunters in the Amazon dip their arrows in to help them kill game.

A lot of research has gone into the study of the health of effects of chemical pesticides. Before the 1960s, pesticides were often sprayed without concern because it was assumed that they had no effect on human health. For example, during World War II, soldiers were routinely sprayed with Dichlorodiphenyltrichloroethane (DDT) to control body lice. Lice are vectors of typhus and they readily spread from person to person in close quarters such as in military encampments. Prior to World War II, typhus often killed more soldiers than battlefield injuries. DDT proved effective in controlling body lice and thus reducing the spread of typhus.

However, especially after the publication of Rachel Carson's book, *Silent Spring*, in 1962, the public became increasingly aware of both the public health and environmental impacts of pesticides. This eventually led to the banning of DDT in the United States in the 1970s.

Synthetic chemicals like DDT can travel through the environmental and there are numerous ways people can be exposed to them. These can include from food and water, airborne transport, workplace exposure, and medical procedures. These chemicals and their effects can also be passed from parent to child in the womb, through breast milk, or through changes in the DNA.

Toxic substances can be classified by the effects on health and the kinds of harm they cause. For example, **carcinogens** cause cancer. Cancer is a condition where cells begin to grow and divide uncontrollably, forming tumors that attack healthy cells. **Teratogens** cause damage to a developing fetus in the womb. **Mutagens** cause mutations, changes to the DNA of living organisms, which can lead to cancer or other abnormalities. **Neurotoxins**, which include many heavy metals, damage the brain and central nervous system. **Allergens** cause an overstimulation of the immune system. The risks of allergens is often hard to gauge as some people may not be affected by them at all while others can have serious, even fatal reactions to them. **Pathway inhibitors** interfere with important biological functions. For example, cyanide disrupts the production of energy in the cell's mitochondria while strychnine, or rat poison, prevents blood clotting, causing animals to die of internal bleeding.

As noted before, **endocrine disruptors** affect the hormone system, either by mimicking hormones or blocking them from functioning. Hormones are the body's chemical messengers. They are released by various glands and travel throughout the body, looking for protein structures on the membranes of cells called receptors. When a hormone binds to a receptor, it triggers a chemical response inside the cell. For example, at the onset of puberty, hormones are produced that trigger the development of secondary sex characteristics, like breasts in women or facial hair in men. Hormone mimics are chemically similar to certain hormones and will also bind to receptors, triggering the same response. Often, these responses are inappropriate or harmful to the organism.

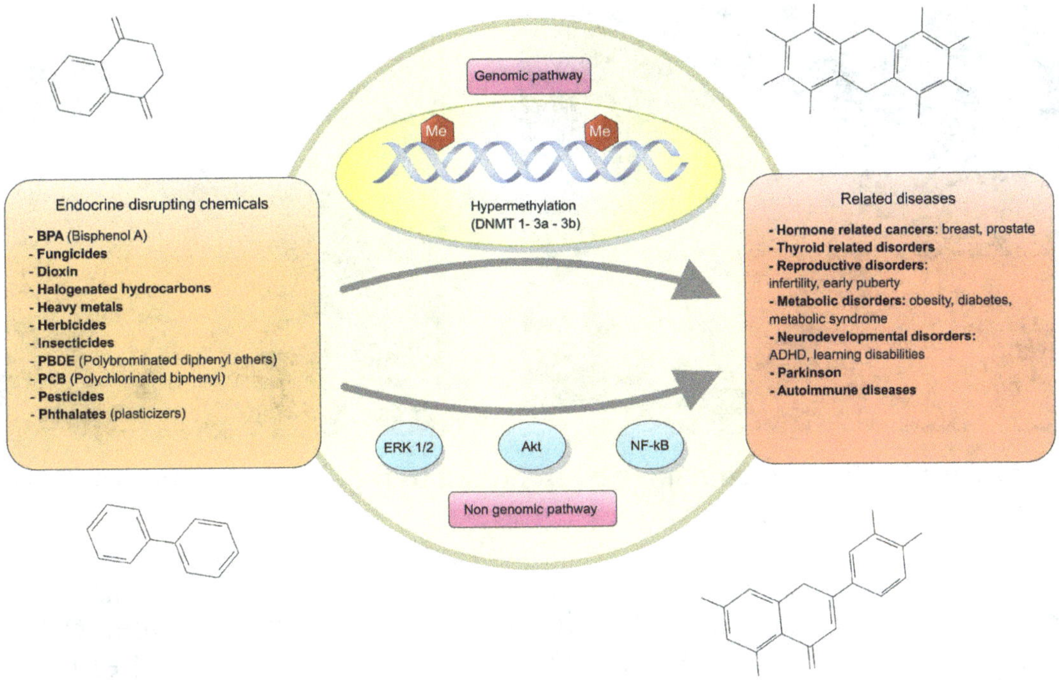

ellepigrafica/Shutterstock.com

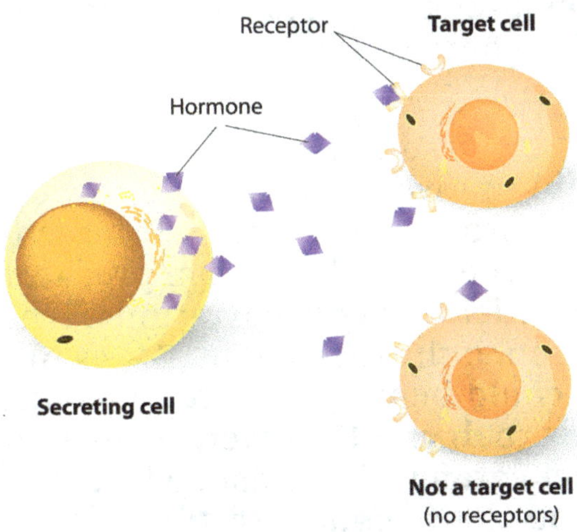

Designua/Shutterstock.com

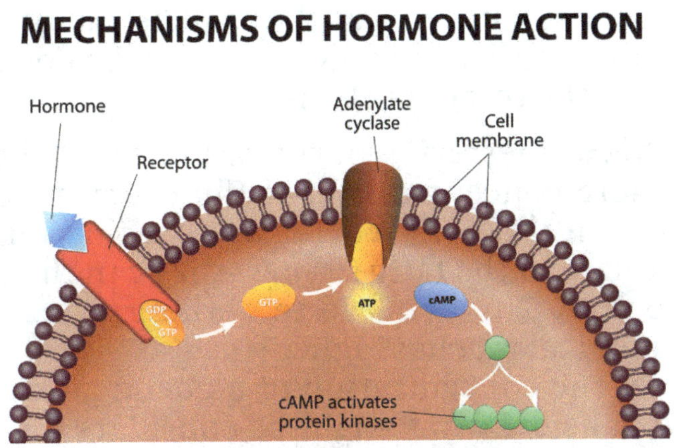

Designua/Shutterstock.com

Louis Guillette is a biologist who studied reproductive abnormalities in alligators, He found that among alligators that were exposed to the pesticide atrazine, the males had low testosterone levels while the females had elevated estrogen levels. His laboratory research found that atrazine triggered the production of an enzyme that converted testosterone into estrogen. Another researcher, Tyrone Hayes, performed similar studies in frogs exposed to atrazine. Hayes found that male frogs were becoming feminized and hermaphroditic after exposure to atrazine.

The study of hormone disruptors is still an emerging field and many in the chemical industry dispute the findings of researchers like Guillette and Hayes.

Stephen Chantzis/Shutterstock.com

Kurit afshen/Shutterstock.com

Hormone disruptors like Atrazine have been linked to reproductive abnormalities in species such as alligators and frogs.

Organisms have involved many defenses against hazardous chemicals in the environment. Skin, hair, feathers, and scales all serve as barriers to keep these chemicals from entering the body. However, these chemicals can still enter through ingestion or inhalation. Once inside, organs like the liver have metabolic pathways that can break down and neutralize these chemicals or make them more water-soluble so that they can be excreted through the kidneys. If the chemicals cannot be broken down, they can be deposited into fatty tissues and sequestered from the rest of the body.

This sequestering in the fatty tissues, however, can lead to **bioaccumulation**, where concentrations of the chemical can build up over time and then overwhelm the body's defenses, causing long-term harm. As substances bioaccumulate in the fatty tissues, they can then move up the food chain from predator to prey. This can lead to **biomagnification**, in which the concentration of the chemical reaches higher levels as it moves up through the food chain. For example, the concentration of a chemical like DDT could be very minute in a lake, well below what might be considered hazardous to drink. However, the plankton in the lake absorbs some of the DDT, who get eaten by little fish, who are in turn eaten by larger fish. At each level in the food chain, concentration of DDT increases as more and accumulates in the tissues of each predator. Finally, the large fish are eaten by birds of prey like eagles or osprey. As result, the eagles may accumulate enough DDT in their tissues to cause problems like the softening of their eggshells.

A tragic example of biomagnification occurred in the 1950s in Minamata Bay Japan. A battery factory was discharging mercury into the bay. Even though the levels of mercury in the water were not in excess of what was considered unsafe for drinking, as it moved up the food chain, hazardous levels of mercury accumulated in the fish. Problems were first noticed in cats that had eaten the fish. As the mercury damaged the cats' nervous systems, they began to have seizures. Initially, they called the problem "dancing cat syndrome" because of the seizures. However, people were also eating fish from the bay and eventually, over 2,200 people were sickened from mercury poisoning, many of whom died.

Once released to the environment, hazardous chemicals can be transported by a number of means. For example, wind currents could carry them across the globe. As a result of this windborne transport, we have found traces of certain pesticides in the tissues of Arctic polar bears and Antarctic penguins, even though these animals are thousands of miles away from where these pesticides were used. Chemicals can also be transported by water. For example, water running off farmland can carry fertilizers and pesticides into rivers and streams. Chemicals can also leach into the soil and contaminate groundwater supplies. Wastewater treatment plants can also release minute amounts of pharmaceuticals or disinfectant products.

Once released to the environment, they can cause declines in the populations in affected species. Some chemical pesticides are suspected to be responsible for the global collapse in amphibian populations.

Many chemicals break down into simpler compounds overtime. The breakdown rate depends on the chemical nature of the substance as well as factors like temperature, sunlight exposure, and moisture. The **breakdown products** of these chemicals could be less hazardous, but not always. Some breakdown products are just as or even more toxic than the original chemical. For example, DDT breaks down into DDE, which is also hazardous.

PongMoji/Shutterstock.com

DDT spraying in Thailand.

Some chemicals resistant breakdown and very persistent in the environment. Plastics and other synthetic materials are very resistant. That is part of what makes them so useful. No one wants to wake up in the morning and find out a fungus has eaten their tires. This persistence, however, means we may be dealing with these substances for a long time.

12.3 Hazard Characterization

Identifying the toxicity of a particular chemical is a complex science. Sensitivity to certain chemicals can vary from one individual to another. Factors that can influence sensitivity include age, gender, genetics, and overall health with infants and small children being the most vulnerable to chemicals. As a rule, organisms with smaller body weights and younger organisms with still-developing organs are more vulnerable than larger and mature organisms.

The EPA has based most of its hazard assessments of adults and then extrapolated down for the smaller sizes of children. Critics argue that this may not take into account the added vulnerabilities children face.

The impact of hazardous chemicals also depends on the extent of the exposure. A **chronic exposure** happens when an individual receives minute dosages over an extended period of time. Habitual smoking is an example of a chronic exposure.

On the other hand, an **acute exposure** occurs when an individual receives a large dose over a short period of time. Acute exposures usually occur from a discrete event, such as the Deepwater Horizon oil spill, which gave numerous organisms acute exposures to petroleum. Because of this, acute exposures are easy to identify and easier to determine a cause and effect relationship between exposure and health impacts. Chronic exposures, in contrast, are more difficult to measure and determine cause and effect relationships. Most regulations are based on acute exposures and may not take into account the effects of chronic exposures.

The most reliable method for determining cause and effect is through a manipulative experiment where the researchers can control the exposure each test subject receives. However, manipulative experiments involving humans are difficult and unethical, so animals such as mice and rats are often used in place of humans. Results are then scaled up by body to determine the effects on humans. However, using animals as proxies are not a perfect solution, as there may be biochemical differences between humans and the test animals that may result in different responses.

lanatoma/Shutterstock.com

There are, however, two methods that can be used to study the health effects of chemicals on humans. One is the **case study approach**, where researches seek out individuals who have certain ailments and then look for common exposures among the population. While case studies can identify possible chemicals of concern, they cannot establish cause and effect the way manipulative experiments can. The other approach is to conduct an **epidemiological study** in which researchers conduct long-term studies on a population and looking for differences in illnesses or death in the test population compared to the general population. For example, if researches want to conduct an epidemiological study on the effect of a certain pesticide, they would look for a population that is already being exposed to the pesticide, such as farm workers, and track them for any differences in health compared to people who are not being exposed. The drawback to epidemiological studies is that, because it can take many years for health effects to become noticeable in the population, they are very time-consuming.

Manipulative experiments involving rodents, however, can be conducted in a much shorter time scale and can achieve better cause and effect relationships despite their limitations. Commonly, a manipulative experiment will involve different populations of rodents being exposed to different dosages and compare their health effects to the control (unexposed) group. The results are then plotted on a **dose–response curve**, which describes how the effect measured changes with dose. One factor often determined in from a dose–response curve is the LD_{50} (lethal dose) or ED_{50} (effective dose). The LD_{50} or ED_{50} is the dose at which 50% of the test population is expected to die or experience some other effect. LD_{50}/ED_{50}

are used because individuals may have a different response and some may be able to metabolize or excrete the chemical better than others.

Some dose–response curves are linear and start at zero, with the effect rising in direct proportion with the dose. Others may be linear but with a **threshold dose**, in which there is no effect below the threshold dose and then effects begin to become noticeable in the population in dosages above the threshold. Not all dose–response curves are linear, however. Some may be J-shaped, U-shaped, or even inverted-U-shaped. In these cases, the effect may initially rise with dose but then eventually start to fall again at higher dosages. Inverted-U-shaped curves are common with endocrine disruptors as organisms are adapted to responding to very tiny amounts of hormones and higher dosages may actually interfere with themselves.

Kirill Kurashov/Shutterstock.com

Lab rat

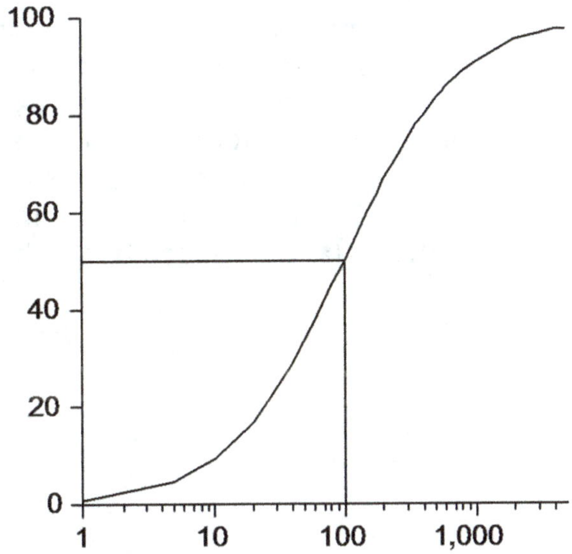

Source: https://toxtutor.nlm.nih.gov/resources/toxtutor_dose_response_curve_text_version.pdf

A dose–response curve.

Of course, unlike lab rats, humans are not exposed to just one chemical throughout their lifetimes. Some chemicals can sum or interfere with each other's effects. For example, DDE, the breakdown product of DDT, can either promote or inhibit sex reversal in alligators, depending on what other chemicals are present. Other combinations of chemicals may multiply their effects, causing a greater effect in combination than the sum of their individual effects. This is called a **synergistic effect** and the reasons of many of them are not well understood. With hundreds of thousands of synthetic chemicals on the market, it is not possible to conduct adequate research on every possible combination.

12.4 Risk Assessment and Risk Management

We defined a hazard as anything that has the potential to cause harm. However, since not every individual will have the same response to the same hazard, we cannot speak with any certainty what the effects of a given hazard will be for a single individual. Instead, we speak in terms of the probability or **risk** of a harmful outcome.

While we try to risk our individual risk as much as possible, our perception of risk may not always match up with reality. For example, we may assign a smaller a risk over things we have control over and a larger risk to things we see as out of our control. So, people tend to underestimate the risk of behaviors like overeating while overestimating the risks of things like nuclear power or flying. Other factors that may affect how people perceive risk is the degree to which the risk is fairly shared among people and whether they have an understanding or familiarity with the activity.

Because our perception of risk does not always match up with reality, we may tend to waste resources combating things that have a low risk while ignoring other factors that actually pose a much higher risk. **Risk assessment** is the process by which scientists attempt to quantify risk. This is important because risk managers need accurate information about the actual risks in order to make a better allocation of resources in reducing risks.

The steps of risk assessment include:

1. **Hazard Identification**: Identifying which substances have the potential to cause harm.
2. **Risk Analysis**: This includes all efforts to determine the degree of hazard a substance may pose and may include a dose–response analysis.
3. **Risk Evaluation**: The process of determining what the actual risk is for the population, including how often and to what degree with the population may be exposed to the chemical.
4. **Risk Management**: The final step in which reasonable efforts are made to reduce the risk posed by the hazard.

Petr Vaclavek/Shutterstock.com

Risk management involves any efforts to reduce risk. Like any other policy decision, risk management, especially relating to government policy, can be influenced by not only science, but also citizens and private interests through lobbying, campaign donations, and voting. In the United States, there are several federal agencies that are involved in risk management. This includes the following:

1. Environmental Protection Agency (EPA): Regulates pesticides and other chemicals not covered by other agencies.
2. Centers for Disease Control and Prevention (CDC): Works to reduce the risk of disease and injury.
3. Occupational Safety and Health Administration (OSHA): Regulates workplace hazards.
4. Food and Drug Administration (FDA): Regulates food additives, drugs, cosmetics, and medical devices.

There are two broad approaches to risk management. The first is called **innocent until proven guilty**. This assumes that any new product is not harmful until it has been proven to cause some specific harm. The advantage of innocent until proven guilty is that it promotes innovation and economic development of new products. The disadvantage is that products that cause harm are usually only identified after they have already injured people.

The other approach is the **precautionary principle**. This assumes any new product could be hazardous until rigorous testing demonstrates that they are safe. The advantage of the precautionary principle is that it identifies hazards before they can cause harm. The disadvantage is that it slows the advancement of technology and economic development.

Many standards in the United States have leaned toward the innocent until proven guilty approach. However, with the passage of the Frank R. Lautenberg Chemical Safety for the 21st Century Act in 2016, the EPA has been directed to adopt more stringent testing requirements for new industrial chemicals, which may lead to a more precautionary principle approach.

On the international stage, the European Union implement its REACH (Registration, Evaluation, Authorization, and Restriction of Chemicals) program that adopts a strong precautionary principle approach and shift much of the burden to industry to prove the safety of their products. In addition, the Stockholm Convention of Persistent Organic Pollutants (POPs) aims to phase out the "dirty dozen" of the most POPs. These are all chemicals that tend to bioaccumulate and biomagnify in the environment and can be transported long distances. The dirty dozen POPs are in three major categories as follows:

1. **Pesticides**: aldrin, chlordane, DDT, dieldrin, endrin, heptachlor, hexachlorobenzene, mirex, toxaphene
2. **Industrial chemicals**: hexachlorobenzene, polychlorinated biphenyls (PCBs)
3. **By-products**: hexachlorobenzene; polychlorinated dibenzo-*p*-dioxins and polychlorinated dibenzofurans (PCDD/PCDF), and PCBs

The Stockholm Convention has been ratified by 172 countries, however, it has not been ratified by the United States as yet.

12.5 Discussion Questions

1. Describe the four major categories of hazards.
2. Why are infectious diseases like measles making a comeback in some industrialized nations?
3. What is Bisphenol A used for why are some governments restricting its use?
4. Why did removing asbestos insulation sometimes make the problem worse?
5. Describe the health effects of carcinogens, mutagens, teratogens, neurotoxins, allergens, and pathway inhibitors.
6. How do endocrine disruptors cause health effects?
7. Describe the processes of bioaccumulation and biomagnification.
8. What is the difference between an acute exposure and a chronic exposure? Which one does the federal government use to determine exposure risk?
9. Describe the case study and epidemiological study approaches to identifying toxic substances in the environment.
10. Why is people's perception of risk often different from the reality?
11. Describe the four basic steps in risk assessment.
12. What is the difference between the "innocent until proven guilty" and the "precautionary principle" approaches to risk management?

12.6 Key Terms

Acute exposure
Allergens
Asbestos
Bioaccumulation
Biological hazards
Biomagnification
Bisphenol A
Breakdown products
By-products
Carcinogens
Case study approach
Chemical hazards
Chronic exposure
Cultural hazards
Dose–response curve
Endocrine disruptors
Environmental health
Epidemiological study
Hazard Identification
Industrial chemicals
Infectious disease
Innocent until proven guilty
Mutagens
Neurotoxins
Noninfectious disease
Phthalates
Pathway inhibitors
Pesticides
Physical hazards
Polybrominated diphenyl ethers
Precautionary principle
Risk
Risk Analysis
Risk assessment
Risk Evaluation
Risk management
Synergistic effect
Teratogens
Threshold dose
Toxicant
Toxicity
Toxicology
Vector

Chapter 13

WATER RESOURCES AND WATER POLLUTION

Chapter Topics

In 2003, China completed construction on the largest dam project in the world, the Three Gorges Dam across the Yangtze River. The hydroelectric dam created a reservoir 600 km long and has a generating capacity of 22,500 megawatts. It also displaced a record of 1.2 million people and submerged 13 cities, 140 towns, and 1,350 villages. It cost US$31.765 billion to construct.

Impacts from the Three Gorges Dam include pollutants from many of the factories, mines, and dumps that were submerged, landslides caused by erosion of the reservoir and downstream, and loss to fisheries in the South China Sea. The dam is also trapping sediments in the reservoir and is depriving downstream ecosystems of water and nutrients.

This dam represents one of the most significant impacts humans have ever made on the hydrologic system in history. National Aeronautics and Space Administration (NASA) even estimates that the redistribution of water will even lengthen the day by 0.06 microseconds and will make the Earth slightly rounder in the middle and flatter at the poles.

PRILL/Shutterstock.com

13.1 Human Impacts on Aquatic Systems

Human activities can affect aquatic systems in three main ways: (1) Altering the flow of water through the construction of levees, canals, and dams; (2) Withdrawing water from surface and groundwater sources faster than they can be replenished; and (3) Discharging pollution.

In terms of water usage, the vast majority (70%) of water use is for Agriculture. Industrial use accounts for 20% and residential usage takes up the remaining 10%. Our use of water can be defined as either consumptive or nonconsumptive use. **Consumptive use** involves the removal of water from a source and not returning it. Examples of consumptive use include irrigation and drinking. **Nonconsumptive use** is any use the does not remove the water from the source or only temporarily removes it, such as hydroelectric dams, washing, and flushing the toilet. Nonconsumptive may not deplete a water source, but it can alter its quality through pollution.

D. Hammonds/Shutterstock.com

REDPIXEL.PL/Shutterstock.com

Overdrawing surface water systems has resulted in several major river systems, including the Colorado River, the Rio Grande, which forms the border between Texas and Mexico, and China's Yellow River, being depleted to the point where the run dry before reaching the ocean. This alters the ecosystems, for example, organisms are deprived of an aquatic habit and migratory fish like salmon are no longer able to swim upstream. River deltas are also deprived of sediments that replenish the nutrients needed for coastal wetlands.

Water mining, or the removal of water faster than it can be replenished, is a worldwide problem. For example, the Aral Sea borders on Uzbekistan and Kazakhstan and was once one of the largest lakes in the world. However, during the Soviet era, the two rivers that feed into the Aral Sea were diverted to irrigate cotton fields. With an influx of new water, the Aral Sea has dried up and lost over 80% of its original volume, and is now divided into two smaller lakes. This has resulted in the loss of 60,000 fishing jobs and the winds have picked up pesticide-laden dust from the dried seabed. The pesticides have in turn destroyed most of the cotton production.

The world's largest aquifer, the Ogallala Aquifer is located in the Midwest of the United States under states ranging from Nebraska, Kansas, Oklahoma, and Texas. Drawing water from the Ogallala Aquifer has turned the land in these states into some of the most productive farmland in the world. However, the water is being withdrawn faster that it can be replenished. Once depleted, it may take thousands of years to refill the aquifer, meaning farmers in these states will have to find alternative sources of water. One proposal is to construct aqueducts to transport water down from the Great Lakes.

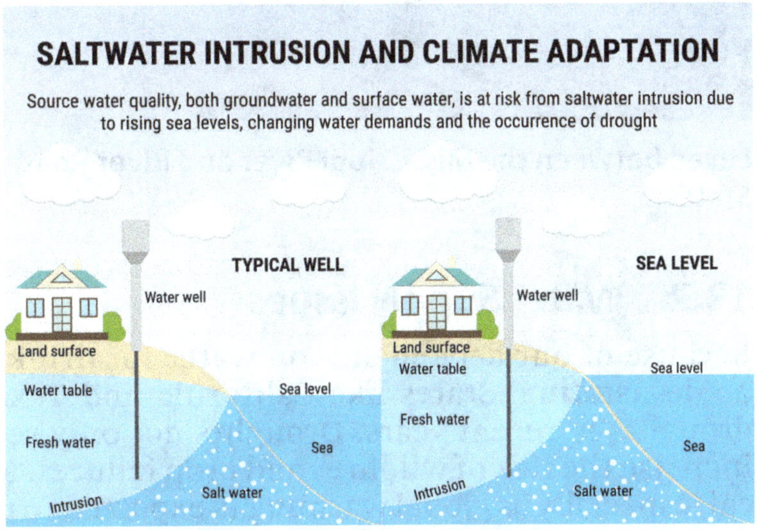

m.malinika/Shutterstock.com

Groundwater depletion is particularly a problem in coastal communities. As groundwater flows, it pushes out into the ocean while the salt water pushes back into the aquifer. Overdrawing the aquifer causes the water table to drop. With less water pushing out into the ocean, the ocean water moves into the aquifer to take its place. This **salt water intrusion** has occurred in California, Florida, India, and various nations in the Middle East. Once this happens, the wells have to be abandoned.

Sheryl Chapman/Shutterstock.com

Another problem of groundwater depletion is that, as the aquifer is drained, it loses some of the ability to support the ground above it. This can result in **subsidence** forming depressions in the ground or even **sinkholes** opening in the ground. Sinkholes will cause buildings and roadways above to collapse, causing millions of dollars in property damage.

Flooding occurs naturally. Some floods are seasonal, as snowpack in the mountains melt in the spring, it can cause a surge in water that causes rivers to flood their banks. Other floods can be caused by heavy storm systems such as hurricanes and tropical storms. To protect property against flooding, **levees**, large earthen or stone mounds, have been built along the river banks. Levees prevent flooding by keeping the water in the river channel. However, as this causes larger volumes of water to travel further downstream, levees can actually make flooding worse in one location even as they prevent it in another.

MDay Photography/Shutterstock.com

Roberto Michel/Shutterstock.com

Levee between the Mississippi River and River Road near Baton Rouge, Louisiana, United States.

13.2 Water Supply Issues

Because of our dependence on water for drinking and irrigation, shortages can be devastating. States like California and Texas have experienced long-lasting droughts in recent years. Droughts not only result in water shortages; they also increase the risk of wildfires and crop failures. Managing our use of water is critical, especially as global warming is expected to result in droughts becoming more frequent and longer-lasting in some parts of the world. Overall, our consumption of water has doubled over the past 50 years.

Angel DiBilio/Shutterstock.com

California aqueduct.

Aqueducts or canals are artificial waterways constructed for transportation or to divert water for human use. For example, huge aqueducts have been dug to transport water from the Colorado River for use in cities such as Denver, Las Vegas, Los Angeles, and San Diego, as well as for irrigation in farmlands throughout the Southwest.

An obstruction built across a river or stream to block the flow of water is called a **dam**. Dams are built to prevent flooding, generate electricity, or to create reservoirs for drinking or irrigation. Dams have a number of advantages and drawbacks as summarized by the table below:

Advantages of Dams	Disadvantages of Dams
Reliable source of water for drinking and irrigation	Millions of people have been displaced by dam projects
New recreational opportunities are created in the reservoir	Recreational opportunities are lost downstream
Generates clean electricity with less greenhouse gases emitted	Sediments accumulate in the reservoir, depriving lands downstream of nutrients and fresh topsoil
Renewable electricity	Habitats are altered upstream and downstream
Controls flooding	Small risk of catastrophic flood if the dam fails
	Fish migration is blocked and a cold-water reservoir is created, resulting in declines in fisheries

brgfx/Shutterstock.com

Many older dams have outlived their usefulness and their costs outweigh their benefits. There is often public pressure to remove older dams. Removing the dam returns the river to its normal flow and restores ecosystems to their previous state. It also restores the migration of fish like salmon and recreational opportunities like fly-fishing and rafting can resume. In addition, sediments start flowing back into the deltas, reviving shellfish and fish habitats. Restoring the natural flood cycles also deposits more nutrients to the soils in the floodplains.

Many rivers cross international boundaries or form the borders between nations. As result, nations must come to agreements to share the water resources in order to avoid conflicts. Disputes over water rights are a major source of conflict among nations in the Middle East. In contrast, the United States has negotiated a treaty with Mexico over rights to the Rio Grande. In addition, India, Pakistan, Bangladesh, Bhutan, and Nepal have formed agreements over several transboundary rivers.

When looking for solutions for water shortages, we can either try to increase the supply or reduce the demand. Reservoirs created by dams can increase the supply, but at the cost of depleting the available water elsewhere.

Desalination/desalinization can increase the water supply by removing the salt from seawater. This can be achieved by **distillation**, in which the water is evaporated and then recondensed. The other method is **reverse osmosis** in which the water is forced through a semipermeable membrane under pressure to separate the salt from the water. Desalination is more expensive than tapping freshwater sources and it produces a very brackish waste. However, as California experiences more and more water shortages, it has invested in several desalination plants to take advantage of its long coastline. Many desert countries like Saudi Arabia get most of their drinking water from desalination as it is the only possible means of meeting their demand.

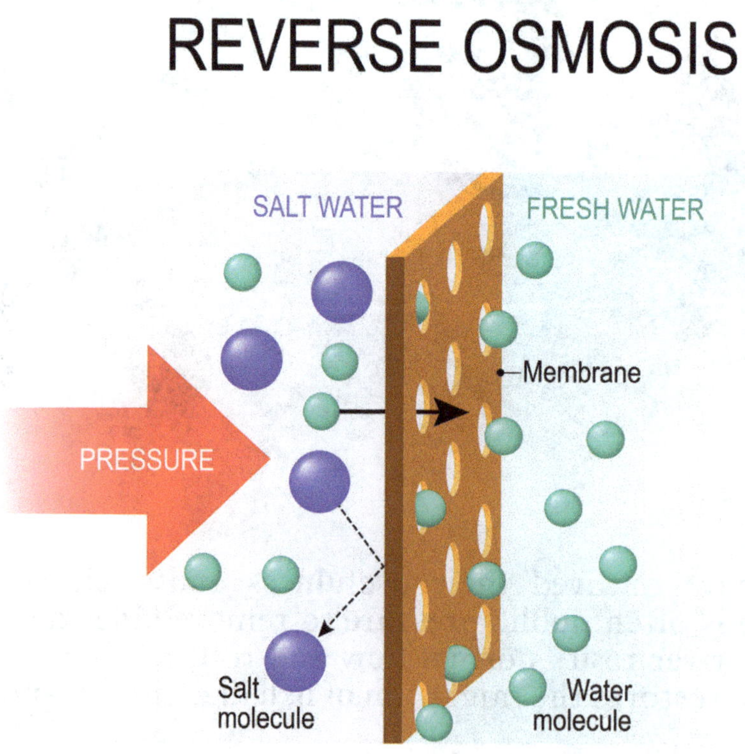

Designua/Shutterstock.com

DESALT SALINE WATER

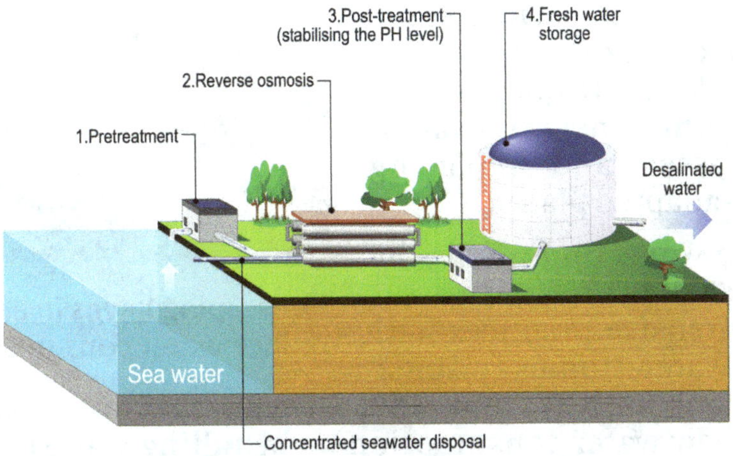

Designua/Shutterstock.com

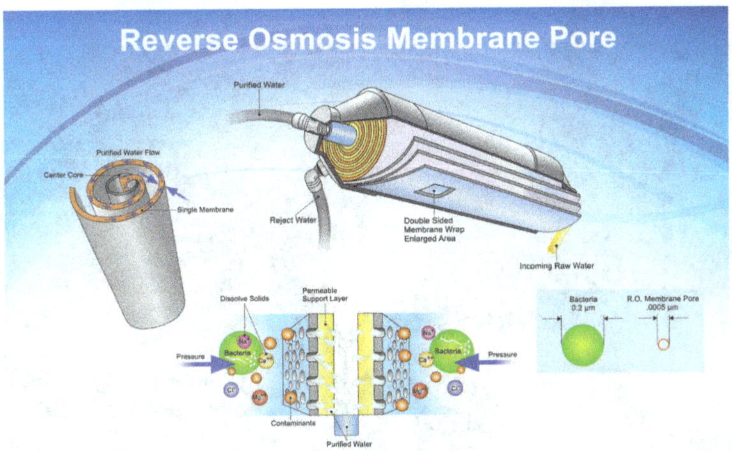

Studio BKK/Shutterstock.com

Desalination is expensive, but is often the only solution to produce an adequate supply of drinking water in arid regions.

Our largest single use of water is in agriculture, so any water savings that can be found there can make a huge difference. Almond orchards in California alone use 10% of the state's total water use. Lining irrigation canals and leveling fields can help reduce runoff. However, switching from wasteful practices like flood-and-furrow to drip irrigation can produce much more significant returns. Not planting water-intensive crops in arid regions can also save a lot of water.

In the home, water usage can be reduced by installing low-flow faucet heads and washing machines as well as using toilets with smaller tanks. Federal rules adopted in the 1990s phased out the 6.5-gallon toilet and now most toilets sold in the United States use about 1.3 gallons per flush. Collecting rainwater for gardening and water lawns can also help. Some states even permit residence to recycle graywater, water from the sinks and washtubs, for use on lawns. Some cities like Phoenix, Arizona, will offer property tax breaks to residents who practice **xeriscaping**, the removal of water-thirsty grasses and replacing them with desert-adapted plants.

Kathryn Roach/Shutterstock.com

Xeriscaping lawns in arid regions can reduce water consumption.

Industry can save water and money by conducting a water audit and find ways to replaced water-intensive practices with ones that use less water. Homes and business should also look for leaks and replace them as soon as possible. Businesses can also reduce their water consumption by installing waterless urinals in their men's rooms.

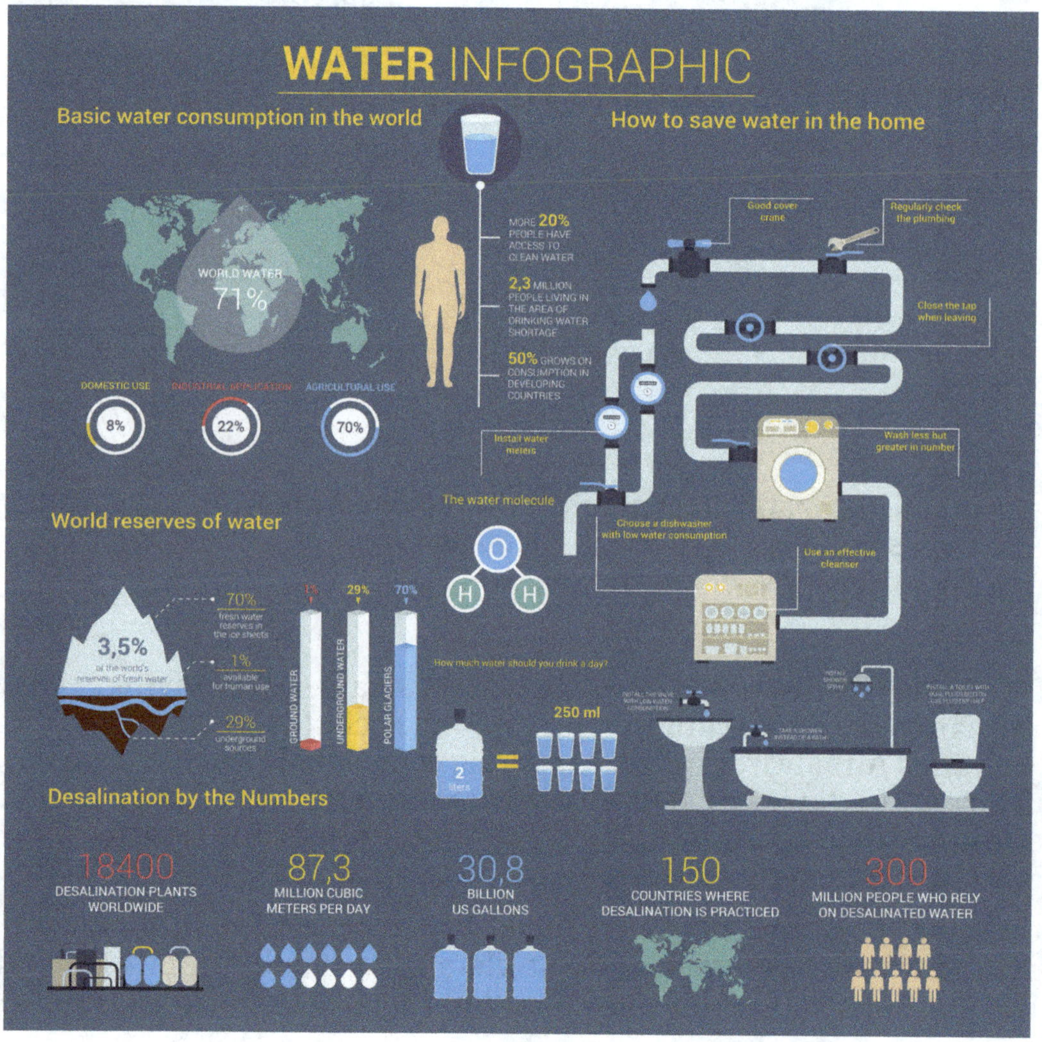

Elegant Solution/Shutterstock.com

13.3 Water Pollution Sources and Categories

In addition to depletion of water supplies, pollution remains a major problem. While much has been done to improve the condition of the world's waterways, the World Commission on Water has rated nearly half the world's major rives as "seriously depleted and polluted, degrading and poisoning the surrounding ecosystems, threatening the health and livelihood of people who depend on them."

Water pollution is the alteration of the chemical, biological, or physical properties of water in ways that threaten public health or damage ecosystems. Essentially, its any change to the water that does harm. While some forms of water pollution are readily apparent, others, such as changes in pH, nutrient concentration, and dissolved oxygen content are not visible and require chemical testing to detect.

Point sources are any single, identifiable source of pollutants, such as a sewage discharge pipe. **Nonpoint sources** are more diffuse, consistently of many smaller sources spread out over large areas, such as runoff from farmlands. The table below summarizes the major point and nonpoint sources of water pollution.

Point Sources	Pollutants	Nonpoint Sources	Pollutants
CAFOs[a]	Nutrients, organic waste, pathogens	Farms, lawns, golf courses	Fertilizers and pesticides
Sewage treatment plants	Nutrients, organic waste, pathogens	Residential streets	Motor oil, road salt, grease, and other chemicals
Waste disposal sites	Toxic chemicals	Construction sites, deforested land, and overgrazed lands	Sediments
Factories and other industrial sites	Toxic chemicals	Abandoned mines[a]	Acid drainage
Oil spills	Petroleum		

[a]Depending on how diffuse they are, CAFOs and mines can either be point or nonpoint sources.

Pathogens are the most significant pollutants in terms of their effect on human health. The United Nations estimates that 3,800 children die each day from waterborne illnesses like cholera and dysentery. The biggest of pathogen pollutants is untreated sewage that makes its way into the drinking water supply. Millions of people in developing countries lack adequate sewage treatment and safe drinking water, making these diseases one of the biggest causes of death, especially among children. Pathogens can also come from CAFOs.

Toxic chemicals can enter the waterways from farm runoff, leaching landfills, industrial facilities, and oil spills. This include pesticides, heavy metals like lead, mercury, and cadmium, petroleum and petroleum products, and acids from mine drainage. As water infiltrates old mines, it can pick up sulfur compounds that will lower the pH. Acid drainage can kill fish and can leach out toxic metals like aluminum from the soil, poisoning plant life. Other toxic chemicals are carcinogens and neurotoxins.

Nutrient pollution comes from fertilizer runoff, sewage discharge, and CAFOs. These nutrients promote algae growth, which leads to eutrophication and hypoxia (Chapter 2). In addition to depleting the dissolved oxygen content, nutrients in the Gulf of Mexico can lead to **red tides**. Red tides are reddish-colored algae that release toxins that can make swimming the water dangerous.

Alfred Rowan/Shutterstock.com

Red tides can imperil aquatic life and make swimming in these waters hazardous for people.

In addition to nutrients, **biodegradable wastes** can also deplete the dissolved oxygen content of water as bacteria break down. Biodegradable wastes include human waste, pulp mill waste, yard waste, and any other organic material.

Wastewater is water contaminated by human activity, such as from toilets, dish washers, washing machines, and bathtubs. Wastewater can contain biodegradable waste, nutrients, and pathogens.

Any activity that removes the vegetation from the land, such as deforestation, construction, mining, or overgrazing, results in soil erosion and **sediment pollution**. Sediments block sunlight in waterways, preventing photosynthesis. They also clog gills and feeding structures of aquatic organisms.

Oil pollution from major spills is very dramatic, but even small sources from boat engines or natural seepage from the ocean floor can be significant. In 1989, the oil tanker Exxon Valdez ran aground off the coast of Alaska, causing what was at the time the worst oil spill in history. In response, Congress passed the Oil Pollution

Act in 1990. This law and its subsequent amendments established a fund for cleanup of spills and required oil tankers to have double hulls by 2010. The good news is that oil tanker spills have declined sharply since the Exxon Valdez spill. The bad news is that in 2010, the **Deepwater Horizon** oil rig exploded, causing a spill that surpassed the Valdez spill.

Breck P. Kent/Shutterstock.com

Oil from the Deepwater Horizon spill, Louisiana.

James Jones Jr/Shutterstock.com

Plastic debris from discard fishing nets, fishing line, plastic bags, bottles, and balloons can entangle and drown sea turtles and dolphins as well as be a choking hazard for a variety of aquatic species. Prevailing currents have concentrated the plastic debris into what is known as the **Great Pacific Garbage Patch**. In fact, there are actually two garbage patches one in the Eastern Pacific and the other in the West Pacific. Each one is larger than Texas. Because they are concentrated in international waters, no nation is claiming responsibility for them, however, a number of nonprofit NGOs are attempting to find ways to clean them up.

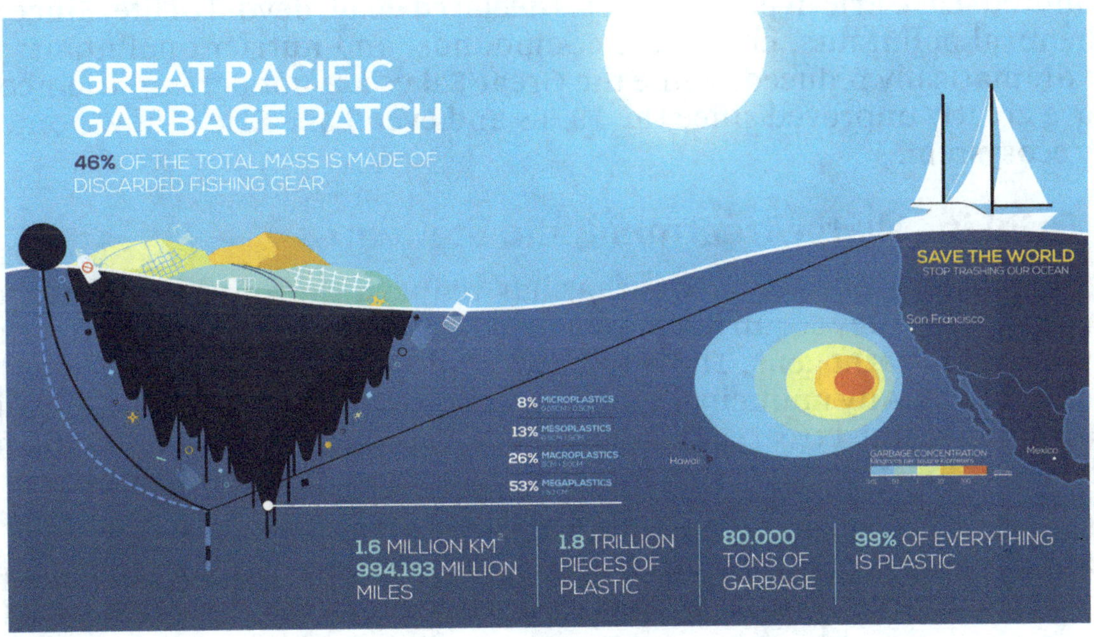

Pro_Vector/Shutterstock.com

Ostariyanov/Shutterstock.com

While most pollution control efforts are aimed at surface water pollutants, groundwater pollution is also a serious problem. Often, it is called the "invisible crisis" since it cannot be seen. Due to the lack of sunlight exposure and microbial action, groundwater pollutants do not decompose as readily as they would in the surface. They also don't disperse the way surface water pollutants do, so they remain in the ground much longer until they finally do decompose. While anthropogenic pollutants are serious, there are naturally occurring pollutants that can be picked up from the soil by the groundwater. These include sulfur compounds, fluorides, nitrates, arsenic, and aluminum.

One of the biggest sources of groundwater pollution is leaking underground storage tanks. Fuel tanks are stored underground to reduce the danger of explosions. However, this makes leaks more difficult to detect. Federal regulations require underground storage tanks be inspected and replaced regularly. Ironically, one of the most contaminated groundwater sites in the United States is a federal facility, the Hanford Nuclear Reservation in Washington State. Over half the underground tanks at Hanford are leaking. Pesticides, solvents, and animal wastes from CAFOs can also be significant sources of groundwater pollution.

In response to water pollution, Congress passed the **Clean Water Act** in 1977, which set standards for pollution discharges, provided funding for sewage treatment plants, and funded research. One of the biggest successes of the Clean Water Act has been the Great Lakes System. In the 1960s, the Great Lakes were all seriously polluted. Lake Erie in particular was declared dead, devoid of life. Since then toxic chemical pollutants, chlorinate compounds, and nutrient pollutants have all been dramatically reduced. While the Great Lakes are not perfect, their conditions have greatly improved since the 1970s and even Lake Erie is now home to a thriving ecosystem.

13.4 Drinking Water Treatment and Standards

One of the most significant factors that has reduced the spread of waterborne illnesses has been rules requiring public drinking water to be treated to remove pollutants and kill pathogens. These rules were adopted under the **Safe Drinking Water Act** of 1974. Any municipal agency or private company providing drinking water to the public must treat the water to federal and state standards for disinfection. The water must also be tested for 90 pollutants for which the EPA has established **maximum containment levels** (MCLs).

It's important to note that these rules only apply to public drinking water suppliers. Private wells on residential properties are not subject to these standards and any treatment done to private well water is at the discretion of the property owner.

Treatment of drinking water begins with removal from the reservoir, whether a surface or groundwater source. If the pH of the water is low, lime may be added to raise it to 7. This is what should have been done with water from the Flint River before being distributed to the residents of Flint, Michigan.

Picsshots/Shutterstock.com

A drinking water treatment plant.

Next, **alum** is added. Alum is a coagulant, which means it causes sediment particles to stick together. Next, the water is sent to **coagulation and flocculation**, where the water is first stirred and then settled to let the **floc**, or sediment particles that are sticking to each other settle out and be collected from the bottom. Next, the water is filtered through sand, gravel, and/or charcoal to remove the tiniest sediment particles and trapped parasites that cause illnesses like giardia and cryptosporidium. After **filtration**, chlorine is added for disinfection. Federal and state rules require that drinking water must have a chlorine residual as it travels through pipes underground. This is in case any pathogens enter through the pipes. The chlorine will then kill them. After disinfection, the water is distributed to residents.

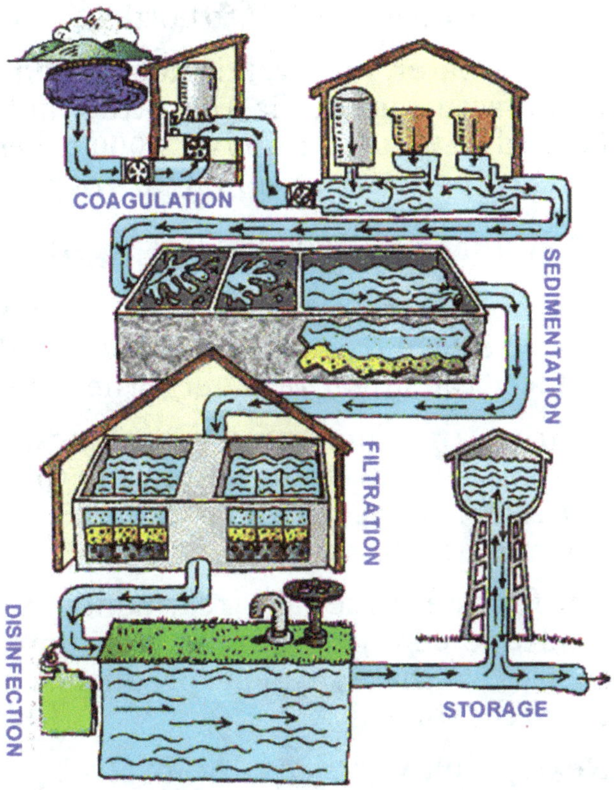

Image Source: cdc.gov.

Drinking water treatment plant.

What about **bottled water**? Many people like to drink it. Bottled water is convenient, portable, and often tastes better than tap water because it lacks the chlorine after taste. Nearly every food and beverage company owns at least one brand of bottled water, making it a $160 billion a year industry. But is it safer than tap water from a public drinking supplier?

Bottled water is considered a food product and is therefore regulated by the FDA instead of the EPA. However, the FDA has adopted by reference the same MCL standards that EPA has for public drinking water. The one exception is that since bottled water is packaged and doesn't travel through underground pipes, it is not required to have a chlorine residual. Other than that, the standards are the same.

In fact, a lot of bottled water brands actually are tap water from a public drinking water supplier. Some bottled water brands advertise themselves as "spring water" and the FDA only allows them to use the term "spring water" on their labels if it actually comes from an artesian spring. Other brands will be labeled as "filtered water" or "purified water." In those cases, the bottlers are taking water from a public drinking water source and filter it with charcoal to remove the chlorine before packing it into bottles.

Somphop Nithi/Shutterstock.com

Bottled water uses more energy than public tap water and generates tons of plastic waste.

Bottled water, however, is not very environmentally friendly. Taking the costs of manufacturing the plastic bottles and transporting them by truck into account, bottled water uses 1,000 to 2,000 times as much energy as public drinking water. It also produces 1.5 million tons of plastic waste each year.

13.5 Wastewater Treatment and Biosolids

Sewage contaminated water is still a big problem in communities that lack adequate sewage treatment. In developed countries, sewage treatment has greatly reduced the spread of waterborne illness.

In urban and suburban communities, homes are connected to a sewage collection system that transports wastewater to a sewage treatment plant. There, the organic material is removed and pathogens are killed before the water is discharged to a river or into the ocean.

The first step in sewage treatment is **preliminary treatment**. This involves some kind of screen and grit removal system to capture much of the very solid materials, such as eggshells, hygiene products, and any small toys that the kids have flushed down the toilet.

Next, **primary treatment** involves putting the water into the **primary clarifier** or sedimentation tank. A **clarifier** is a large circular tank where solids are allowed to settle to the bottom while oils and grease float to the surface and are skimmed off. **Secondary treatment** involves aeration that promotes the growth of bacteria to digest the organic material in the wastewater. Aeration can be achieved in an **aeration tank**, where water is bubbled in to provide the bacteria with oxygen. Next, the water is sent to the secondary clarifier where the digested solids are settled out again. Some of the solids are sent back to the aeration tank to provide bacteria feedstock to continue digestion. The rest is collected as sludge and further treated. The aeration tank may also include **biological nutrient removal (BNR)** in which specialized bacteria will solidify the phosphorus, so it can be settled out while the nitrogen is released back into the air. BNR is often done in areas where eutrophication is of high concern.

An alternative to an aeration tank is the **trickling filter**, where the water is sprayed onto a pile of rocks or seashells. Bacteria growing on the rocks then digest the organic material so that it can be settled out in the secondary clarifier.

Scuall/Shutterstock.com

Aerial view of a wastewater treatment plant. The clarifiers are the circular tanks. The large elliptical tanks are aeration tanks.

After secondary treatment, the water is filtered through sand or charcoal and then disinfected. Chlorine has long been used to disinfect wastewater; however, the water must be chlorinated before the treated **effluent** can be discharged to a body of water in order to produce the health of aquatic life. Chlorine gas is toxic and for safety reasons, many treatment plants have switched to using ozone or ultraviolet to disinfect the effluent. These have the advantage that they do not leave a residual that must be removed prior to discharging the effluent.

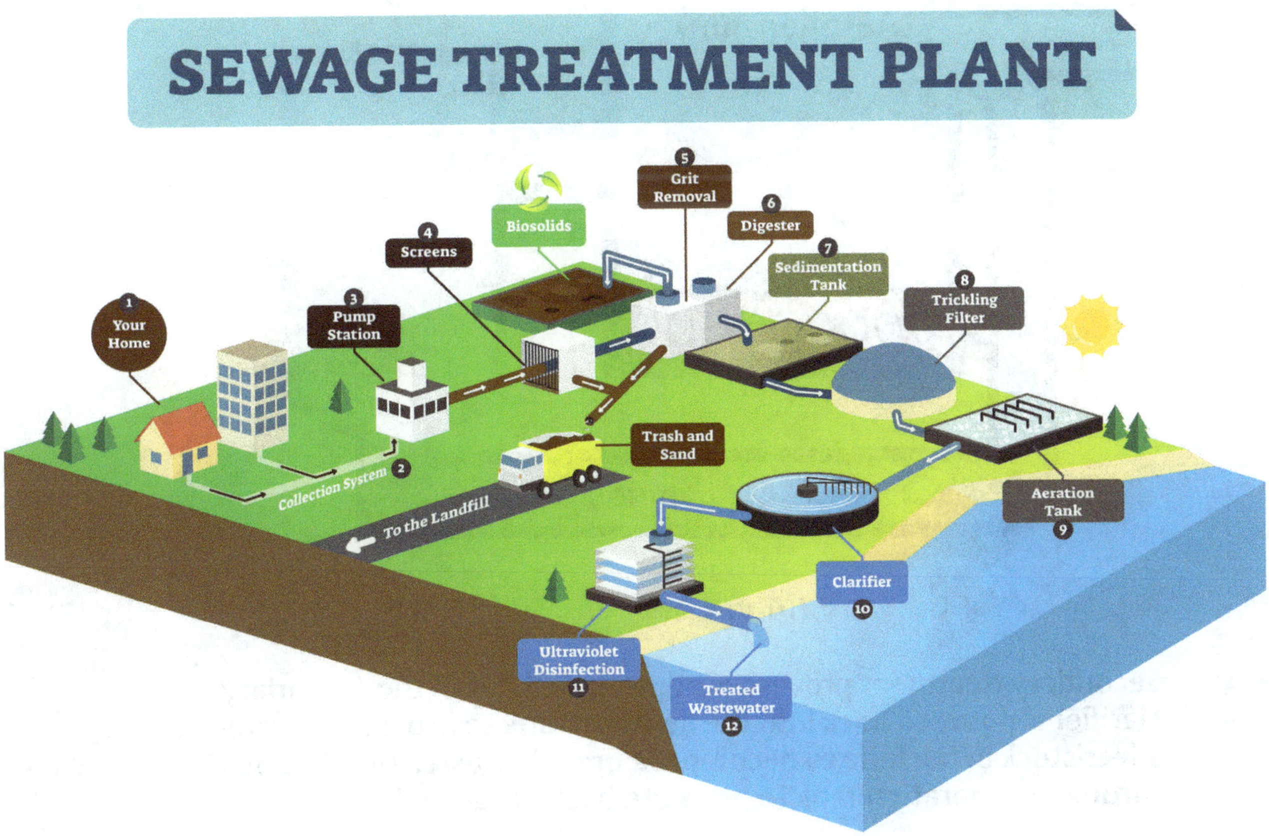

VectorMine/Shutterstock.com

Overall schematic of the sewage treatment process.

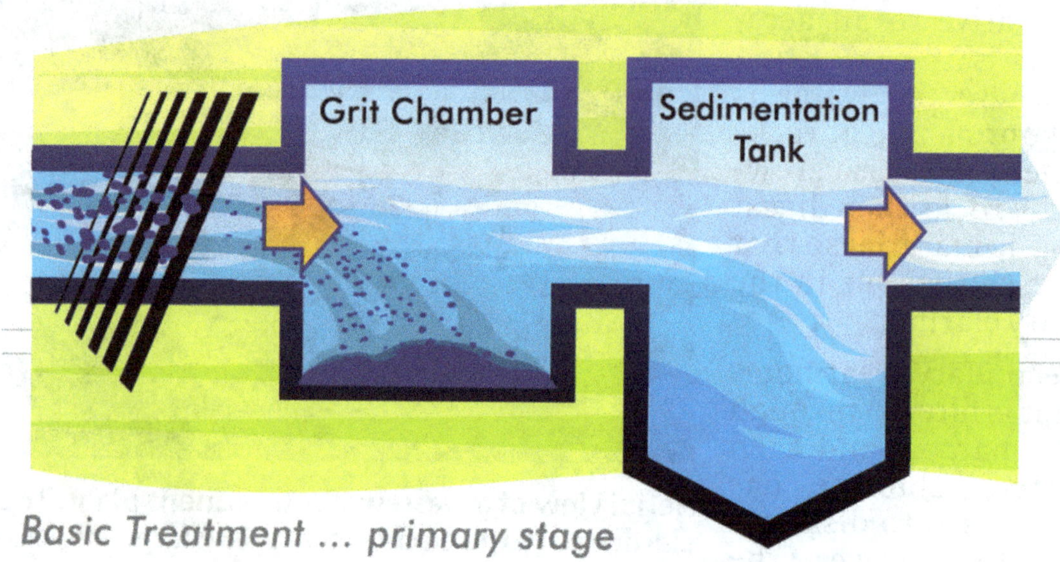

Preliminary and primary treatment, including the bar screen, grit removal chamber, and primary clarifier/sedimentation tank.

Image Source: epa.gov.

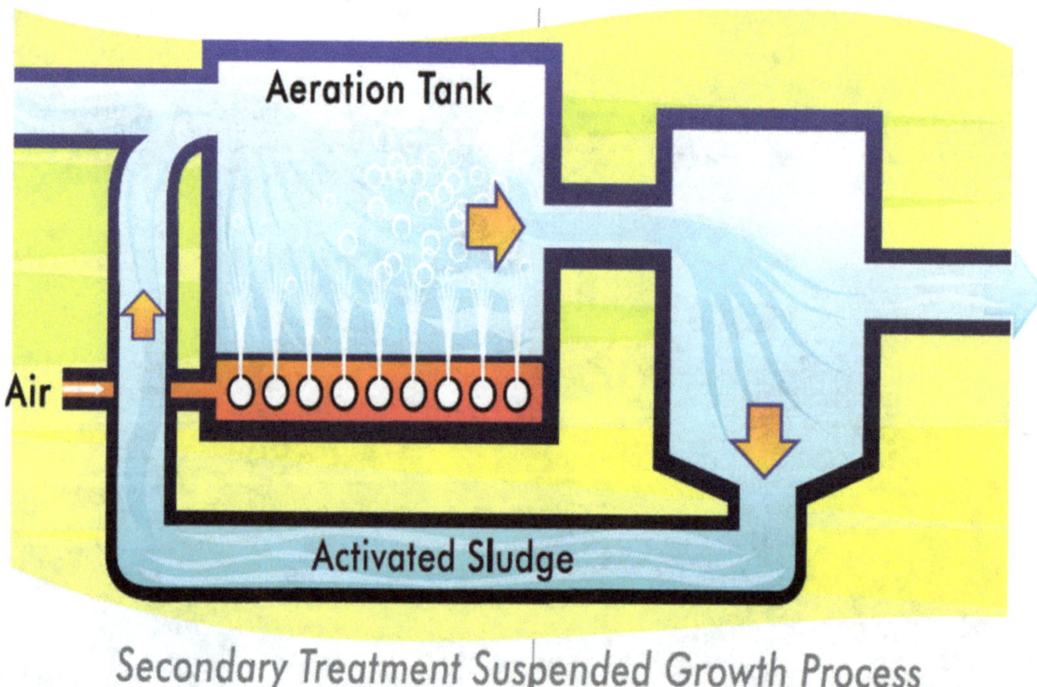

Secondary treatment process. Sludge settled out of the secondary clarifier is pumped back into the aeration tank to provide the tank with a feedstock of bacteria to decompose organic waste. Air is bubbled through the aeration tank to promote bacteria growth.

Image Source: epa.gov.

RODIKOVA TATIANA/Shutterstock.com

Ultraviolet light used for disinfection. Many wastewater treatment plants are switching from chlorine to ultraviolet light or ozone disinfection.

The solid material collected from the clarifiers, called **sludge**, must then be further treated. First, it is run through belt presses or centrifuges to dewater it. Then, it is treated to kill pathogens. This treated sludge or **biosolids**, is then landfilled, incinerated, and used as a fertilizer. Methods to kill pathogens in biosolids include

Lime stabilization: Lime is added to raise the pH to 12, which kills most pathogens.

Pasteurization: The sludge is heated and dried into a solid, pellet-like material.

Anaerobic digestion: Anaerobic bacteria thrive in an environment with little or no oxygen. They will further digest the sludge and kill the pathogens.

Composting: The sludge is mixed with wood chips or other water-absorbing material and bacteria decompose the material.

hans engbers/Shutterstock.com

Anaerobic digesters for treatment of sewage sludge into biosolids.

Composting of organic waste into biosolids.

Once treated to federal and state standards, biosolids rich in nutrients can be used as fertilizers, mostly for growing hay or other forms of animal feed. Regulations in place require testing of biosolids for pathogens and pollutants. There are also rules limiting where it can be applied, including setback distance keeping it away from features like property lines, water sources, and sinkholes. There are two categories of biosolids. Class A biosolids have more stringent treatment standards for pathogens but fewer restrictions on where they can be applied. Conversely, Class B biosolids have less stringent treatment standards but more restrictions on where they can be applied.

In rural communities where it is not economically feasible to hook homes to a centralized collection system, on-lot sewage systems are used. **Septic tanks** are underground storage tanks. Wastewater flows from the house into the tank where solids settle into the bottom of the tank. The water is then allowed to flow through a series of perforated pipes in the drainage or leach field. The nutrients are absorbed by plants in the leach field and pathogens are filtered through sand before the water enters the aquifer.

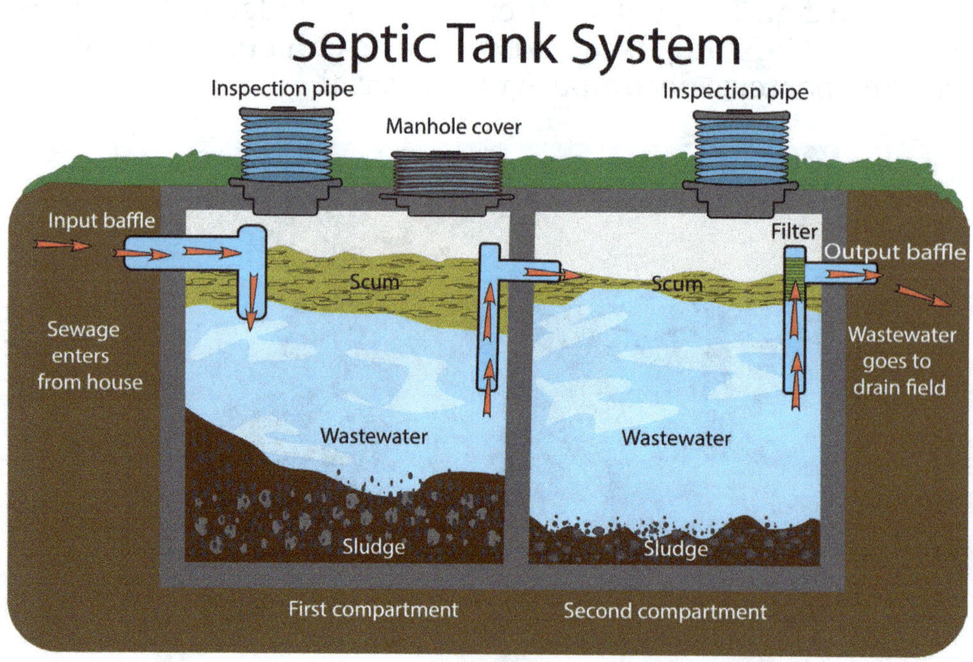

KajaNi/Shutterstock.com

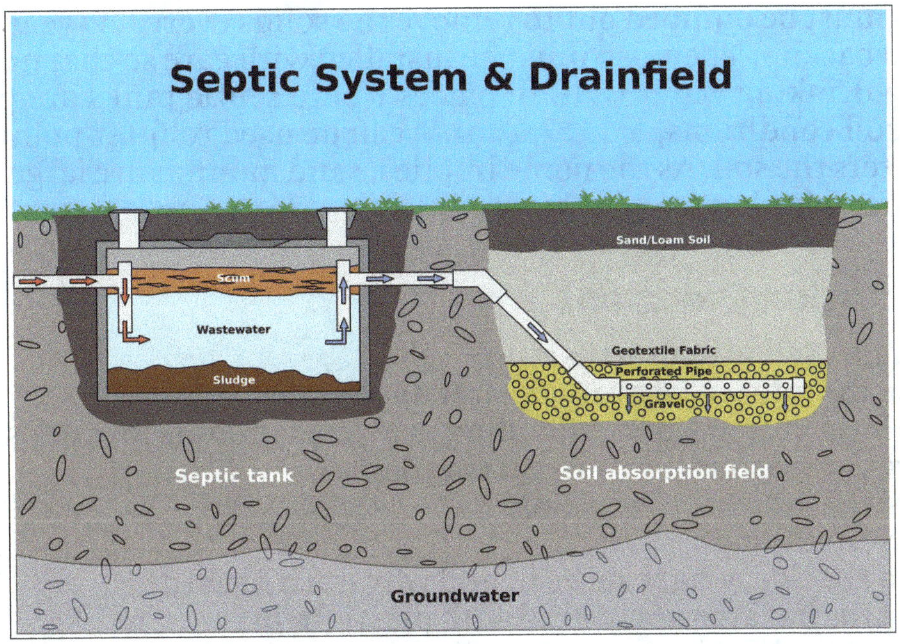

Slave SPB/Shutterstock.com

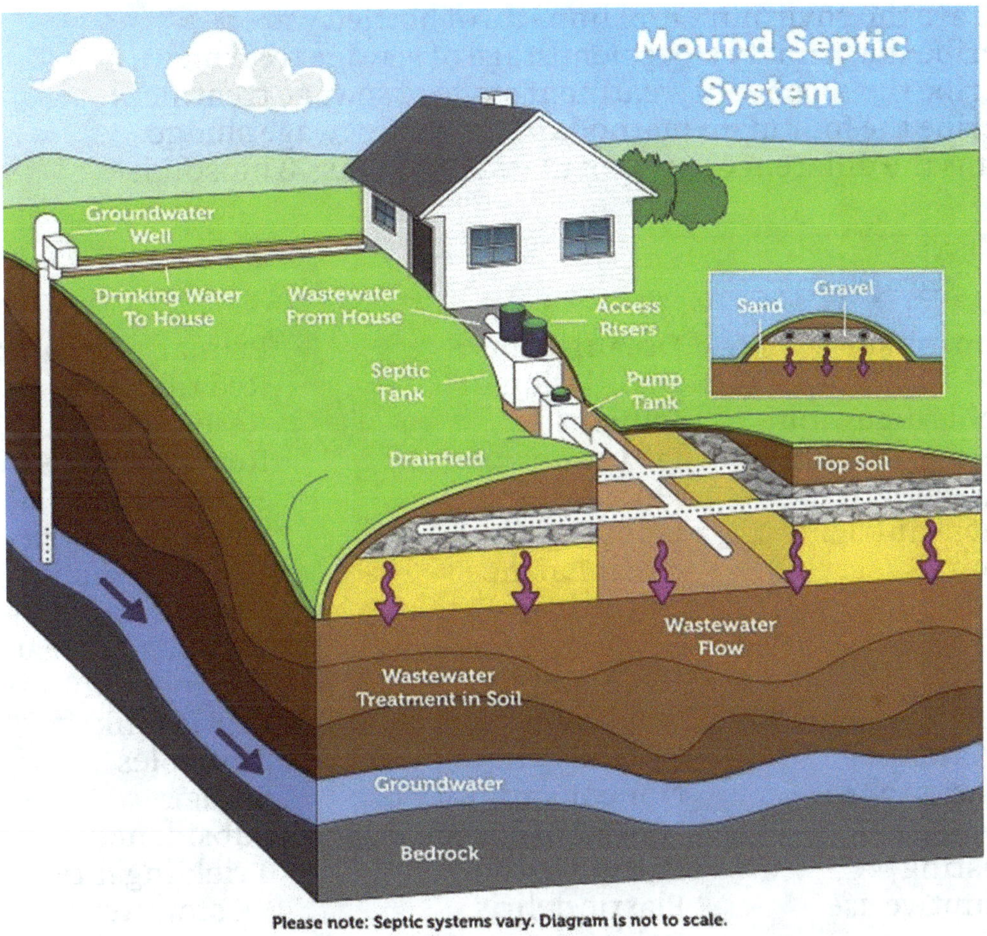

A sand-mound septic system.

Image Source: epa.gov.

Septic tanks must be pumped out to remove the solids every few years or they will overflow and backup. They also must be installed with care so that none are above grade from a drinking water well. In places where septic tanks are prone to failures due to soil conditions, **sand mounds** can be used to filter pollutants before the water enters the soil. As the name implies, sand mounds are large piles of sand that make effective filters.

13.6 Discussion Questions

1. What causes salt water intrusion?
2. Describe the impacts of groundwater depletion.
3. How do levees often worsen flooding problems in some locations?
4. What are the reasons for constructing dams?
5. What are the drawbacks to dams?
6. What is xeriscaping?
7. What is a red tide, what causes it, and why is it a hazard?
8. Describe the hazards associated with plastic debris.
9. Why do public drinking water suppliers have to add a chlorine residual?
10. What are the environmental impacts of bottled water?
11. Describe the primary treatment stage of sewage treatment.
12. Describe the secondary treatment stage of sewage treatment.
13. Describe the four main methods of treating sewage sludge.
14. What is the difference between Class A and Class B biosolids?

13.7 Key Terms

Aeration tank
Alum
Anaerobic digestion
Aqueducts
Biodegradable waste
Biological nutrient removal
Biosolids
Bottled water
Canals
Clarifier
Clean Water Act
Coagulation and flocculation
Composting
Consumptive use
Dam
Deep Water Horizon Spill
Desalination
Distillation
Effluent
Filtration
Floc
Flooding
Great Pacific Garbage Patch
Levee
Lime stabilization
Maximum contaminant level
Nonconsumptive use
Nonpoint source
Pasteurization
Plastic debris
Point source
Preliminary treatment
Primary clarifier
Primary treatment
Red tide
Reservoir
Reverse osmosis
Safe Drinking Water Act
Salt water intrusion
Sand mound
Secondary partner
Secondary Treatment
Sediment pollution
Septic tank
Sinkholes
Sludge
Subsidence
Trickling filter
Wastewater
Water mining
Water pollution
Xeriscaping

Chapter 14

THE ATMOSPHERE AND AIR POLLUTION

Chapter Topics

14.1 Structure and Layers of the Atmosphere
14.2 Atmospheric Conditions, Weather, and Climate
14.3 Outdoor Air Pollution
14.4 Stratospheric Ozone Depletion
14.5 Acid Deposition
14.6 Indoor Air Pollution
14.7 Discussion Questions
14.8 Key Terms

Air pollution problems have been around since the dawn of the Industrial Revolution. In London, homes were heated with coal and the city was plagued with "killer fog" events. The last such event was one of the worst. In 1952, a killer fog hit the city, killing 12,000 people. Pittsburgh became synonymous with coal and steel and its air quality reflected it. Los Angeles is infamous for its traffic jams as it seems every resident of the city is on the road 24 hours. The city also sits in valley in a warm climate, which creates the conditions for a temperature inversion that traps the pollutants over the city. Mexico City has similar conditions as Los Angeles: warm climate, heavy vehicle traffic, and mountains surrounding it. More recently, Beijing has had a reputation for some of the worst air quality in the world. Beijing's air has been so bad that when the Olympics were held there in 2008, many athletes were hesitant to go there out of concern that the pollution would affect their performance.

Slowly, however, solutions to air pollution issues have emerged. Londoners no longer heat their homes with coal and congestion fees are collected to support mass transit. While Los Angelinos have yet to embrace mass transit, catalytic converters and vehicle inspections, have reduced its smog levels. Mexico City has adopted similar measures and has reduced the sulfur content of diesel, banned leaded gasoline, closed an old oil refinery. With the decline in the steel industry, Pittsburgh has transitioned into an academic hub for the robotics and biotech industries. Pittsburgh has also engaged in a sand-blasting program to clean decades of soot from its buildings. After years of denying the problem, China finally began tackling the air pollution in Beijing. It has focused on reducing coal consumption and putting emission limits on auto emissions in the city. In 2012, the city's air was so bad many experts considered the city uninhabitable. Today, while Beijing still has a long way to go, its air quality has improved considerably. Despite many of these improvements, urban sprawl and increasing amounts of traffic still contribute to air pollution problems in Los Angeles, Mexico City, and Beijing.

14.1 Structure and Layers of the Atmosphere

The **atmosphere** consists of the layer of gases that surround the Earth. It serves to retain some of the energy the Earth receives from the Sun to keep the surface temperature habitable and provides us with the oxygen we need to breathe. It also protects us from meteor impacts, harmful radiation, and serves as an important reservoir that helps transport and recycle water, nitrogen, and carbon.

The Earth is the only planet in our Solar System, indeed, the only planet that we know of, that has a breathable atmosphere. Our two neighboring planets have radically different atmospheres, neither of which could support life as we know it. Venus has an ultra-dense atmosphere where the surface pressure is 90 times that of the Earth at sea level. In addition, its atmosphere is mostly carbon dioxide, creating a runaway greenhouse effect that traps so much heat that the temperatures are hot enough to melt lead on the surface. In contrast, Mars has a very thin atmosphere that retains very little heat and doesn't support liquid water, making it a cold desert. So, Venus is too hot, Mars is too cold, but Earth, like the bowl of porridge Goldilocks ate, is just right. At least for now.

The Earth's atmosphere is mostly nitrogen (71%) and oxygen (21%) with the remaining 1% being a mixture of gases including carbon dioxide, water vapor, argon, and methane. Human activity has been altering the quantities of some of these gases, such as carbon dioxide, through various processes.

Nasky/Shutterstock.com

The atmosphere consists of four main layers. The lowest layer, the **troposphere**, starts at sea level and extends to about 10 km above the surface. It contains the air we breathe and the prevailing weather patterns that we experience. Temperatures drop with elevation in the troposphere, which is why tall mountains are snow covered throughout the year and high-flying aircraft have to be concerned with icing of the wings. The top edge of the troposphere, the **tropopause**, serves as a barrier that prevents little vertical mixing of gases between the troposphere and the stratosphere. As a result, most pollutants we release into the troposphere tend to stay in the troposphere and do not cross over into the stratosphere. One of the few excerpts at the CFCs that damage the ozone layer.

The **stratosphere** extends from 10 to about 50 km above sea level. Unlike in the troposphere, temperatures rise with elevation in the stratosphere. Compared to the troposphere, air in the stratosphere is drier and less dense and there is very little vertical mixing of gases. The stratosphere contains the **ozone layer**, a region of high ozone concentration that shields us from incoming ultraviolet radiation. Without it, we would have many more cases of sunburn and skin cancer.

Temperatures again fall with elevation in the **mesosphere**, which extends between 50 and 85 km above sea level. Finally, the **thermosphere or ionosphere** goes up about 500 km above sea level, where it thins out into the vacuum of space.

As the Sun's rays pass through the atmosphere, they are absorbed by the Earth's surface and reemitted as infrared radiation. This warms the air near the surface. Because warm air is less dense than colder air, it rises. As it reaches the upper regions of the troposphere, it cools and then sinks back down to the surface. In addition, warm air holds more water vapor than cold air, so the air near the surface picks up water vapor, carries up into the upper troposphere and, as it cools, the water vapor condenses into clouds. This regular motion of rising and falling air forms **convection currents** that drive much of the prevailing climates patterns.

CRStocker/Shutterstock.com

VisualCurriculums/Shutterstock.com

14.2 Atmospheric Conditions, Weather, and Climate

Weather refers to the short-term atmospheric conditions over a limited area. **Climate**, on the other hand, is defined as the long-term average atmospheric conditions over a larger region. Weather can vary from day to day, but climate averages out over the course of a year or longer. It's important to emphasize the difference between the two, especially as we discuss climate change and global warming in Chapter 15.

Convection currents are divided into large groupings called cells and have a big influence over the climate. For example, in the tropical region, a pair of convective cells called the **Hadley cells** are driven by solar radiation. At the equator, warm, moist air rises into the upper troposphere where it cools. The water contained in this air condenses into clouds and bring heavy rains. These create conditions for tropical rainforests along the equator in places like the Amazon basin, Central Africa, India, and Malaysia. Meanwhile, the cooler, drier air migrates and then sinks down at about 30 degrees north and south latitude. This dry air creates deserts in the American Southwest, Chile, the Gobi Desert in Central Asia, the Sahara Desert, and so on.

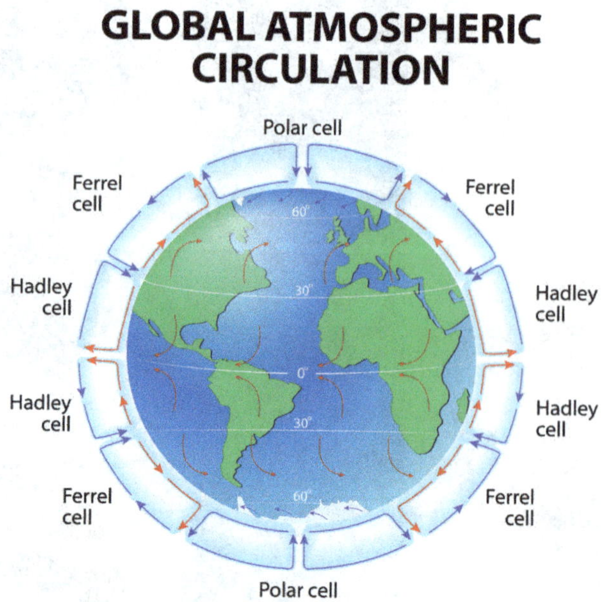

Designua/Shutterstock.com

The **Ferrell Cells** create similar conditions between 30 and 60 degrees north and south, bringing heavy rains at the 60 degrees latitudes, which is about where the temperate rainforest of the Pacific Northwest resides. Finally, the **Polar Cells** run between 60 degrees and the poles. As a result, both the North and South Poles receive very little precipitation. It may not seem like it since it's covered in ice, but Antarctica receives so little precipitation that it qualifies as a desert. However, in recent years, this has begun to change as the central region of Antarctica has begun to see an increase in precipitation in the form of snow.

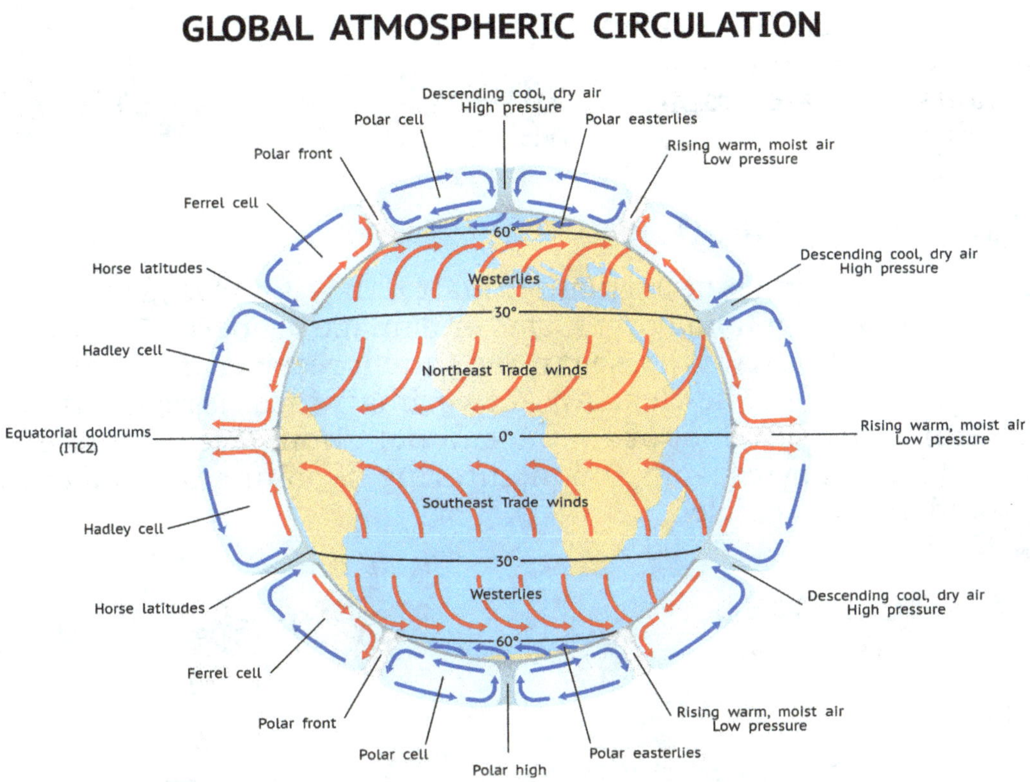

Jamilia Marini/Shutterstock.com

Of course, these cells do not just move air north and south. The Earth rotates underneath them, creating a twisting and deflection of the air currents due to the **Coriolis effect**. This effect deflects air currents to the right in the Northern Hemisphere and to the left in the Southern Hemisphere. This creates the curving wind patterns that we refer to as the **trade winds**. During the Age of Exploration, when sailing the ships were the primary means for crossing the oceans, the trade wins played an important role in enabling these crossings. Meanwhile, at the equator, the **doldrums** result in very few winds.

One atmospheric phenomenon that can influence the levels of pollution over a city is what is known as a **temperature inversion**. Under normal conditions, when the temperature falls with elevation, warmer, less dense air rises. As a result, this rising air carries pollutants with it, dispersing them into the upper troposphere. During an inversion, a colder air mass is trapped underneath a warm air mass.

As a result, there is less vertical mixing of gases and pollutants are trapped. Inversions usually occur in valleys, where the mountains create a shadow that blocks the light from the morning Sun. This generates a cooler air mass that cannot rise and escape the valley. This is why the smog in Los Angeles is usually the worst during the early morning rush air, when lots of cars are on the road emitting pollutants that are trapped around the city.

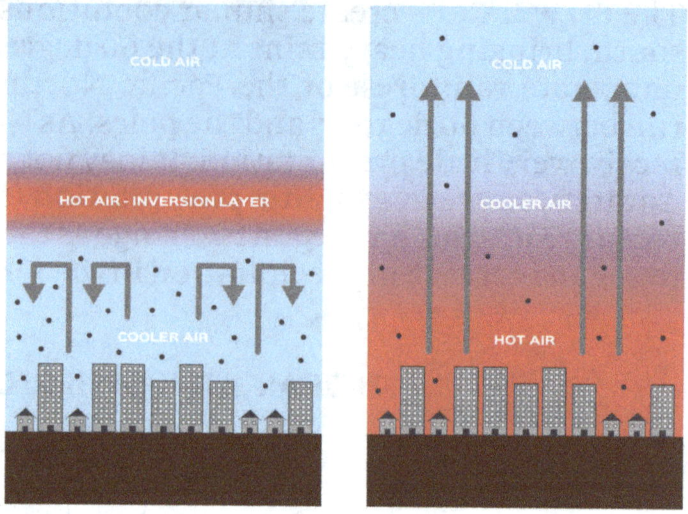

gritsalak karalak/Shutterstock.com

14.3 Outdoor Air Pollution

Air pollutants are any gases or particulate matter that is released into the air that has the potential to harm living organisms or damage property. The first version of the **Clean Air Act** (CAA) was passed in 1963 and it received major amendments in 1970 and again in 1990. The Clean Air Act's primary focus is regulating and controlling the emission of outdoor or ambient air pollution. It sets emissions standards, provides funds for research and monitoring of pollutants, and encourages emission standards for vehicles.

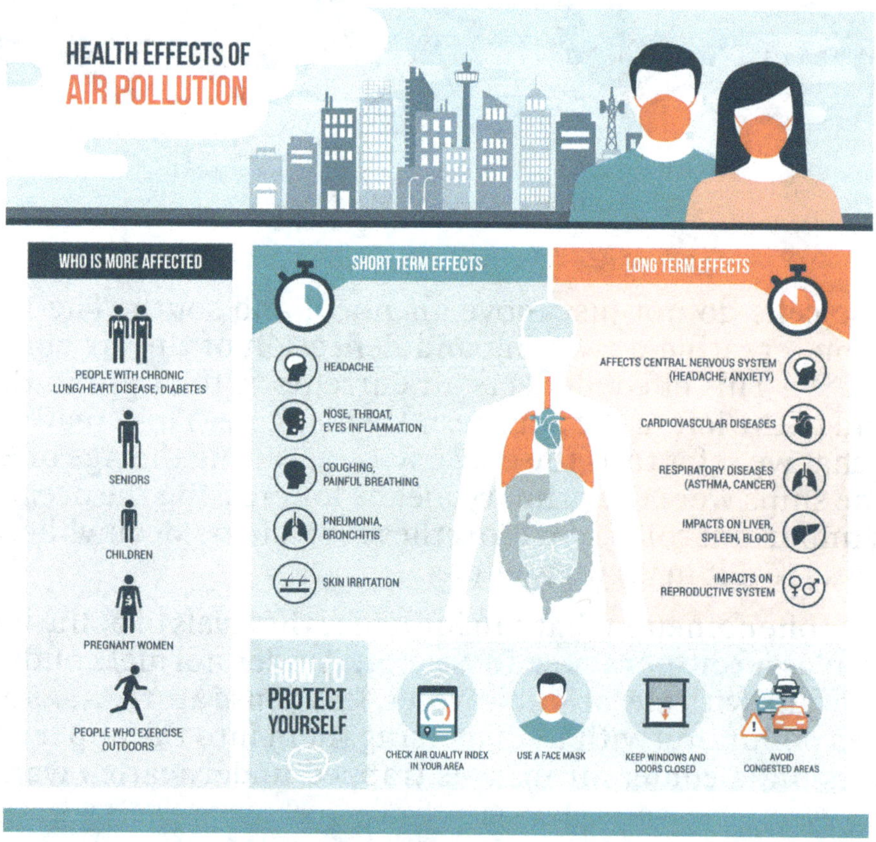

elenabsl/Shutterstock.com

There are natural sources of air pollutants. For example, volcanic eruptions emit particulate matter and sulfur dioxide into the air. These eruptions can ground local air traffic by blocking visibility. The particulate matter can also damage the lungs and property. In the longer term, the sulfur dioxide droplets can cause a cooling effect by reflecting some of the Sun's rays away from the Earth's surface.

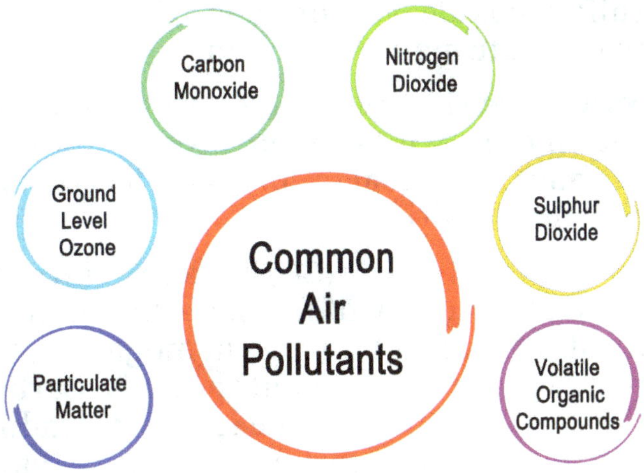

arka38/Shutterstock.com

Forest fires also emit particulate matter. While fires occur naturally, human activity can worsen them by suppressing fires where the fire is part of the natural ecosystem (Chapter 6). Winds can also pick up dust particles as they blow across drylands like the Sahara or during the Dust Bowl in the Midwest. Desertification can worsen the movement of dust, as can practices that remove the vegetation from the land like deforestation and overgrazing.

Primary pollutants are emitted directly from the source such as a factory smokestack or the exhaust pipe of a vehicle. Ash from a volcanic eruption and carbon monoxide from the burning of fossil fuels are examples of primary pollutants. In contrast, **secondary pollutants** form when primary pollutants react with other pollutants, sunlight, or gases in the atmosphere. Ozone and the chemicals that produce acid deposition are secondary pollutants.

One important factor about pollutants is the amount of time that remains in the atmosphere before breaking down or settling to the ground or the **residence time**. The longer the residence time, the further a pollutant is able to travel in the air. Therefore, pollutants that have short residence times, measured in hours or days, tend to have more local impacts. In contrast, pollutants with residence times measured in decades or centuries will have global impacts. Metals, large particulates, and sulfur dioxide tend to have short residence times, while carbon dioxide and CFCs have much longer residence times.

Until the Clean Air Act, the EPA sets standards for the emission of air pollutants and turns the enforcement over to the state agencies. Each state is required to develop a State Implementation Plan (SIP) and submit it for approval by the EPA. The goal of each SIP is to identify major sources of air pollutants and outline strategies for reducing emissions. If the EPA determines that a state's SIP is inadequate, it may take over direct enforcement.

bunnyphoto/Shutterstock.com

The CAA requires states to monitor the emissions of six major pollutants. These **emission pollutants** are sulfur dioxide, carbon monoxide, nitrogen oxides, volatile organic compounds, particulate matter, and lead.

Sulfur dioxide (SO_2) is a colorless gas with a very strong odor. If you've ever smelled rotten eggs, you've smelled sulfur dioxide. The major source of sulfur dioxide is the burning of coal to generate electricity. Sulfur dioxide is the largest contributor of acid deposition.

Carbon monoxide (CO) is a colorless, odorless gas that is formed from the incomplete combustion of fuels. Carbon monoxide is a major health threat because of its affinity for binding to hemoglobin, the protein in red blood cells that transports oxygen throughout the body. Carbon monoxide will push the oxygen out of the hemoglobin and take its place, leading to oxygen deprivation. Car exhaust is a major source of carbon monoxide, which is why you should never run your car in an enclosed garage. Carbon monoxide can also be produced by burning fuels like oil, natural gas, and kerosene, so for homes that use these fuels for heating or cooking, a carbon monoxide detector can be a good investment. Symptoms of carbon monoxide poisoning include dizziness, nausea, and loss of consciousness. Often, the early stages of carbon monoxide poisoning can be mistaken for the flu.

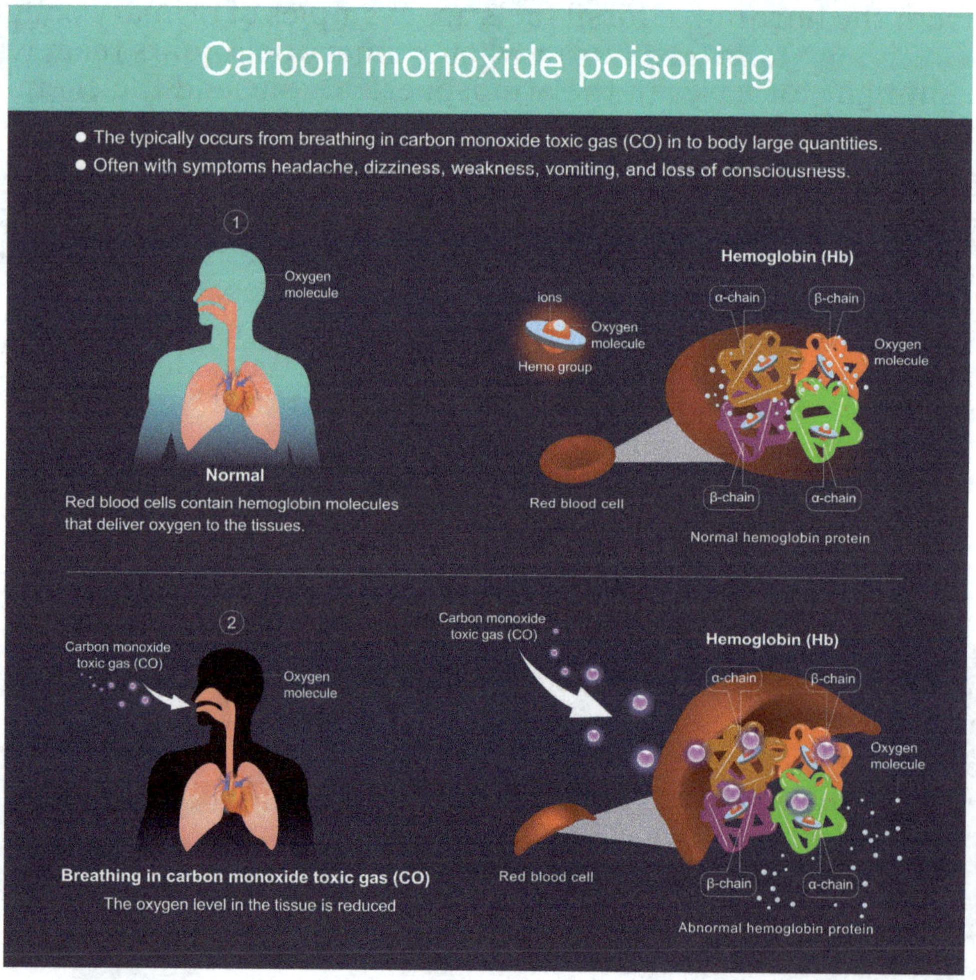

Studio BKK/Shutterstock.com

Lemon Tree Images/Shutterstock.com

Nitrogen oxides (NO_s) are a group of gases that form when high-temperature combustion causes nitrogen and oxygen in the air to react with each other. NO_x contributes to smog, acid deposition, and stratospheric ozone depletion. Many are also serious lung irritates.

Volatile Organic Compounds (VOCs) are a large group of carbon-containing compounds that evaporate readily. They can include emissions from paints, solved, pesticides, and adhesives. They can also include fuels such as methane and propane, as well as the combustion products of many fuels including gasoline and diesel. Many VOCs are carcinogens and toxic and contribute to smog.

Particulate matter consists of solid particles or droplets that become suspended in the air. Many irritate the lungs or can transport toxic chemicals into the lungs. Originally, the CAA regulated particulates under a standard called PM_{10}, which focused on all particles with a diameter of 10 μm or less. After the 1990 amendment to the CAA, a new standard, $PM_{2.5}$, was adopted. Now, the standards focus on particles that are 2.5 μm in diameter or smaller. The reason for the switch is that the smaller particles travel the deepest into the lungs and cause the most damage. They are also harder to capture, so by focusing on the tiniest particles, regulators can be assured that the larger particles are being removed anyway.

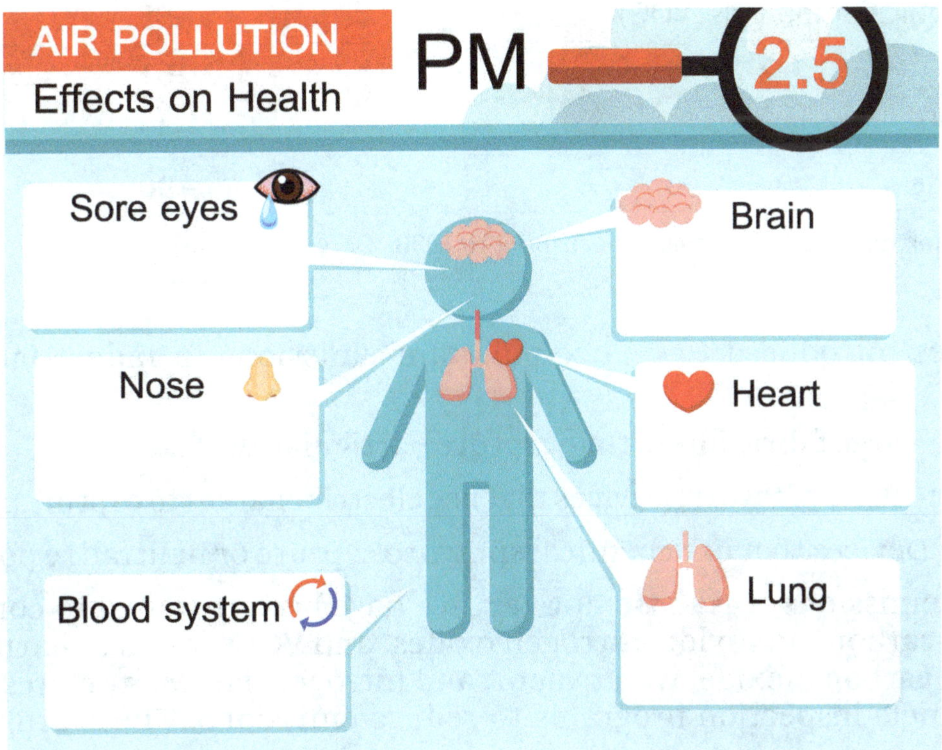

aririnstory/Shutterstock.com

Kevin T Brett/Shutterstock.com

Lead (Pb) is a toxic heavy metal. While it is a particulate, it was given its own category as an emission pollutant due to its history and its effects on health. Lead is a neurotoxin that can bioaccumulate and cause damage to the central nervous system, including causing permanent learning disabilities in children. Lead was once added to gasoline to make engines run quieter. Phasing leaded gasoline out in the 1970s has resulted in a 99% drop in lead emissions. Today, the only significant source of lead emissions is from smelting operations.

The CAA began a coordinated effort to reduce the emission of these pollutants. The following table shows the percentage change of the emission of each pollutant from 1980 to 2018.

Changes in Pollutant Emissions: 1980–2018

Pollutant	Percent Change in Emissions
Carbon Monoxide	−73
Lead	−99
Nitrogen Oxides (NO_x)	−62
Volatile Organic Compounds (VOC)	−55
Direct PM_{10}	−61
Direct $PM_{2.5}$[a]	−30
Sulfur Dioxide	−90

[a]$PM_{2.5}$ range is for 1990–2018 due to lack of data prior to 1990.

Source: epa.gov

Technologies introduced to reduce pollutant emissions in industrial facilities include:

Baghouses: Large fabric filters that capture particulates.

Electrostatic Precipitators: Devices that use electric charges to capture pollutants.

Scrubbers: Devices that use chemical sprays to capture or neutralize pollutants.

For vehicle emissions, **catalytic convertors** have been installed to convert pollutants like carbon monoxide, nitrogen oxides, and VOCs into less harmful compounds like carbon dioxide, water vapor, and nitrogen gas. Many states have also adopted vehicle inspection programs to reduce emissions. These programs can

keyfoto/Shutterstock.com

Baghouses, scrubbers, and electrostatic precipitators have reduced the emission of pollutants from industrial facilities like factories and power plants.

vary from state to state. For example, the state of Pennsylvania uses a three-tier system. Vehicles in the largest urban areas of Philadelphia and Pittsburgh have the most stringent testing requirements. The smaller urban counties like Dauphin, York, Lehigh, and so on, have a less stringent standard while the primarily rural counties do not require any emissions inspections. Catalytic convertors and vehicle inspections, as well as the removal of lead from gasoline have reduced the emissions from vehicles despite the fact that the number of vehicles on the road have increased significantly over the past few decades.

In addition to the six emission pollutants, the EPA also regulates the emission of 187 air toxics or hazardous air pollutants (HAP). HAPs are pollutants that cause cancer, reproductive health effects, or other serious health effects.

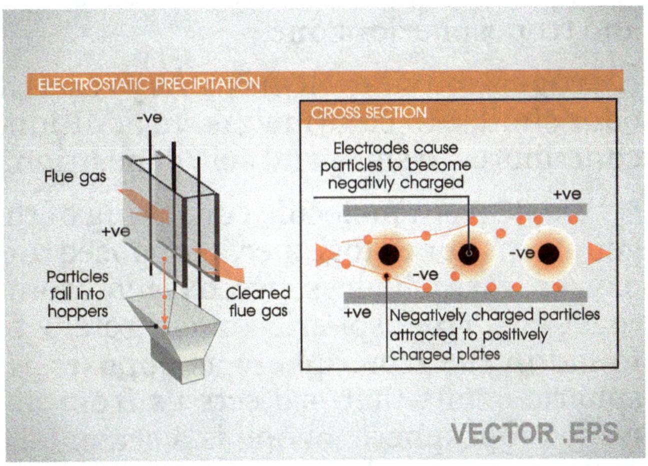

Studio BKK/Shutterstock.com

Mark Winfrey/Shutterstock.com

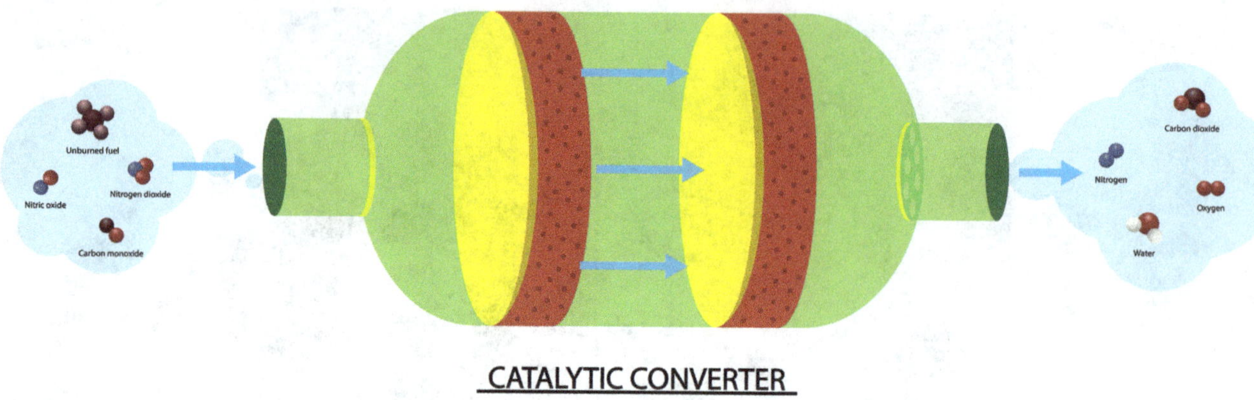

Studio BKK/Shutterstock.com

Finally, the EPA also has standards for the **National Ambient Air Quality Standards** (NAAQS) that monitor and grade the quality of the air. For these standards, **six criteria pollutants** are monitored through daily or hourly samples taken across the country. This data is compiled and used to grade the quality of the air by region in each state. This information is available on the web at https://www.epa.gov/outdoor-air-quality-data/air-quality-index-report. Four of the six criteria pollutants are the same as the emissions pollutants: carbon monoxide, sulfur dioxide, particulate matter, and lead. The other two are nitrogen dioxide and tropospheric ozone.

Nitrogen dioxide (NO_2) is a very reactive, reddish-brown gas that has a pungent odor similar to chlorine gas. In addition to being a powerful lung irritant, it also contributes to smog and acid deposition.

Ozone (O_3) is a molecule consisting of three oxygen atoms joined together, as opposed to ordinary oxygen gas, which is two oxygen atoms joined together. Tropospheric ozone refers to ozone found in the troposphere as opposed to stratospheric ozone that protects us from ultraviolet rays. Tropospheric ozone is a secondary pollutant that forms when sunlight triggers reactions between NO_x and VOCs. A highly reactive gas, ozone is a powerful lung irritant.

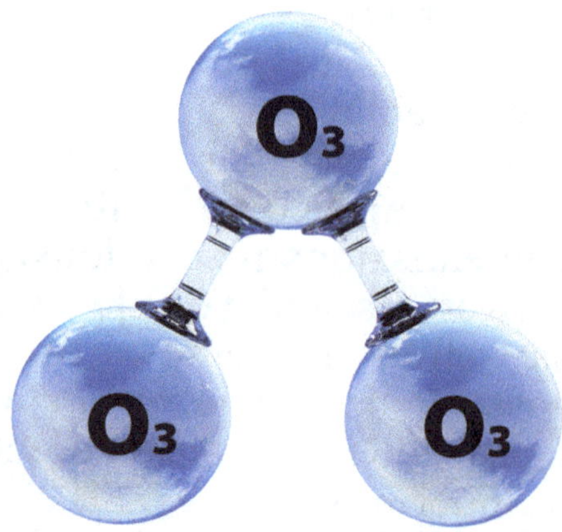

Victor Josan/Shutterstock.com

Since the passage of the CAA, the concentrations of the criteria pollutants have all gone down significantly. However, tropospheric ozone is the one that still most frequently exceeds the current EPA standards. Ozone alerts can become incorporated into the daily weather reports. When an ozone alert has been issued for a region, the elderly and people with asthma or other respiratory conditions are advised to limit their physical exertions as this could endanger their health. Overall, however, the air quality even in the largest urban areas has improved steadily since 1980.

The following table summarizes the percent change in air quality for the criteria pollutants from 1980 to 2018.

Percent Change in Air Quality: 1980–2018

Pollutant	Percent Change in Air Quality
Carbon Monoxide	−83
Lead	−99
Nitrogen Dioxide (annual)	−65
Nitrogen Dioxide (1-hour)	−61
Ozone (8-hour)	−31
PM_{10} (24-hour)[a]	−26
$PM_{2.5}$ (annual)[b]	−39
$PM_{2.5}$ (24-hour)[c]	−34
Sulfur Dioxide (1-hour)	−91

[a] PM_{10} data is from 1990 to 2018.

[b] $PM_{2.5}$ data is from 2000 to 2018.

Source: epa.gov

Quality Stock Arts/Shutterstock.com

Smog is a portmanteau of smoke and fog and is a mixture of air pollutants that accumulate over a large area. There two main types of smog.

Industrial smog results from the burning fossil fuels, especially coal. It contains carbon monoxide, carbon dioxide, mercury, and sulfur. It results in a grayish mass of soot hanging over the city. Industrial smog is the kind of smog that caused the killer fog events in London. One of the worst industrial smog events in the United States occurred in 1948 in Donora, Pennsylvania. An industrial accident released a large quantity of pollutants into the valley where Donora is located, trapping the pollutants. The smog was so thick that they had to turn the street lights on at noon. Scrubbers and the reduction in coal burning have reduced the prevalence of industrial smog.

Photochemical smog is the more common kind of smog that forms in cities with heavy vehicle traffic like Los Angeles. Sunlight reacting with vehicle exhaust to produce a variety of pollutants produces a yellowish or brownish haze that reduces visibility. Secondary pollutants formed in photochemical pollutants, include ozone, Paroxyacyl Nitrates (PANs), and aldehydes.

Efforts to reduce photochemical smog include catalytic convertors, vehicle inspection programs, and reducing emissions from oil refineries and power plants.

LukeandKarla.Travel/Shutterstock.com

14.4 Stratospheric Ozone Depletion

While ozone in the troposphere is harmful, stratospheric ozone absorbs a lot of the Sun's ultraviolet rays that can cause skin cancer. Unfortunately, certain chemicals called halocarbons have been among the most significant contributors to the depletion of the ozone layer.

Halogens are a group of elements that include chlorine, fluorine, and bromine. **Halocarbons** are a class of hydrocarbons in which one or more of the hydrogen atoms have been replaced with a halogen. A particular group of halocarbons, called **chlorofluorocarbons** (CFCs), contain chlorine, fluorine, and carbon. CFCs have a low-boiling temperature and appeared to be chemically inert. This made them useful for a lot of applications such as refrigerators, air conditioners, and fire extinguishers. They were also used in the manufacture of Styrofoam and as propellants in spray cans. Their chemical inertness meant they didn't react with other compounds. This is good for use in spray cans. You don't want the propellant to alter the color or adhesive properties of a spray paint.

However, CFCs are able to migrate into the stratosphere. Once there, sunlight causes them to break apart, releasing chlorine atoms, which in turn, react with ozone molecules to convert them into ordinary oxygen.

Kerckweb/Shutterstock.com

CFCs were widely used as spray propellants and refrigerants.

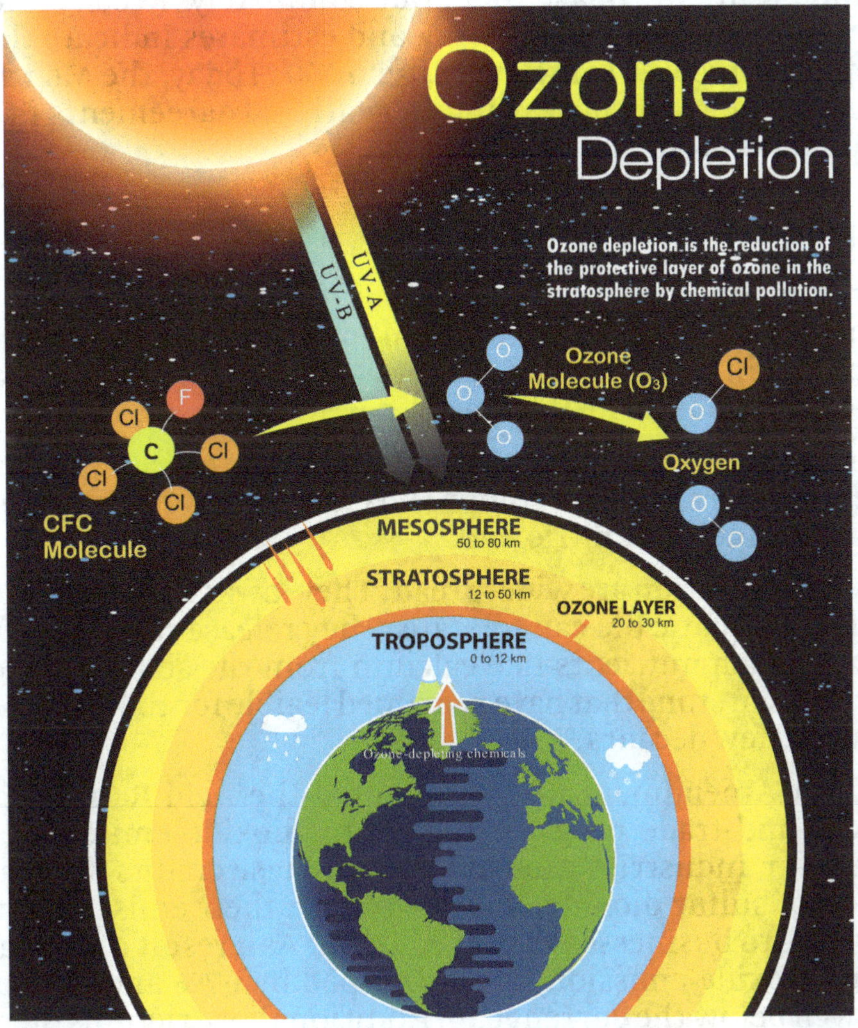

Nagendra_Yadav/Shutterstock.com

In the 1980s. scientists began to discover the damage being done to the ozone layer and by 1985, they discovered the **hole in the ozone layer** (actually, a region where the ozone concentration is the thinnest) over Antarctica. The "hole" formed when stratospheric clouds containing nitric acid accelerated the breakdown of CFCs and concentrated the chlorine in the spring over the south polar region. This concentration was the result of the Earth's rotation and many prevailing wind currents.

Narin C/Shutterstock.com

The hold in the ozone layer was discovered in the 1980s. In recent years, it has shown signs of stabilizing and may be getting smaller.

To combat the problem of ozone depletion, nations signed the **Montreal Protocol**, a treaty to phase out the production of CFCs. Countries eventually agreed to cut CFC production in half by 1998 and eventually phase them out completely. Recent evidence shows that the **ozone hole** has stopped growing and estimates indicate that the ozone layer will completely recover sometime after 2060. Today, the Montreal Protocol is held up as one of the most successful international agreements in addressing a global environmental problem.

14.5 Acid Deposition

Acid deposition is the release of acids or acid-forming compounds from the air onto the Earth's surface. The most common form of acid deposition is **acid precipitation**, whether in the form of rain, snow, or fog. Dry deposition can also occur when the acid compounds deposit from dry particles.

Acid deposition begins when the burning of fossil fuels releasing sulfur dioxide and nitrogen oxides into the air. These compounds mix with water vapor and condense into clouds as sulfuric acid and nitric acid.

The effects of acid deposition are widespread. They lower the pH of lakes and ponds below the levels the aquatic life can tolerate. They release toxic ions from the soil, killing plants and leach nutrients like calcium from the soil. It also erodes stone structures. Writings in stone that have remained legible for thousands of years have been worn away in a few decades.

The Clean Air Act Amendment of 1990 created the Acid Rain Program, which included the cap-and-trade program for sulfur dioxide emissions (Chapter 8). Under this program, industries were able to purchase credits up to a specific cap for the emission of sulfur dioxide. As they reduce their emissions, they can sell their excess credits to businesses that need them. As a result of the cap-and-trade program, sulfur dioxide emissions have dropped by 67% and the current emission levels are well below the current cap. Additional restrictions on power plants have also reduced nitrogen oxide emissions. As a result, the pH of rainwater in the United States has been rising. Normal rainwater has a pH of about 5.6, making it slightly acidic. So, what we refer to as acid rain is actually any rainwater with a pH below 5.6. In 1990, the average pH of rainwater in the Northeastern United States was less than 4.1. By 2015, it had risen to close to 4.9. It is hoped that further reduction in sulfur dioxide and nitrogen oxide emissions will continue to reduce acid deposition.

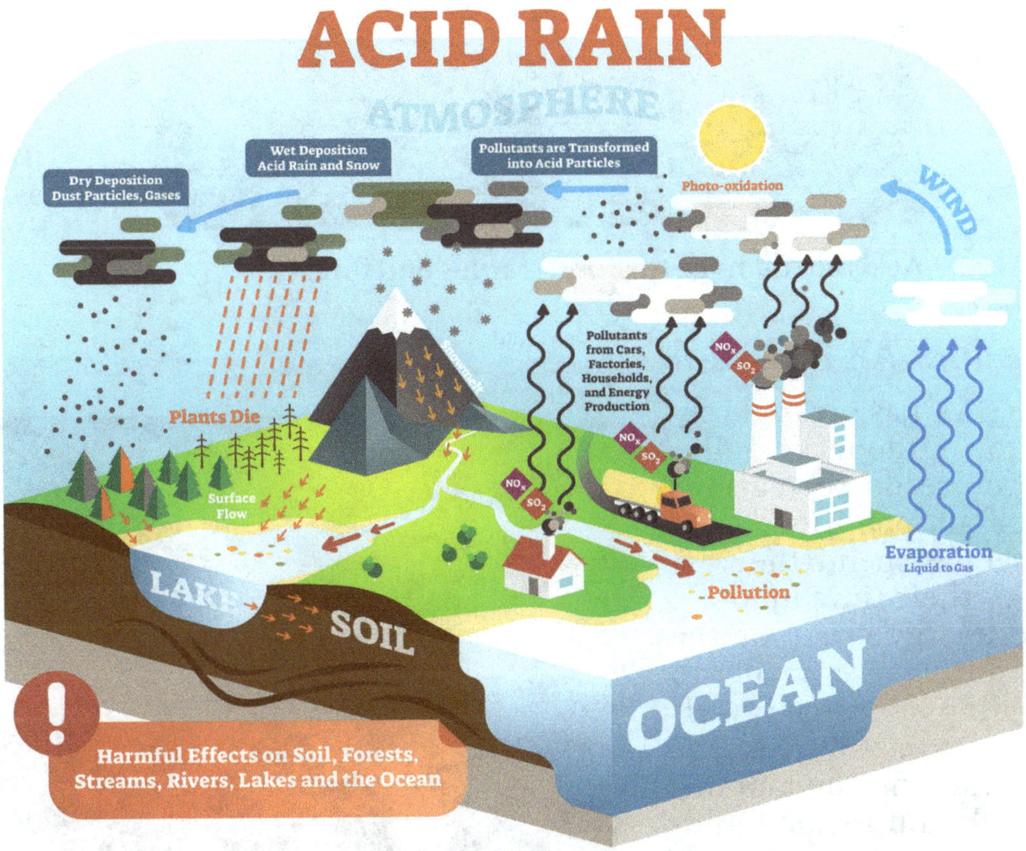

VectorMine/Shutterstock.com

Vandathai/Shutterstock.com

Mary Terriberry/Shutterstock.com

Acid rain damage to stone statues and forests.

14.6 Indoor Air Pollution

While the CAA has created regulations for outdoor air pollution, there are no regulations for indoor air pollution, even though people spend the vast majority of their time indoors. After the energy crisis of the 1970s, efforts were made to make buildings more energy-efficient. However, this had the result of reducing the amount of air circulation inside buildings, with less fresh air from the outside coming in and pollutants becoming trapped inside.

piyaphong/Shutterstock.com

In the developing world, the main sources of indoor air pollution are the burning of fuels for heating and cooking indoors. Fuels like firewood, charcoal, dung, and crop waste are often burning inside homes with little or no ventilation. Studies have found that the particulate matter concentrations in some developing nation homes are 20 times the EPA standards for outdoor pollutants.

Francesco Scatena/Shutterstock.com

Indoor air quality can be improved in developing countries by drying fuelwood or switching to less polluting fuels like natural gas. Also, they can install chimneys or other sources of ventilation or simply move the cooking outdoors.

In the developed countries, the two most significant indoor pollutants are cigarette smoke and radon, both of which are a link to lung cancer. Many public buildings have become smoke-free, either by choice or by state regulations. Homes can also be installed with systems that can ventilate the radon to the outside.

VOCs are emitted indoors from plastics, oils, paints, perfumes, adhesives, cleaning fluids, and so on. New carpets and furniture that are made with adhesives will off gas VOCs like formaldehyde (this includes that "new car smell" everyone loves). While these emissions fade over time, other sources of VOCs like copier and printers will emit them every time they are used. Health effects of these compounds are not well understood as the concentrations are minute. Individual reactions can vary. Many people might not have any complaints from them while others will complain that the copier gives them a headache.

Other indoor air pollutants include pet dander, asbestos, lead particles from paint, carbon monoxide, and various chemicals emitting from household cleaning products, fuels, and paints stored in the basement.

Organisms can also contribute to indoor air pollution. Legionella is a bacterium that grows in water the condenses from air conditioning systems and then travels through the ducts, giving people respiratory infections. Molds and other fungi can grow in areas that have leaky plumbing or other sources of moisture. Many mold species release allergens or toxins that can cause reactions in some individuals.

Often, the exact source of people's ailments is unknown. The symptoms people complain of, such as sneezing, coughing, headache, congestion, and so on are often non-specific to any disease. As a result, the term **sick building syndrome** is applied.

WHYFRAME/Shutterstock.com

These indoor problems can be reduced by avoiding smoking and testing buildings for radon and mold spore. Preventing leaks and repairing water damage can also reduce the growth of mold. Moving printers and other devices that emit VOCs to separate, ventilated rooms can reduce complaints. Finally, increasing the flow of fresh air into the building can also disperse these pollutants and improve the quality of the air.

Francesco Scatena/Shutterstock.com

14.7 Discussion Questions

1. Describe the troposphere and stratosphere and how they differ in temperature and composition.
2. What is the difference between weather and climate?
3. How do the Hadley cells influence the climate between the equator and 30 degrees north/south?
4. What is a temperature inversion and how does it affect pollution in a city?
5. What is the difference between primary and secondary pollutants?
6. Name the six emissions pollutants and describe the health and environmental impacts of each of them.
7. Why did the federal government switch from the PM_{10} standard to the $PM_{2.5}$ standard?
8. What do catalytic convertors do?
9. Name the six criteria pollutants used in the National Ambient Air Quality Standards.
10. Name the two different kinds of smog and describe how each is formed.
11. Where is the ozone layer located and what benefit does it give us?
12. How do CFCs affect the ozone layer?
13. What are the causes of acid deposition?
14. What program did the CAA amendment of 1990 create and how did it reduce acid deposition?
15. What is sick building syndrome?

14.8 Key Terms

Acid deposition
Acid precipitation
Air pollution
Atmosphere
Baghouse
Carbon monoxide
Catalytic convertor
Chlorofluorocarbons
Clean Air Act
Climate
Convection current
Coriolis effect
Doldrums
Electrostatic precipitator
Emission pollutants
Ferrell cells
Hadley cells
Halocarbons
Halogens
Hole in the ozone layer
Industrial smog
Lead
Mesosphere
Montreal Protocol
National Ambient Air Quality Standards
Nitrogen dioxide
Nitrogen oxides
Ozone
Ozone hole
Ozone layer
Particulate matter
Photochemical smog
Polar cells
Primary pollutants
Residence time
Scrubbers
Secondary pollutants
Sick building syndrome
Six criteria pollutants
Smog
Stratosphere
Sulfur dioxide
Temperature inversion
Thermosphere or ionosphere
Trade winds
Tropopause
Troposphere
Volatile organic compounds
Weather

Chapter 15

GLOBAL WARMING

Chapter Topics

Bernhard Staehli/Shutterstock.com

Evidence for climate change is all around us. The ice sheets in Greenland and West Antarctica are melt, as are glaciers in Alaska, the Himalayas, Alps, Africa, and Rockies. The number of glaciers in Glacier National Park has dwindled from 150 to 30. In addition, the Arctic sea ice is also dwindling.

Populations of Adélie penguin in Antarctica have collapsed by 90%. Other species have shifted their ranges. Foxes, butterflies, pikas, and some flowers have moved north or to higher elevations, while species like polar bears that already live on the northern margins are struggling. On the other hand, many pest species like mosquitos and ticks are thriving in the new environment. Outbreaks of the bark beetle, which kill millions of acres of pine and spruce trees, are becoming more frequent.

Precipitation, including rain and now, has increased in many areas around the global. Major storm events like hurricanes have become more intense and more frequent. Meanwhile, other regions like the American Southwest have experienced more frequent and severe droughts, while the number of record high-temperature events in the United States has been increasing. Meanwhile, flooding in Miami has been increasing and the city is spending millions installing pumps to contend with it.

The global sea level has risen by 8 inches in the last century, with most of that rise having occurred in the past two decades. Ocean acidity has increased by about 30%, threatening coral populations and other species that need calcium carbonate to make their shells. Meanwhile, the temperature of the upper 2,300 feet of the oceans has risen by about 0.9° F.

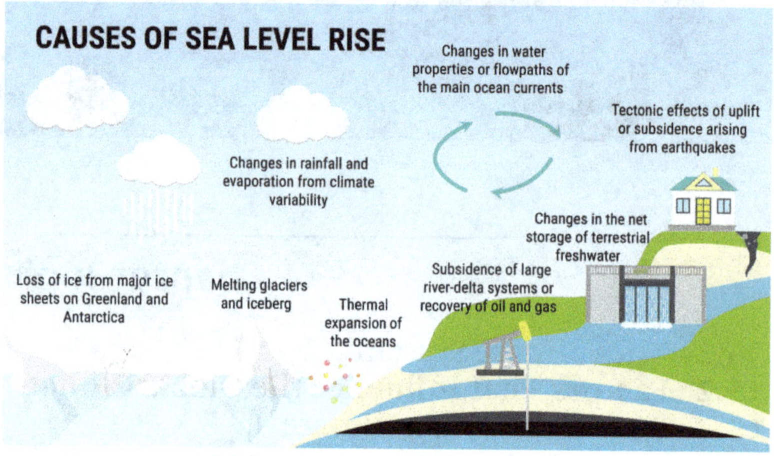

m.malinika/Shutterstock.com

Despite all of this evidence, there are many that deny that global warming is happening or that the changes are cyclical. Or, if they are really happening, it is not being influenced by human activities. However, these arguments are not supported by the evidence.

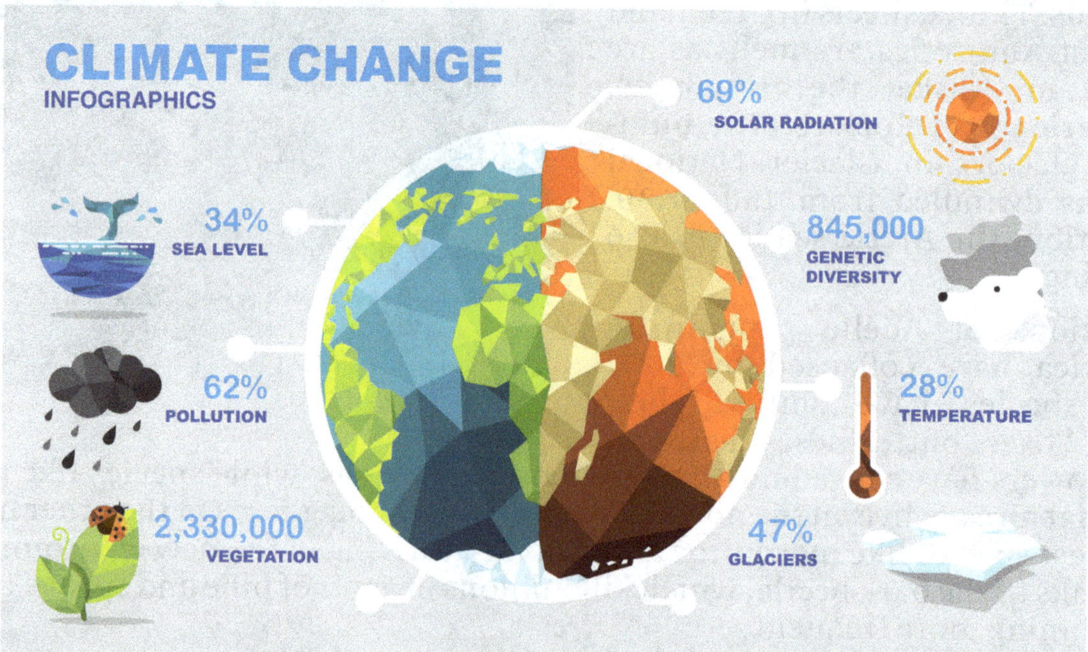

Cube29/Shutterstock.com

15.1 The Climate System

As we noted in Chapter 14, **climate** is the atmospheric condition of large areas averaged over long periods of time. This is different from weather, which is the local, short-term atmospheric condition. Climate measures a number of conditions, including

- Temperature
- Precipitation
- Wind
- Humidity
- Barometric pressure
- Solar radiation

Global climate change refers to any alterations in the overall climate of the planet. We know from geologic evidence that the climate has varied over time. It has been warmer than it is today and it has also been cooler than it is today at various times in the past. Atmospheric concentrations of gases like carbon dioxide have been one of the many factors that have altered the climate in the past. Other factors include variations in the Earth's motion, shifts in the ocean currents, changes in the configuration of the continents, and volcanism have all also influenced the climate of the past. **Global warming** is a specific subset of climate change that involves an increase in the average global temperature. While the global temperatures have shift in the past, these changes have occurred over periods of thousands of years or more. The current-accelerated rising temperatures is being driven in large part by human activities, in particular, the emission of greenhouse gases.

15.2 Factors That Influence the Global Climate

There are three major factors that influence the Earth's climate: the sun, the oceans, and the atmosphere. As we noted in Chapter 7, the oceans store a lot of heat energy and transport it around the world through the currents as part of the thermohaline circulation.

The sun emits electromagnetic radiation, the majority of which that passes through our atmosphere is visible light, which is why our eyes are adapted to seeing this narrow band of wavelengths. The energy from this light warms the surface of the Earth, which reemits the energy as infrared radiation, which has a longer wavelength. There are many gases that are transparent to visible light but tend to absorb infrared radiation and trap it near the Earth's surface. These are referred as **greenhouse gases** because a greenhouse is a structure designed to trap heat. Greenhouse gases include water vapor (H_2O), ozone (O_3), carbon dioxide (CO_2), nitrous oxide (N_2O), methane (CH_4), and halocarbons. This warming of the lower troposphere from the trapping of infrared radiation is referred to as the **greenhouse effect**.

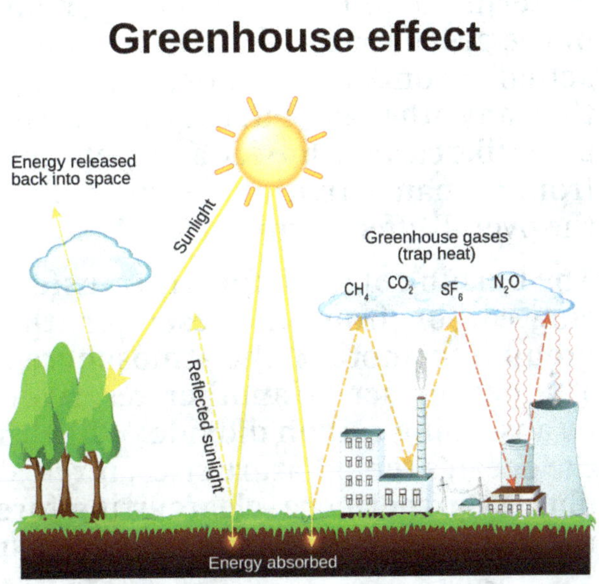

Designua/Shutterstock.com

Generally, the greenhouse effect is beneficial for life. Without it, the Earth would be much colder, similar to Mars. What we are concerned with is the anthropogenic (human-caused) increase in the greenhouse effect by the emission of greenhouse gases. Different gases have different capability to treat heat, its **global warming potential**. For comparison, carbon dioxide is given a global warming potential of 1 to serve as a baseline. The other gases are then measured in terms of how much more they contribute to global warming on a molecule-per-molecule basis. The table below summarizes the global warming potential of some significant gases.

Relative Global Warming Potential

Greenhouse Gas	Over 20 Years	Over 100 Years
Carbon dioxide	1	1
Methane	84	28
Nitrous oxide	264	265

Data from IPCC, 2013. Fifth assessment report.

While these other gases may be stronger in terms of how much heat each molecule can trap, carbon dioxide is the gas of most concern because we are emitting much more it than any other gas. As indicated by the chart to the right, carbon dioxide accounted for 82% of the total greenhouse gas emissions in the United States.

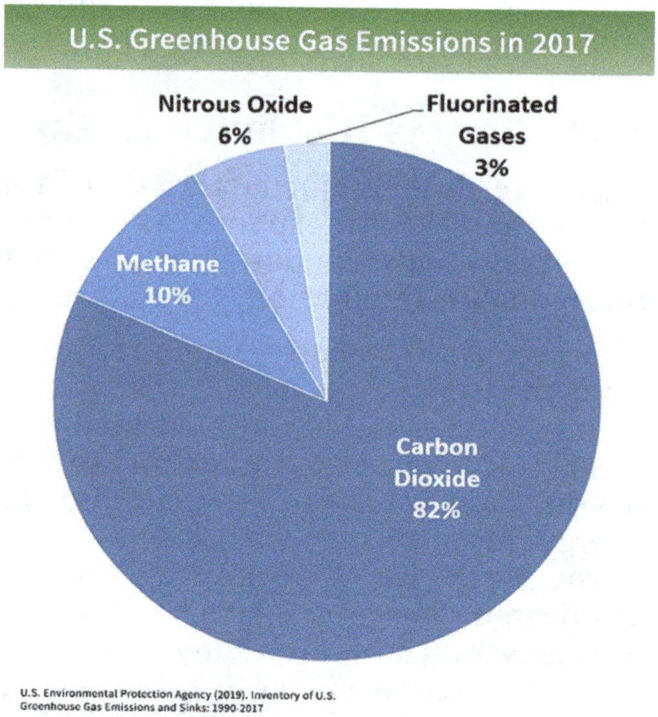

Source: epa.gov

In terms of the natural component of the greenhouse effect, water vapor actually contributes more warming than any other gas, but the increase in gases like carbon dioxide and methane from human activities are increasing the overall effect significantly.

The burning of fossil fuels is transferring carbon from one reservoir (the ground) to another (the atmosphere). The oceans act as another reservoir by absorbing carbon dioxide. While this has offset global warming somewhat, as the ocean's temperature rises, it releases the carbon dioxide back into the atmosphere. Furthermore, clearcutting forests removes another sink for carbon dioxide, so less is removed from the atmosphere through photosynthesis.

Other factors that can influence the global climate include aerosols, microscopic droplets, and soot particles released from combustion or volcanic activities. Soot particles that settle on the surface cause a net warming by absorbing heat. On the other hand, sulfate aerosols suspended in the air can cause a net cooling by reflecting the sun's rays away from the Earth.

Radiative forcing measures the amount of change a given factor can have on the Earth's temperature. A positive force warms the planet, while a negative force cools it. When we take into account all of the phenomena that can influence the climate, such as greenhouse gases, aerosols, and the relative reflectivity of the land, water, and ice, the current radiative forcing comes out to be a net warming factor of 2.3 watts/m^2.

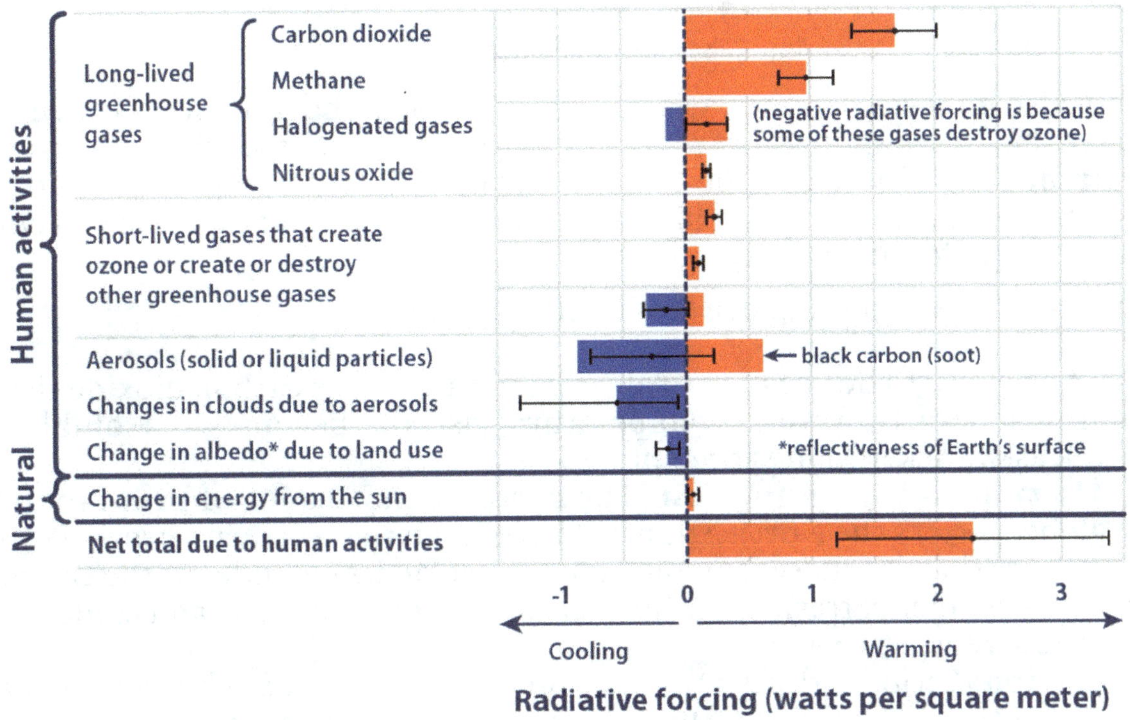

Table Source: epa.gov

Beyond the Earth's surface and atmosphere, variations in the Earth's motion can influence the global climate as well. These changes follow regular patterns called the **Milankovitch Cycles**. For example, the tilt of the Earth's axis varies between 22° and 25°. Also, the Earth's axis experiences precession and revolves like the axis of a spinning top. This means the axis changes direction. Currently, the axis points in the general direction of the star Polaris, making it the North Star as it appears in the north all year round. Over time, the axis will change direction and no longer point toward Polaris. In the future, it will point toward different stars or even have periods when it points to no particular star at all (much like how southern end of the axis does not point to any star today). Finally, the eccentricity, or the degree of how stretched out the Earth's orbit also periodically changes from more circular to more elongated. While all of these cycles can have a profound effect on the global climate, they run on the order of thousands to tens of thousands of years and cannot account for the accelerated warming the Earth is currently experiencing.

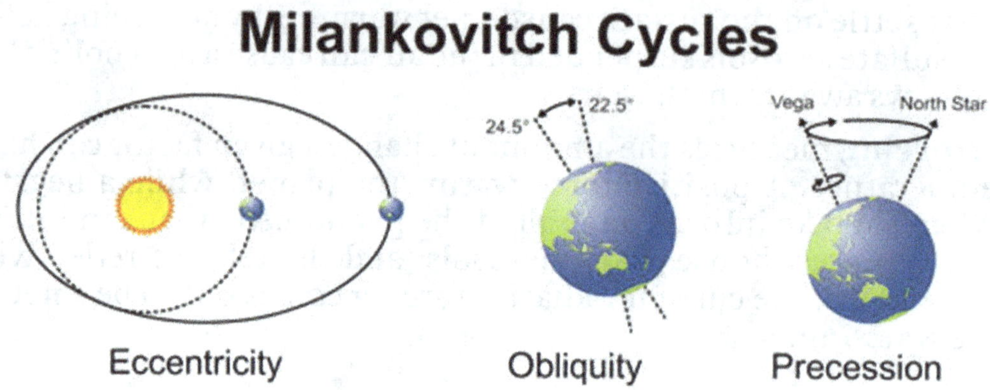

Image Source: https://skepticalscience.com/print.php?r=33

The sun also has cycles in the variation of its output. The most well-known is the 11-year cycle of sunspot activity. Every 11 years, the sun goes through a sunspot maximum, which increases the frequencies of solar flares and other activities. These changes, however, do not have a significant influence on the Earth's climate with a radiative forcing of only 0.05 watts/m^2.

Other factors that can influence that climate include

> **The oceans**: The oceans can absorb 50 times more carbon dioxide than the atmosphere. However, as temperatures rise, CO_2 becomes less soluble and is released back into the atmosphere.
>
> **Thermohaline circulation**: The currents move a lot of heat around the globe, such as the North Atlantic Current, giving Western Europe a warmer climate than it otherwise would have based on its latitude. As the Greenland Ice Sheet melts, freshwater flows into the North Atlantic and could disrupt the current.
>
> **El Niño**: Periodic shifts in atmospheric pressure cause shifts in the Pacific currents. In recent years, these shifts are becoming more frequent and more intense.

15.3 Researching the Climate

We do not have direct data of the climate from the distant past, so we have to rely on **proxy indicators**, indirect measurements that can be used as a substitute for direct measurements.

Ice core samples are one of the most useful proxy indicators. Like sedimentary rocks, ice layers form one on top of the other. So, the deeper one drills through ice in Greenland and Antarctica, the older the ice. Looking at slices from the ice core samples under a micrograph will reveal tiny bubbles. These bubbles are samples of the atmosphere as it was when the ice formed. By measuring the concentration of different gases and isotopes in these bubbles, scientists can calculate the global climate conditions of the past and even the frequency of fires and volcanic eruptions. Data collected from ice cores taken from Antarctica indicate that the current CO_2 levels in the atmosphere are the highest in 800,000 years.

Image Source: https://climate.nasa.gov/news/2616/core-questions-an-introduction-to-ice-cores/

Ice core samples are used to gather information about the climate of the distant past.

Dendrochronology is the study of tree rings. As trees grow, they add a new ring about every year. By making measurements of the rings, scientists can record variations in the climate throughout the tree's life. Scientists have collected tree ring samples from numerous trees that reach back thousands of years.

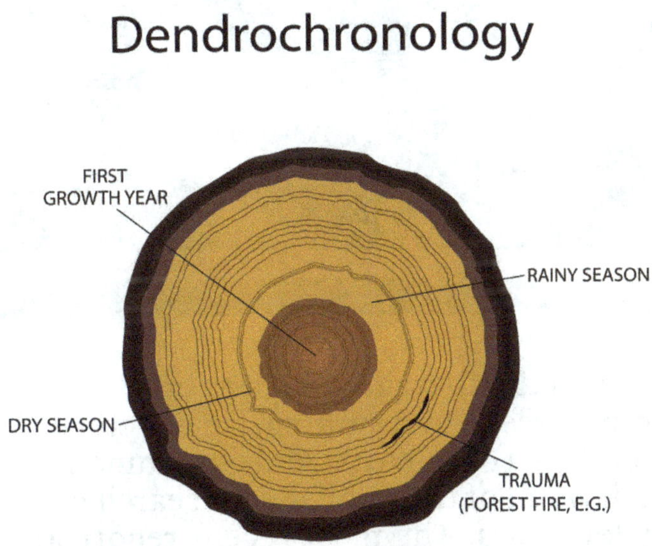

Nasky/Shutterstock.com

Other proxy indicators include fossilized packrat middens, isotopic analysis of coral reefs, and sediment cores from the bottom of lakes and other bodies of water.

We also have direct measurements of carbon dioxide levels for several decades now. The Mauna Loa Observatory has been taking hourly samples and measuring CO_2 concentrations since 1958. From these measurements, the CO_2 levels have risen from 315 ppm in 1958 to over 400 ppm today. So, there is direct data the CO_2 have been increasing.

Climate modeling involves mathematical simulations run by computer programs. Modeling is still not a perfect science as the global climate is complex with lot of factors and moving parts and we still don't understand all of them. Models are tested by entering data from the past and running them to see how well they simulate current conditions, which is then used to further refine them. When models are demonstrated to accurately simulate the present, they can be run to make predictions about the future based on current trends. As our understanding of the climate situation grows, the accuracy of models has been steadily improving.

15.4 Current and Predicted Impacts of Global Warming

We are seeing signs of global warming every day now such as increased flooding in the Miami metropolitan area, increased droughts in California and Texas, and more frequent heat waves and cold snaps across the United States.

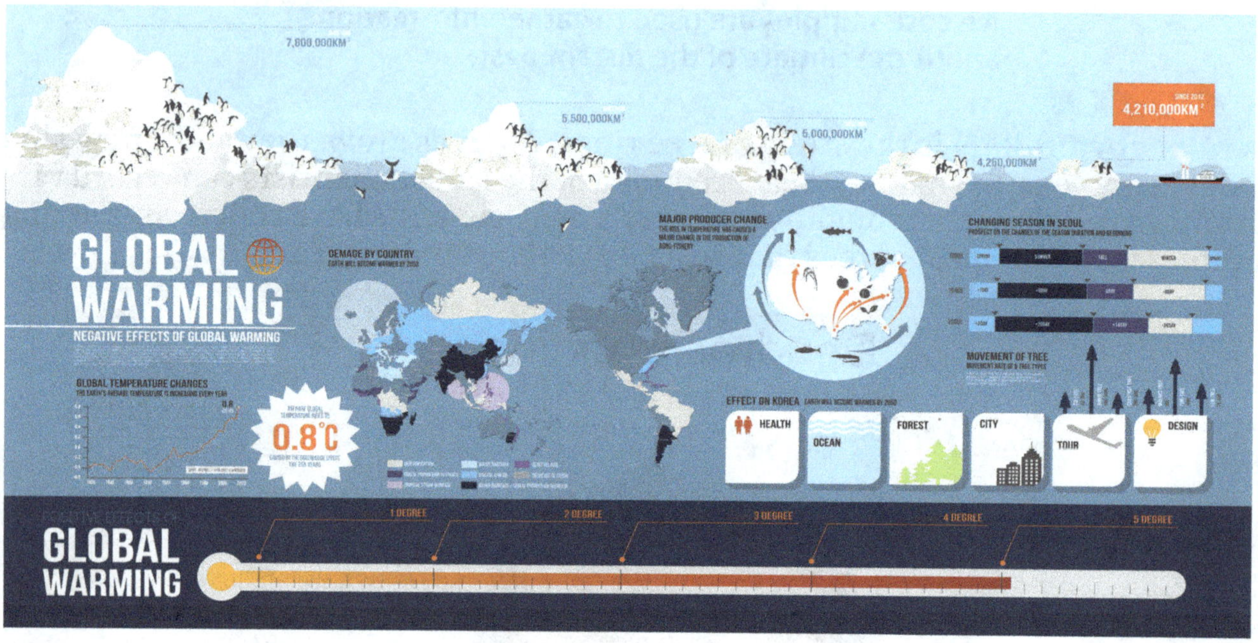

getfile/Shutterstock.com

The United Nations formed the **Intergovernmental Panel on Climate Change** (IPCC) in 1988 to examine, assess, and summarize the current state of knowledge on global warming. Consisting of scientist and government officials from all over the world, the IPCC reviews all of the published research on climate and summarizes in a report every few years. The most recent report, the Sixth Assessment Report, was released in 2018. It covers all of the observed trends in sea levels, storm activity, glaciers and snow cover, droughts, changes in organisms, and other factors. It also reviews possible mitigation strategies and makes predictions for future trends.

According to the Sixth Assessment Report, the global temperature has risen by 1.1° C and the average temperature in the United States has increased by 1° F in the past 20 years.

Further trends in temperature changes are expected to vary by region, with the poles, especially the north pole, experiencing the most net warming.

Warming trends has increased evaporation rates, resulting in a 2% increase in global precipitation. And in winter, that does mean an increase in snowfall, despite what many climate skeptics may say, an increase in snowfall does not mean global warming is a hoax. This increase is actually predicted by climate models. The poles are expected to see the most increase in precipitation (snow) and as a result, the interior of Antarctica is seeing a net increase in ice cover even as the ice shelves along the coasts are disappearing.

Not all regions are seeing an increase in precipitation. Some areas are expected to continue to see a decrease in precipitation and more droughts. These regions include the American Southwest, the Mediterranean Region, Southern Africa, Australia, and Central America.

In the United States, the frequency of extreme weather events such as tornadoes, droughts, floods, and hurricanes, has doubled since 1970. One notable year was 2012 with Superstorm Sandy and in March of that year was plagued with extreme warmth in the east and extreme cold in the west as the jet stream was deflected. This created an atmosphere blocking pattern that prevented the movement of air across the continent. 2017 was a particularly intense year for hurricanes. Storms like Maria, Harvey, and Irma devastated Florida, Texas, and Puerto Rico.

Matty Symons/Shutterstock.com

Worldwide, glaciers are retreating and mountaintops are losing their snow cover. This is serious because the mountain snowpack is the primary source of water to replenish groundwater aquifers that millions of people depend on for drinking water and irrigation. The West Antarctic shelf is on the verge of a major collapse. If it falls, it will increase sea levels by 3 m. The ice sheet in Greenland and the sea ice in the Arctic is also thinning, with many summers in the Arctic Ocean being completely ice-free. This melting of the sea ice is opening up to areas for shipping and oil companies are rushing in to drill in these waters, even as they spend millions of dollars on propaganda programs that sow doubt on the reality of global warming.

In the northern latitudes, permafrost is melting. This is causing buildings and roads to collapse due to subsidence. As permafrost thaws, it releases methane that was trapped in the ground. Since methane is a potent greenhouse gas, this results in a positive feedback loop leading to more warming.

Lorelyn Medina/Shutterstock.com

Michelle Holihan/Shutterstock.com

Barrow, Alaska – June 3, 2012: Cemetery slowing sinking into permafrost.

As surface ice melts, it exposes less reflect surfaces like dry land and liquid water. This in turn absorbs more heat, creating another positive feedback loop that promotes more ice melting.

Chumash Maxim/Shutterstock.com

Maldivian capital from above.

Rising sea levels result in more beach erosion, saltwater intrusion of aquifers, and coastal flooding. More frequent hurricanes and other storm systems result in more storm surges, creating severe, localized flooding. While we can never say with certainty that any one single storm event was caused by global warming, the overall trend in frequency and intensity is indicative of expected changes. Rising sea levels can put many cities at risk. Most of New Orleans is below sea level with only levees protecting it. These levees already failed during Hurricane Katrina and are likely to fail again. The Maldives is an island nation in the Indian Ocean. Most of the land in the Maldives Islands are less than 1 m above sea level and are high risk of flooding. Some islands in Alaska have lost as much as half their land area to sea level rise, putting the indigenous people living on them in jeopardy.

More than half of the US population lives along the coasts and 3.7 million have homes within 1 vertical mile of the high tide line. The region in the United States is South Florida, with 2.4 million people and 1.8 million acres vulnerable to flooding.

Despite this, the state government of Florida has largely been in denial about the threat of rising sea levels.

More CO_2 in the ocean results in ocean acidification and the pH of the oceans has dropped by 0.1 pH unit. Ocean acidification damages coral and other shellfish by dissolving the carbonate ions they need to make shells. Warming waters also promote coral bleaching and increases in storm intensity can damage the reefs.

Temperature-related season events, like the leafing of plants, bird migrations, and insect hatchings are occurring early in the year than before. Many species are shifting their ranges higher in elevation or to more northern latitudes looking for suitable habitats. This puts the species already at the margins at greater risk. For example, pikas are mountainous rodents that are moving higher up the mountains. When they reach the mountaintops and are still unable to find a suitable habitat, they have nowhere else to go.

The loss of Arctic sea ice is also making it difficult for polar bears to hunt during the summer months. Normally, they walk across the ice flows looking for prey, but with the ice, they can't hunt.

Some argue that increases in CO_2 levels will increase growing seasons and agricultural output. While this may be true in some locations, increased droughts will decrease crop yields in other locations. Carbon dioxide may also increase tree growth; the warmer temperatures will increase the frequency of disease and pest outbreaks like the bark beetle.

Major human health impacts will be caused by more frequent heat waves. More photochemical smog will cause more respiratory ailments. Tropical disease like malaria may spread into areas like the United States where they are currently not found. Increased flooding will cause more injuries and drowning as well as increases in disease and sanitation problems.

Economic predictions indicate that the costs of global warming will outweigh any benefits. The Stern Review on the Economics of Climate Change predicts that climate change will cost between 5% and 20% of the global GDP by the year 2200. It recommends that investing 1% of GDP now could avoid those costs. The poorest people, of course, are likely to suffer the worst impacts from global warming.

Climate change is also expected to create humanitarian crises. Droughts, crop failures, and floods will increase human suffering in many developing nations. Many of these people will become climate refugees as they flee their native lands. Others may turn to terrorism. For this reason, the US Department of Defense considers global warming a national security threat.

Despite the overwhelming evidence and scientific consensus, there are many people who doubt the reality of climate change or that human activities like the burning of fossil fuels and deforestation are responsible for it. Many of these "climate skeptics" are funded by corporate interests connected by the fossil fuel industry and conservative think tanks that promote efforts to promote doubt about the global warming. The media often amplifies these voices as it thrives on controversy and seeks to present both sides of every topic, even if one side lacks sufficient evidence to support its claims.

15.5 Adaptation and Mitigation Strategies

Assuming that evidence and scientific consensus are correct and human activity is responsible for climate change, the question becomes one of mitigation or adaptation. **Mitigation** is any effort to reduce the severity of climate change, while **adaptation** refers to strategies to reduce the consequences. At this point, both will likely have to be pursued.

Mitigation strategies include any efforts to reduce greenhouse gas emissions through increases in energy efficiency, switching renewable sources, replanting and preserving forests, recovering methane from landfills, and protecting soil quality. More personal strategies could include eating less meat, reducing commuting times, and cutting back on waste and consumption.

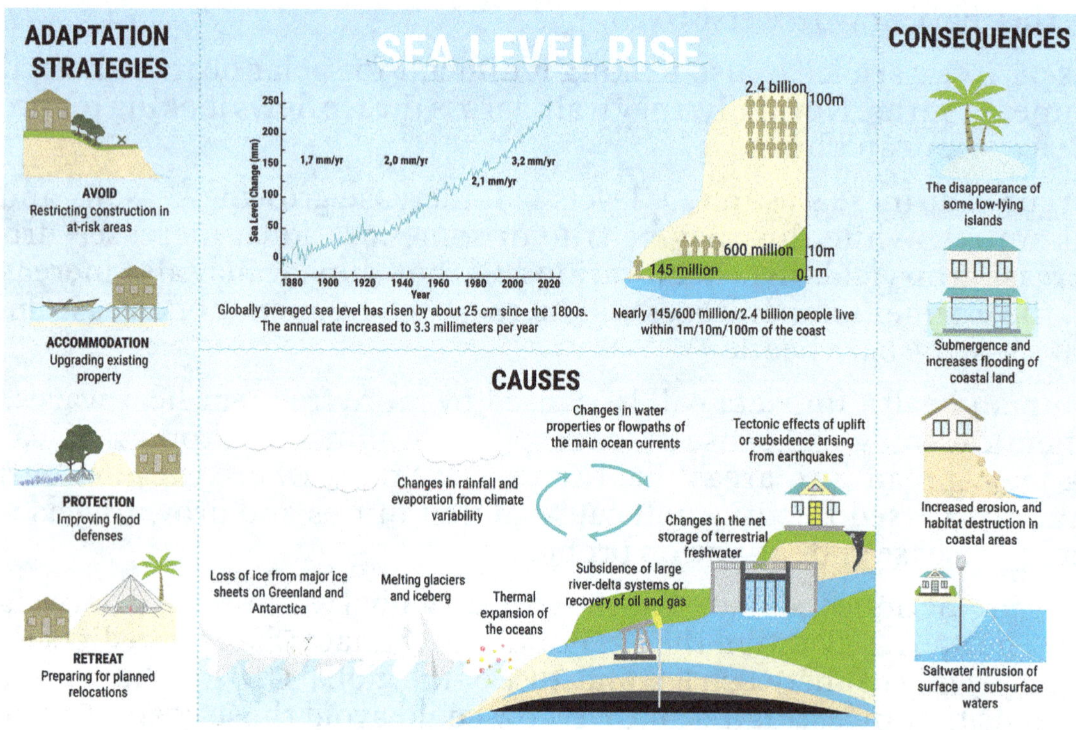

m.malinika/Shutterstock.com

Examples of adaptation include building seawalls to protect against flooding. Already, Miami Beach is installing a system of pumps throughout the city to deal with the increase in flood.

We cannot expect a single "magic bullet" to solve the climate problem, but there are a variety of activities that could reduce greenhouse gas emissions.

One technology being studied is **carbon capture and sequestration**. This involves capturing the carbon dioxide from the burning of fossil fuels and storing it underground in depleted oil and gas deposits or abandoned salt mines. The problem is that despite millions put into research, carbon capture is still not a proven technology. We do not have the technology to capture enough carbon to make a difference and there's no guarantee that will stay underground and not escape. Also, pumping liquified carbon dioxide could dislodge old fault lines and trigger minor earthquakes.

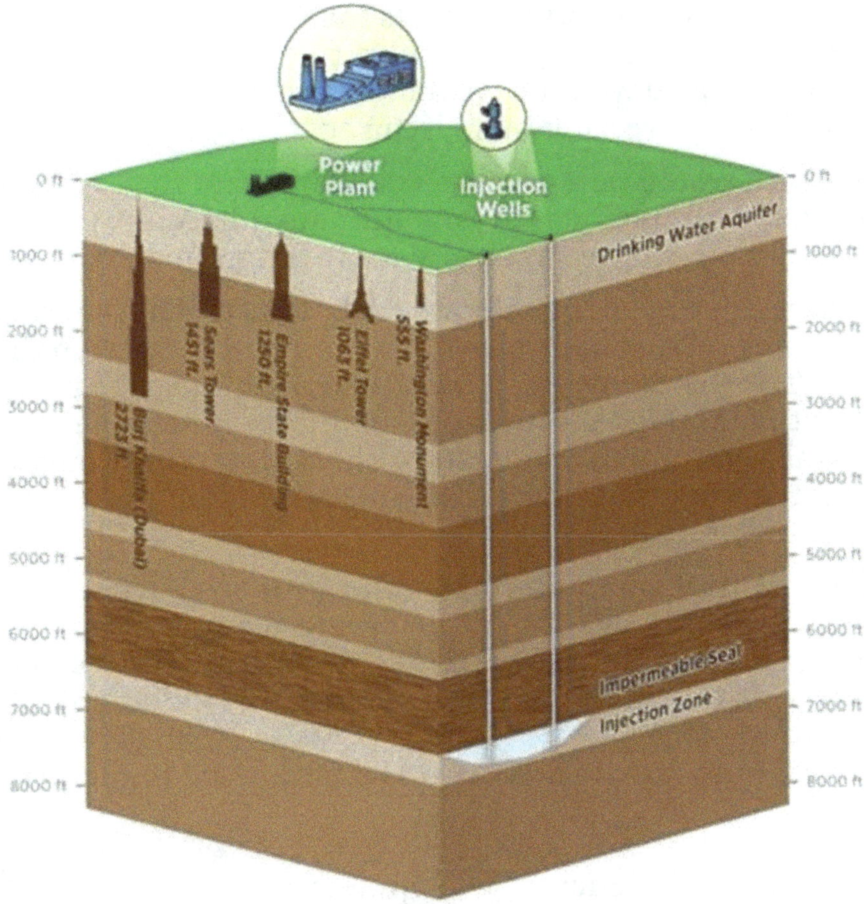

Image Source: epa.gov

Researchers at Princeton University has promoted a strategy that has become known as the "**wedge strategy**." Each wedge represents 1 gigaton in carbon emissions reduced. Seven such wedges would stabilize emissions at their current levels over 50 years. Fifteen of them would significantly reduce emissions this century. Examples of the wedges include

- Incease solar power by 700 times
- Incease wind power by 50 times
- Doubling vehicle efficiency
- Triple nuclear power capacity
- Switch from coal to natural gas in 1,400 plants
- Increase hydrogen fuel production by 100 times
- Increase ethanol production by 50 times
- Adopt carbon capture at 800 coal plants
- Eliminate tropical deforestion
- Adopt conservation tillage on all farms

Another strategy is a **carbon tax**, which is a green tax on the emissions of carbon dioxide. While the theory is that companies would be motivated to reduce carbon emissions to avoid paying the tax, they could also just pass the cost onto their customers. So, many carbon tax schemes include a **fee-and-dividend** where the money collected by the tax would be given to taxpayers as a refund to offset the increased rates. The goal would be to make the carbon tax revenue neutral, where ratepayers would not have a net increase in costs.

Another approach would be a cap and trade program for carbon dioxide, similar to the successful one implemented for sulfur dioxide. Many European nations and several US states have adopted a cap and trade program for carbon.

For several years, there was considerable debate as to whether the Environmental Protection Agency (EPA) had the authority to regulate carbon dioxide under the Clean Air Act. The George W. Bush administration took the position that carbon dioxide did not meet the definition of a pollutant under the CAA. Several states used the federal government and in 2007, the US Supreme Court ruled against the administration, saying that carbon dioxide was a pollutant under the CAA and that the EPA should regulate.

Then, in 2009, the Obama administration proposed a national cap and trade program for carbon dioxide. The bill to create the program passed the House of Representatives but was blocked in the Senate. The next few years saw little action by the federal government as the EPA continued to work on regulations. Then, in 2013, Obama announced his **Climate Action Plan**. Using his executive authority, he ordered the EPA to fast track its regulations and he implemented a number of initiatives to

- Promote renewable energy development
- Modernize the electrical grid
- Finance clean coal and carbon storage efforts
- Improve automotive fuel economy
- Protect and restore forests
- Encourage energy efficiency
- Increase EPA regulation of power plants

These initiatives were reversed by an executive order from President Trump in 2017.

On the international level, the first effort to reduce greenhouse gas emissions was the **United Nations Framework Convention on Climate Change**. Adopted in 1992, it set a goal of voluntary efforts to reduce greenhouse gas emissions to the levels they were in 1990. By 1997, it was clear that the voluntary efforts were not working, so nations got together in and crafter the Kyoto protocol. This set mandatory reductions with a goal of reducing emissions to below that of 1990 by the timeframe of 2008 to 2012. Then President Clinton signed the Kyoto Protocol, but the Senate refused to ratify it. Then in 2001, President George W. Bush pulled the United States out of it. By 2015, the nations that did sign Kyoto had achieved a 12% reduction in greenhouse gas emissions.

Then in 2015, another international conference was held in Paris after President Obama was able to secure commitments from China to reduce emissions under the **Kyoto Protocol**, developing nations were exempt from making any mandatory reductions. Under the Paris Accord, the United States committed to new regulations on coal-fired plants and continue the switch to natural gas. China pledged to cut back on coal production and adopt its own cap and trade program. Brazil promised to end deforestation. Also, Indio promised to reforest previously cleared land, promote renewable energy, and slow emissions growth.

Getting commitments from China and India was key as in the interim between Kyoto and Paris, those countries had both surpassed the United States as the number one and number two emitters of greenhouse gases. The Paris conference was considered the most successful conference to date. Every participating country brought its own ideas to the table and committed to meeting the new set of goals. Even so, the **Paris Accord** would only bring us partly toward the goal of limiting global warming to 2° C by the end of the century and more work would be needed.

Upon taking office in 2017, President Trump announced he was beginning the process of withdrawing from the Paris Accord and is on track to have the United States completely out by November 2020.

One of the objections to efforts to reduce greenhouse gas emissions has been the assumption that it would involve economic sacrifice. However, the United States was able to reduce its emissions by 14% from 2007 to 2016 even as its economy continued to grow. This was achieved mostly by switching from coal to natural gas in many power plants. Natural gas is a cleaner fuel than coal and emits less carbon dioxide. Both China and Germany are forging ahead in developing renewable energy while achieving strong economic growth.

Meanwhile, the Trump administration has been actively trying to increase coal consumption in the United States.

Despite the withdrawal from the Paris Accord, many governors, mayor, university presidents, and corporate leaders have taken the initiative to achieve the goals of the Accord. The most aggressive state has been California, which has implemented a cap and trade program and seeks to reduce greenhouse gas emissions by 25% by 2020.

Many businesses and universities have set a goal of **carbon neutrality**, where no net carbon is released by their activities. One way to achieve this is by purchasing **carbon offsets**. Instead of reducing one's own emissions, a company can pay another party to reduce carbon emissions elsewhere, thus canceling out their own emissions. For example, a power plant could fund a reforestation project.

Finally, some have proposed active efforts to alter the planet's temperature through **geoengineering**. One proposal has been fertilizing phytoplankton to promote photosynthesis. Another is building artificial "trees" to capture carbon dioxide from the air. Others have suggested blocking sunlight by spraying aerosols into the air or using giant mirrors on the surface or in space.

If geoengineering sounds like science fiction, it's because currently, it is. None of these techniques have been proven to work and they may have unforeseen consequences that could be worse that what is happening now. Plus, once we start

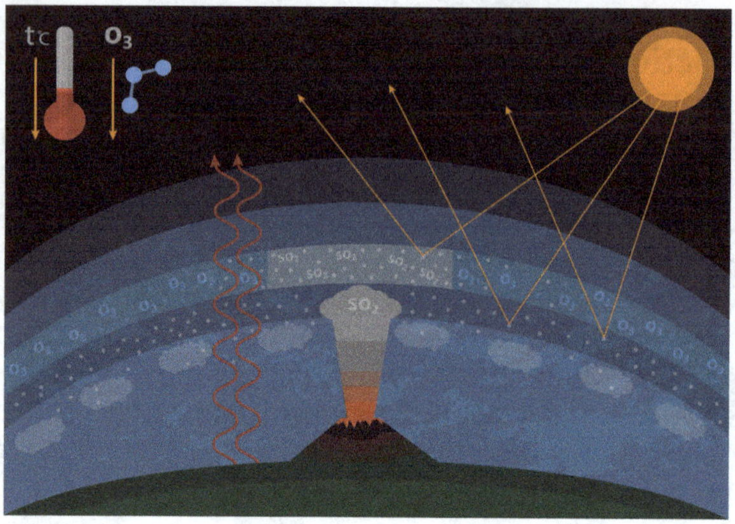

tinkering with the global thermostat, who decides where to set it? What if people in Norway would like the world to be a little warmer than people in India do? Who decides?

15.6 Discussion Questions

1. Define the terms global climate change and global warming.
2. How does the greenhouse effect influence the Earth's climate?
3. What are the Milankovitch cycles and how do they influence the Earth's climate? Can they account for the current warming trend?
4. Name some proxy indicators and describe how they are used to determine the climate of the past.
5. What are some of the expected impacts from the melting of glaciers and permafrost?
6. What are some of the expected impacts of global sea level rise?
7. Describe the expected impacts global warming will have on wildlife.
8. What are the expected impacts of global warming on human health?
9. What is carbon capture and sequestration?

10. Describe the "wedge strategy." What does each wedge represent?
11. What is a carbon tax and how does it incentivize reductions in greenhouse gas emissions?
12. How was the United States able to reduce its greenhouse gas emissions in recent years?
13. What is meant by carbon neutrality and how are carbon offsets meant to help organizations achieve that goal?
14. What is geoengineering? Give some examples and describe how might they be used to offset global warming.

15.7 Key Terms

- Adaptation
- Climate Action Plan
- Carbon capture and sequestration
- Carbon neutrality
- Carbon offsets
- Carbon tax
- Climate
- Climate modeling
- Dendrochronology
- El Niño
- Fee-and-dividend
- Geoengineering
- Global climate change
- Global warming
- Global warming potential
- Greenhouse effect
- Greenhouse gases
- Ice core samples
- Intergovernmental Panel on Climate Change
- Kyoto Protocol
- Milankovitch cycles
- Mitigation
- Paris Accord
- Proxy indicators
- Radiative forcing
- Relative Global Warming Potential
- The oceans
- Thermohaline circulation
- United Nations Framework Convention on Climate Change
- Wedge strategy

Chapter 16

GEOLOGY AND MINERAL RESOURCES

Chapter Topics

In 1962, the borough managers in Centralia, Pennsylvania decided to clean up an old-tire dump by setting it on fire. Unfortunately, the dump was located in an abandoned mine pit and the fire ignited the coal seams beneath it. After years of failed attempts, the state and the federal government eventually abandoned any effort to put it out. The fire continued to burn and the ground below the town reached 900°F in some places. The residents eventually learned to live with it.

Then, in 1981, a 12-year-old boy fell into a sinkhole that opened up in his grandmother's yard. Though he was pulled to safety by his cousin, the incident spurred the government to take action again. Congress voted to allocate funds to buy up the residents' homes and pay them to relocate. Most people moved out, but a small number remained. In 1992, the Pennsylvania government went to court to force the remaining residents to move while the Postal Service revoked Centralia's zip code. The residents found back in court and seven were allowed to remain. Most other buildings were torn down and the highway was rerouted to take traffic around Centralia.

Today, Centralia is all but a ghost town. Smoke pours out of fissures in the ground. The old roadway into Centralia is now known as "Graffiti Highway" due to all the graffiti painted on it. The seven remaining residents are barred from selling or passing on their property to their heirs. Once they die, the last remnant of a once-thriving mining town will be gone.

halfofmoon/Shutterstock.com

Once a thriving mining town, Centralia, PA has become a virtual ghost town due to an underground mine fire that's been burning since 1961.

Yet the fire is expected to burn for another century or more.

Centralia remains a cautionary tale about how we manage our resources and the hazardous of abandoned mines if they are not managed properly.

16.1 Geology and Tectonic Forces

The word **geology** is derived from Greek and literally means "study of the Earth." It is the study of the physical features and the forces that shape them. Geologists study the rock layers, the formation of mountains, and weathering the brings them down. They also help locate deposits of valuable mineral resources and fossil fuels.

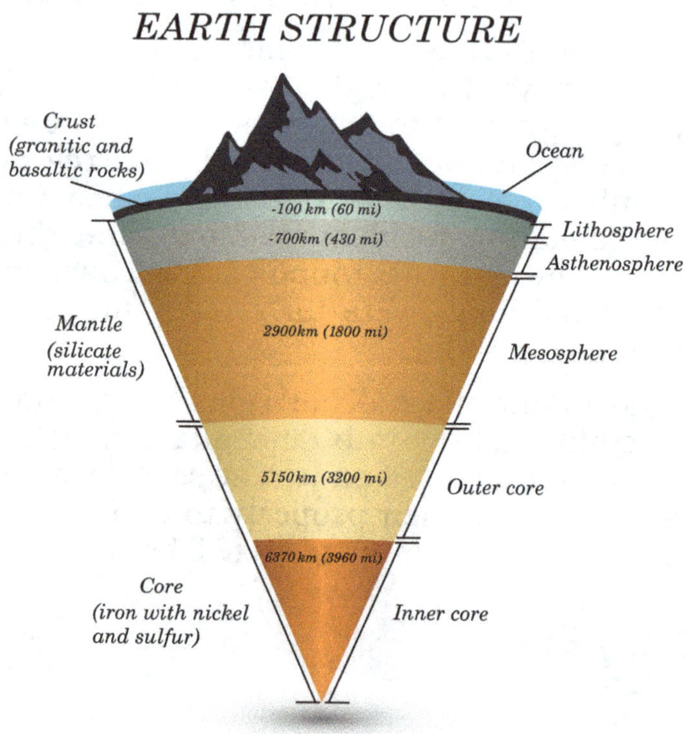

Ellen Bronstayn/Shutterstock.com

The Earth consists of a number of layers. At the center is the solid **inner core** made of iron and nickel. The **outer core**, made of molten iron and nickel surrounds the inner core. Surrounding the outer core is a layer of less dense, elastic rock called the **mantle**. The uppermost layer of the mantle is called the **asthenosphere** and contains a layer of especially

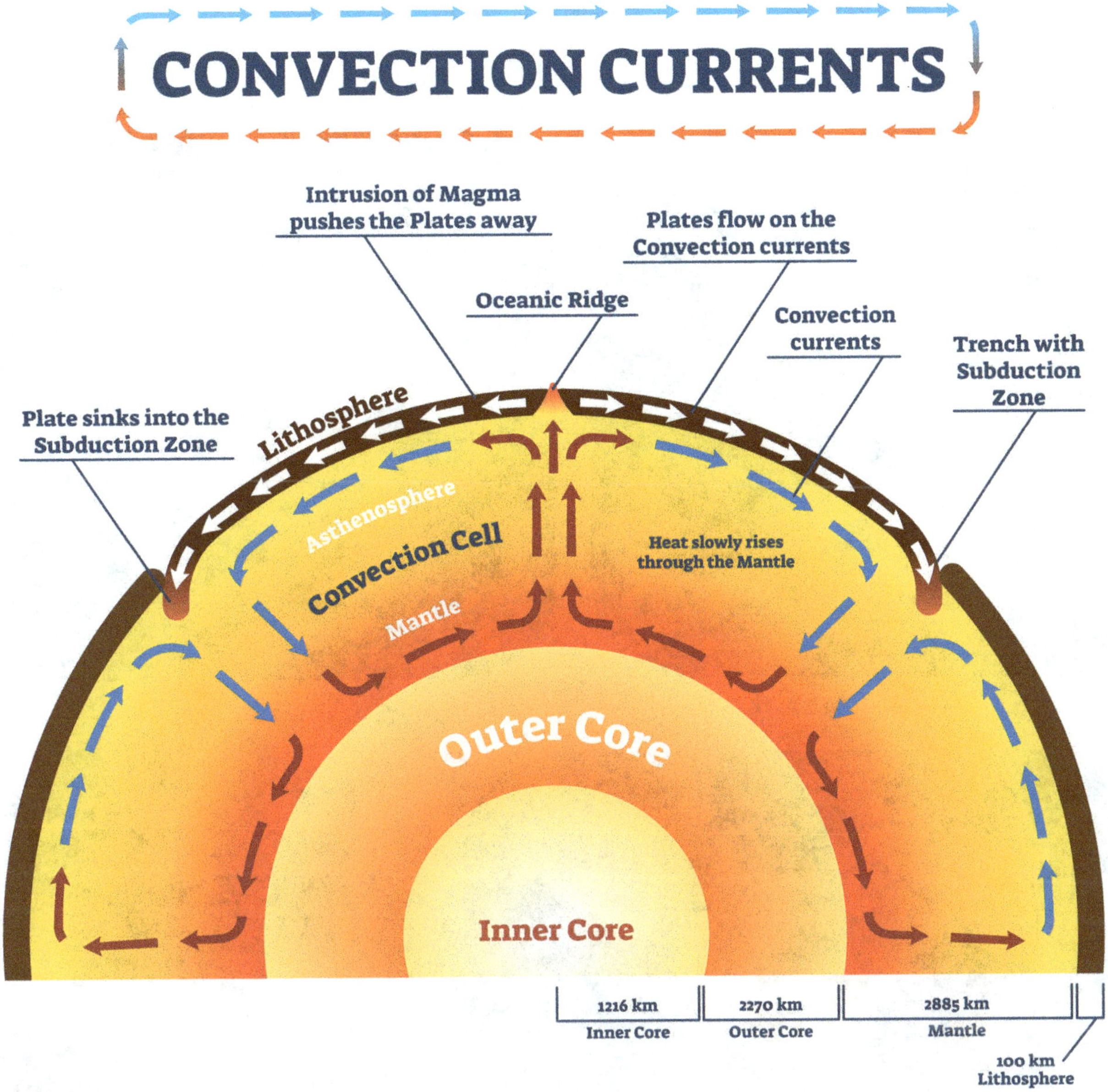

VectorMine/Shutterstock.com

soft and partially molten rock. Finally, a brittle, low-density layer called the **crust** covers the mantle.

The crust is not one single piece, but is broken up into several pieces called **tectonic plates**. Convective forces in the asthenosphere push and drag the various plates, moving them around at a rate of about 2 to 15 cm a year. If you look at a map of the world, it's easy to see how South America and Africa could fit neatly together if one could push them toward each other. Indeed about 335 million years ago, the two continents were touching. In fact, all of the major landmasses were joined together into the supercontinent, Pangea. About 175 million years ago, **plate tectonics**, the movement of the plates, caused Pangea to break up. Then, about 140 million years ago, South America and Africa split and have been moving apart from each other ever since.

VectorMine/Shutterstock.com

The Major Tectonic Plates.

There are three types of plate boundaries, each defined by the relative motion of the plates. At a **divergent plate boundary**, magma rises up from the mantle, pushing the plates apart. As the magma reaches the surface, it cools, forming a mountain ridge. The world's longest mountain range, the Mid-Atlantic Ridge forms at the boundary between the African and South American plates. Divergent boundaries create new crust material as magma forces its way to the surface. As a result, the crust at the bottom of the Atlantic Ocean is some of the youngest crust on Earth.

Two plates slide next to each other at a **transform plate boundary**. These plate boundaries are also called **strike-slip fault lines** as frictional forces can build up in these boundaries. These forces eventually get released, creating a shockwave that travels across the surface that we call an earthquake. The San Andreas fault is a transform plate boundary between the Pacific Plate and the North American Plate and is the reason why earthquakes are common on the West Coast.

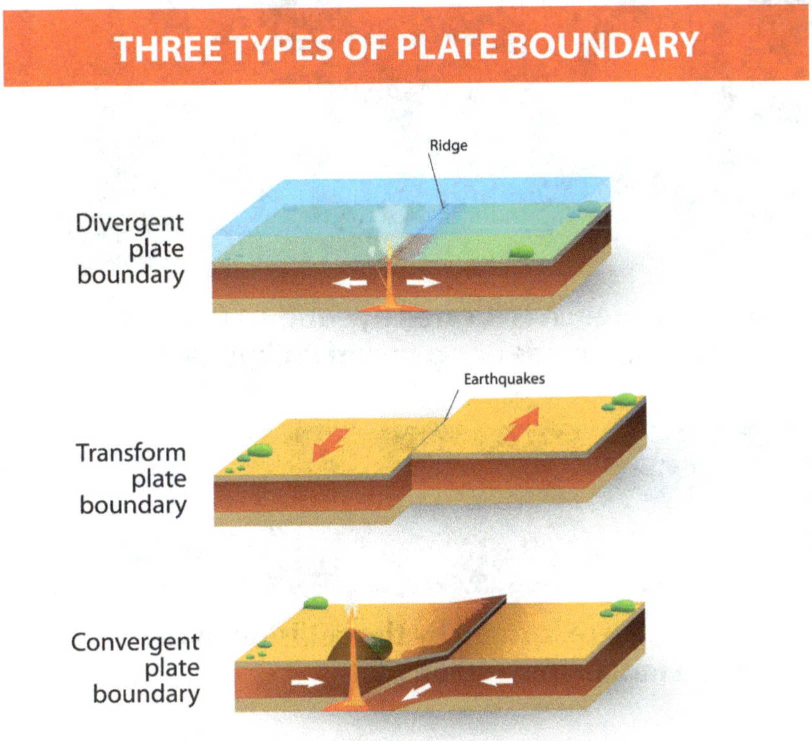

Designua/Shutterstock.com

The three types of plate boundaries: Divergent, Transform, and Convergent.

VicPhotoria/Shutterstock.com

The Mid-Atlantic Ridge forms as magma upwells from the mantle, pushing the two plates apart.

Breck P. Kent/Shutterstock.com

The San Andreas Fault in California, a transform plate boundary that is responsible for periodic earthquakes on the West Coast.

Oliver Denker/Shutterstock.com

The Mariana Trench is formed by subduction that also forms the Islands of Japan and Philippines.

Daniel Prudek/Shutterstock.com

The Himalayas were formed by the collision of the Eurasian and Indian–Australian plates.

A **convergent plate boundary** is where two plates come together. Two things can happen at a convergent plate boundary. One is **subduction**, which occurs where one plate beneath the ocean slides underneath another. The deepest trench in the world, the Mariana Trench in the Pacifica Ocean, is caused by subduction. At subduction, magma is forced upward through the crust, creating chains of volcanic Islands. Japan, the Philippines, and the Malay Archipelago were all formed through volcanoes driven by subduction. The other thing that can happen is **continental collision**, as two continents ram into each other. Continental collision causes the edges of the plates to bend upward, creating **uplift mountains**. The Himalayan Mountains in Asia were formed by the collision of the Indian–Australian Plates with the Eurasian Plate.

Plate tectonics play a major role on the surface of the Earth. They raise up mountains, separate the oceans, create islands, and trigger earthquakes and volcanos. They also play a role in speciation by isolating different populations. For example, after the breakup of Pangea, South America, Africa, Antarctica, Australia, and India were still joined together in a continent called Gondwanaland. During this time, species were free to migrate across these lands. Ancestors of modern monkeys lived in both South America and Africa. As the two continents split, the populations diverged into the two major families of monkeys: Old world monkeys in Africa and New world monkeys in South America. However, during this time, most mammals in South America were marsupials, pouched mammals that give birth to young that are less developed who then migrate into the pouch to suckle and further develop. In contrast, placental mammals nourish their young during gestation through a placenta and are able to carry them longer. Their young are then born more fully developed than those of marsupials. Eventually, South America joined up with North America through the Isthmus of Panama. Then, placental mammals from North America migrated south, outcompeted, and displaced the marsupials in South America. Today, there is only one marsupial found in all of the Western Hemisphere: The opossum.

Benny Marty/Shutterstock.com

This red kangaroo is carrying her young, a joey, in her pouch. Marsupials were outcompeted everywhere by placentals except Australia, where they could thrive due to geologic isolation.

Meanwhile, Australia remained isolated from the rest of the continents. There, marsupials did not have to compete with placentals. They came to dominate Australia as both herbivores (kangaroos and koalas) and predators (Tasmanian devil and the now extinct Tasmanian tiger).

16.2 Rocks and the Rock Cycle

A **rock** is defined as a solid aggregation of minerals. A **mineral** is an element or inorganic compound that has a crystalline structure and its own distinct chemical and physical properties. Note that fossil fuels like coal, being organic, are not minerals. Over time, rocks are subjected to a variety of forces. They are heated, cooled, weathered, dissolved, broken down, compacted, and subjected to tremendous pressures. These forces shape and transform rocks through the **rock cycle**.

Magma is the hot, molten form of rock. Magma that has been released to the surface by volcanos is called **lava**. As magma or lava cools and solidifies, it forms into **igneous rocks**.

Wind and water action breaks rocks down into tiny particles through a process called **weathering**. These particles form sediments and as sediments pile up, they can be subjected to tremendous pressure that solidifies them into **sedimentary rocks**. Fossils are found in sedimentary rocks as fossils form when the remains of an organism get buried in sediments and then leave an imprint behind.

Underground, rocks experience great pressure and temperatures. These forces can alter the crystal structure of the rocks, turning them into **metamorphic rocks**.

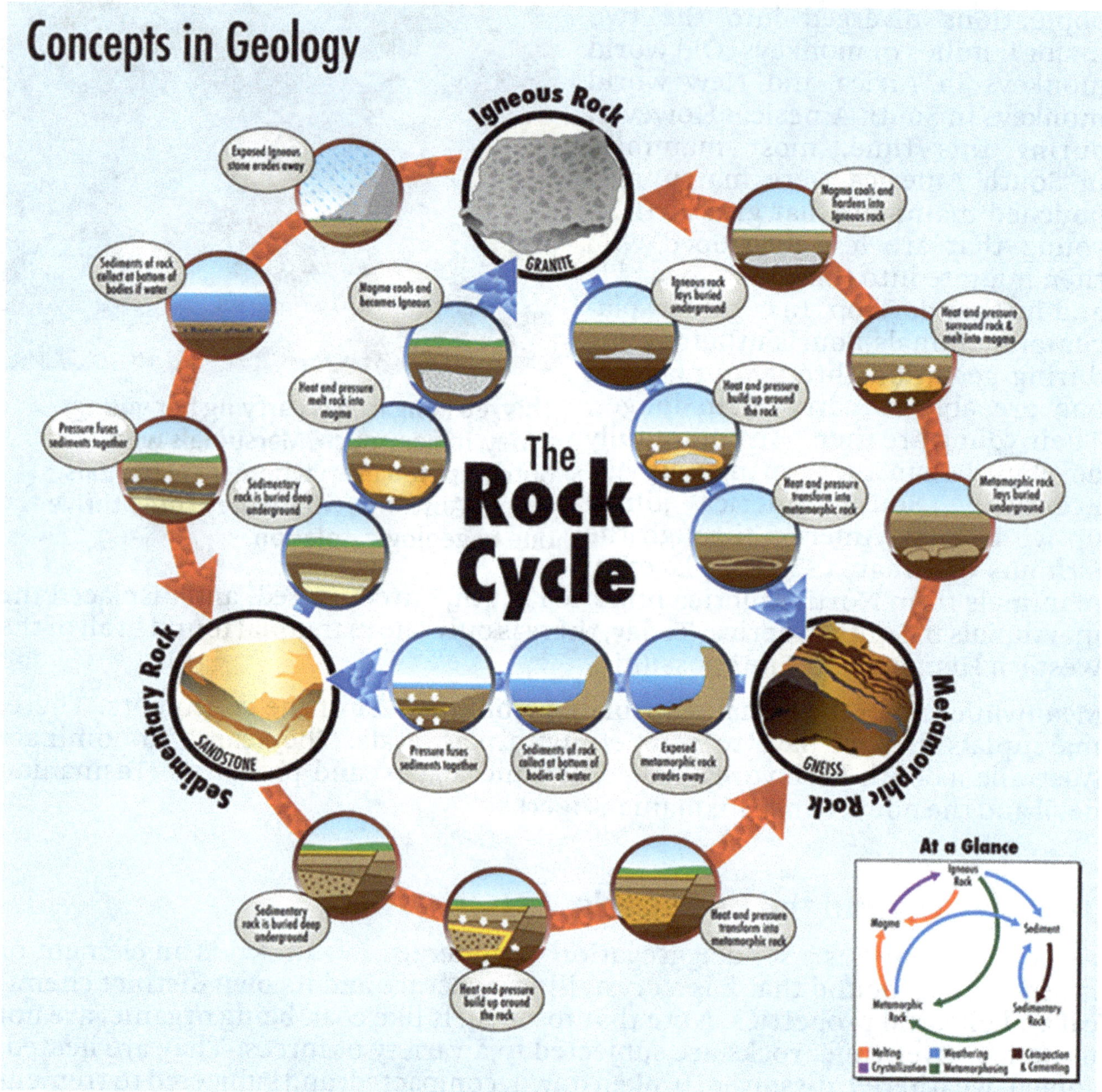

Nerdist72/Shutterstock.com

16.3 Geologic Hazards

An **earthquake** is a sudden release of energy along a tectonic plate boundary, sending a shock wave across the surface. Earthquakes can also occur in the interior of plates due to internal forces acting on the plates. Earthquakes can endanger lives and destroy property. Engineering can make buildings more earthquake-resistant and building codes in the United States require earthquake-resistance in areas of high risk for earthquakes. Poorer countries, however, lack these building codes and often, the construction there is of much lower quality. This can make an earthquake even more devastating, such as the earthquake that hit Haiti in 2010 which claimed about 230,000 lives.

Christoph Hilger/Shutterstock.com

Volcanic basalt rocks of Rubini Rock, Franz Josef Land, Russia.

Leene/Shutterstock.com

Sedimentary rock layers.

LesPalenik/Shutterstock.com

Gneiss (pronounced "nice"), a metamorphic rock, forms when granite is subjected to strong pressures and high temperatures.

arindambanerjee/Shutterstock.com

Earthquake damage in Port-au-Prince, Haiti. The damage of this quake was particularly devastating due to a lack of construction standards.

A **volcano** is a location where molten rock, ash, and gases are released to the surface. Volcanos can release lava in a gradual flow or it can erupt suddenly, release a large amount of ash and gases in what is called a **pyroclastic flow**. Volcanos can occur at rifts and subduction zones at plate boundaries. They can also form over **hot spots**, areas where the mantle is usually hot, creating pressure that can force molten rock up through the surface. The Hawaiian Islands were formed from a hot spot in the middle of the Pacific Ocean, far away from any plate boundaries.

Some notable volcanic eruptions include:

- The Yellowstone Caldera: This "super-volcano" in Wyoming last erupted 640,000 years ago. If it were to erupt today with the same force it did then, it could bury all of North America in ash.
- Mt. Vesuvius: This volcano in Italy erupted in 79 CE and buried the city of Pompei in a pyroclastic flow.
- Mt. Tambora: Erupted in 1815 in Indonesia. It threw so much ash into the atmosphere that it temporarily cooled the Earth, turning 1816 into the "Year Without a Summer." It killed at least 70,000 people.
- Mt. Krakatoa: Also in Indonesia, this eruption in 1883 was heard 3,000 miles away and killed 36,000 people.
- Mt. Saint Helens, Washington: Erupted in 1980. Fortunately, early warnings enabled people to evacuate and casualties were low. Killed 57 people.
- Mt. Kilauea, Hawaii: Has been continuous erupting with a low of lava since 1983.
- Mt. Pinatubo, Philippines: Erupted in 1991 and threw so much ash in the air that it offset global warming by about 0.5°C.
- Mt. Eyjafjallajökull, Iceland: Erupted in 2010 and its ash cloud grounded air traffic in Europe for almost a week.

Maridav/Shutterstock.com

Continuous lava flow from Mt. Kilauea in Hawaii.

BlackMac/Shutterstock.com

The eruption of Mt. Vesuvius buried the residents of Pompeii so quickly that their remains were mummified.

The boundaries of the Pacific Plate, known as the **Circum-Pacific Belt** or the **Ring of Fire**, are particularly active. Ninety percent of all earthquakes and volcanic eruptions occur in the Ring of Fire.

There is not anything we can do to prevent earthquakes or volcanic eruptions. We can avoid them by not living areas that are prone to them. Otherwise, all we can do is prepare through earthquake-resistant construction and early warning systems that give people an opportunity to evacuate. People in areas that are prone to earthquakes or volcanoes should have a "go bag" full of extra clothes, food, and other supplies that can be grabbed when an evacuation order is given.

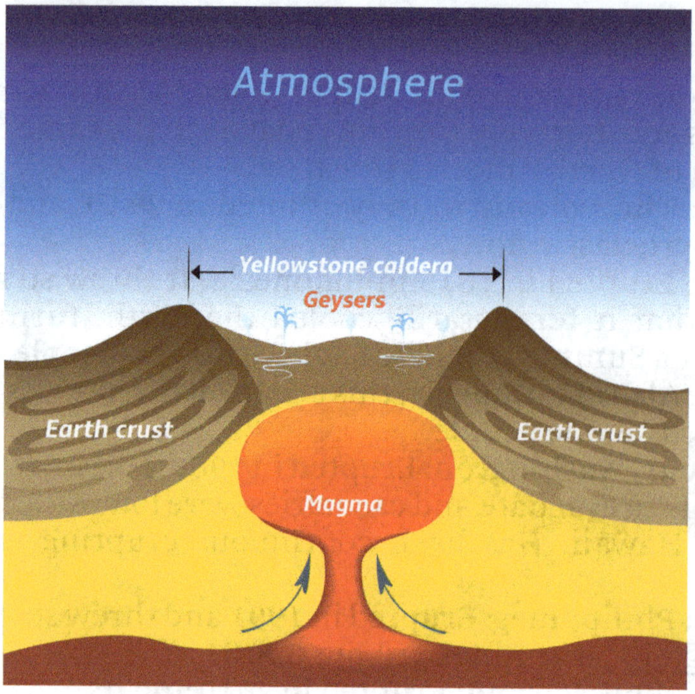

Agil Leonardo/Shutterstock.com

The Yellowstone Caldera. This super-volcano measures 35 by 45 miles in size and is one of the largest volcanoes in the world.

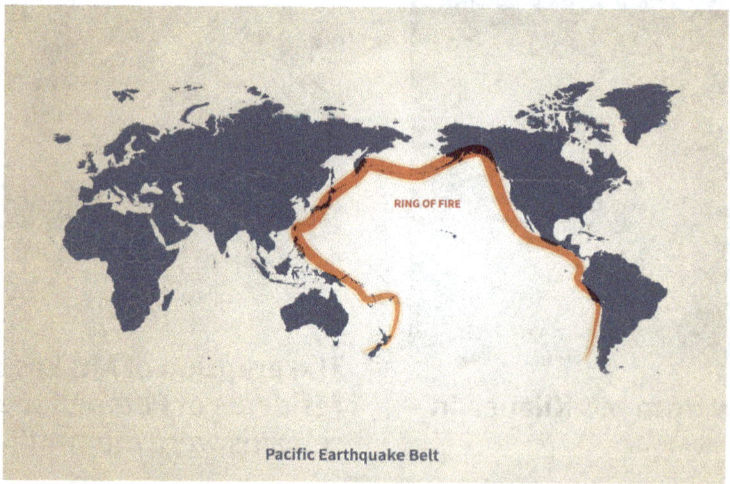

tunasalmon/Shutterstock.com

The Ring of Fire runs along the edges of the Pacific Ocean; 9 out of 10 earthquakes and volcanic eruptions happen in these areas.

Landslides or **mass wasting** happen when a large amount of rock or soil comes loss and falls downhill. Mudslides consist of wet soil and are often triggered by rains over mountains that have been cleared of vegetation. A lahar is a mudslide triggered by a volcanic eruption that melts snow. Clear-cutting mountain forests and other practices that disturb the soil along slopes can increase the risk of landslides.

Alexlky/Shutterstock.com

Lucky Team Studio/Shutterstock.com

Landslides can destroy property and endanger lives.

A **tsunami** is a sudden displacement of water. These are usually triggered by other geologic hazards like earthquakes, volcanos, or landslides. In 2011, an earthquake triggered a tsunami, causing a combined loss of life of about 20,000 people. In addition, the tsunami flooded the Fukushima Daiichi nuclear power plant, causing Japan's worst nuclear accident. In 2004, a devastating tsunami hit several nations along the Indian Ocean, killing about 228,000 people. Since then, nations have been investing in better early detection and warning systems. Tsunamis cannot be prevented, so advanced warning and evacuation are key in preventing deaths. Also, restoring coastal ecosystems like mangrove forests and coral reefs can lessen their impacts.

The 2011 earthquake in Japan triggered a tsunami. Together, they caused at least 20,000 deaths.

16.4 Mineral Resources and Mining

We make use of a variety of mineral resources. These include many useful metals like iron and copper and precious metals like gold and silver. Salt is mined as a food additive and potash is used to make fertilizers. We make glass from silica and wallboard from gypsum. Many rare metals like indium and tantalum are used in electronics. Sand, gravel, and limestone are important for construction projects. Fossil fuels like coal, oil, and natural gas, while not minerals, are also mined from the Earth's crust. Also, uranium is mined for use in nuclear power.

The discovery of gold in California triggered the Gold Rush of the nineteenth century. Many people flocked to California in hopes of striking it rich.

These diamonds are one example of gemstones that are mined for jewelry and industrial use.

Discovod/Shutterstock.com

Salt is mined for use in food.

MarcelClemens/Shutterstock.com

Uranium ore.

Mining, the process of removing soil and rock in order to extract limited resources, is practiced all over the world in order to obtain these resources. The mineral resources that are of economic value are found in only limited quantities. As a result, geologists conduct surveys in order to locate substantial deposits from which we can extract the ones we need. Once located, mining involves the removal of a lot of rock and earth and all mining activities have environmental impacts.

There are several mining methods. These methods are chosen based on the material mined, the nature of the deposits, and what is the most economic method.

Strip mining involves the removal of large amounts of rock and soil, called **overburden**, in order to access the material of interest. The overburden is removed using heavy machinery. Usually, strip mining is practiced to access shallow, horizontal deposits of materials like coal. Once they have removed all of the accessible material, they replace the overburden.

Strip mining has a lot of environmental impacts. Whatever communities and ecosystems were on the surface are destroyed. Also, as water infiltrates the mine, it

Jason Benz Bennee/Shutterstock.com

Strip mining employs a heavy machine to remove rock and soil to access useful minerals or coal.

Acid mine drainage can poison waterways and turn streams red by releasing iron oxides.

can pick up sulfur compounds, forming sulfuric acid. This **acid mine drainage** can poison bodies of water. Often, acid mine drainage releases iron oxides from the soil, causing streams to turn a reddish or orange color. Acid mine drainage also frees up toxic ions like aluminum, which can poison trees.

Subsurface mining involves digging shafts and tunnels underground to access deep deposits. While subsurface mining has a smaller footprint than strip mining, it is more dangerous. Improper use of explosives and tunnel collapses can kill or trap workers underground. In 2010, a Chilean mine collapse, trapping 33 miners underground. It took a month for rescuers to dig them out. Fortunately, all of them survived. Pockets of natural gas can also cause explosions or suffocate miners. In order to warn the miners that the tunnel was filling up with gas, miners would bring canaries into the mine. If the canary stopped singing and died, that was their signal to evacuate. Today, mines use electronic gas detectors instead of canaries. Coal dust inhalation can also cause problems like black lung disease, which has shortened the lives of many miners.

A subsurface mine involves digging deep shafts underground.

Phatthanun Kaewsuwan/Shutterstock.com

An open-pit mine in Thailand.

Subsurface mines are also prone to acid drainage after they've been abandoned. And old mines can collapse, creating sinkholes.

Similar to strip mining, **open-pit mining** involved using heavy machinery to dig a massive hole in order to obtain materials that are widely distributed. The pits are terraced to enable vehicles to enter and exit. Like strip mining, open-pit mining involves the removal of a lot of rock and soil. The environmental impacts of open-pit mining are also similar to that of strip mining. They destroy ecosystems and are prone to acid drainage. Open-pit mines used to mine for sand, gravel, or clay are often called **quarries**.

Mountaintop removal involves the use of explosives to blast the top of a mountain in order to access seams of coal inside. It almost exclusively practiced in the Appalachian Region for coal mining. The practice of **valley filling**, in which the blasted off overburden is dumped into the local valley, results in clogged streams and acid drainage. Mountaintop removal also produces shockwaves that damage buildings and releases a lot of dust that can cause health problems for local residents. Mountaintop removal also increases erosion and flash floods.

For years, practices like valley filling were exempted from laws like the Clean Water Act. The Obama administration adopted some rules to mitigate the damage, but those rules were later repealed by the Trump administration.

Malachijacobs/Shutterstock.com

A mountaintop removal mine in West Virginia.

Placer mining involves using flowing water to sift through sediments for small amounts of minerals. During the California Gold Rush, miners often looked for gold by "panning" along stream beds. Gemstones and some metals like tantalum ores are mined using this method. Placer mining promotes streambank erosion and can result in hazardous chemicals being released into streams.

HildeAnna/Shutterstock.com

Placer mining or panning for gold.

Solution mining is the practice of drilling a hole underground and then injecting water or acid underground to dissolve the desired material. The dissolved material is then pumped up to the surface and separated from the fluid. While solution mining has a smaller footprint than other mining methods, it can also result in hazardous chemicals contaminating surface and groundwater sources. Salt, potash, and uranium can be mined using solution mining.

Russ Heinl/Shutterstock.com

Potash solution mine in Canada.

While seawater contains many valuable minerals, we currently lack the technology to mine them economically. Another possibility involved dredging the seafloor for minerals, including ball-shaped ores of manganese called nodules. However, these methods are also not economically viable.

JUN3/Shutterstock.com

Millions of manganese nodules exist scattered all over the seafloor.

Mining activities can often create short-term economic boons for a community. However, once the mine has been exhausted, it is closed and the mining company moves on, taking its jobs with it. In some places, mining operations involve human trafficking and human rights problems. For example, farmers in the Democratic Republic of Congo have been forced to work at placer mines for tantalum as slave laborers. Money earned from the sale of tantalum is used to finance an ongoing civil war in the region. Diamonds and silver are also mined in places throughout the world using slave labor.

In the United States, coal mining jobs have been in decline for years. While this decline is often blamed on new air pollution regulations, the largest reasons for the decline are economic. Hydraulic fracturing has reduced the price of natural gas and many power plants have converted from burning coal to the cheaper natural gas. In addition, mining companies have shifted toward practices like strip mining and mountaintop removal that require fewer mine workers than subsurface mining.

Reclamation is the practice of restoring the land to some semblance of the pre-mining conditions. Any mining structures are torn down and the overburden is replaced and covered with fresh topsoil. New vegetation is planted. However, ecosystems are complex and even the best reclamation plans often result in creating a community that is less diverse than the original one. Overtime, however, the previous community can eventually return through the process of succession.

Previously, many mining companies would avoid the cost of reclamation by declaring bankruptcy after they have exhausted a mine. To get around this problem, the **Surface Mining Control and Reclamation Act of 1977** requires mining companies to post a bond to covering reclamation costs before receiving their permit. Another important federal law relating to mining is the **General Mining Act of 1872** that was passed in response to the Gold Rush. This law allows any citizen or company to file a claim for the rights to mine on federal land that is open for mining. The fee for filing a claim, set in the 1872 Act, was set at $5.00 an acre. Critics say that this amounts to the virtual giveaway of public resources while proponents say it promotes a domestic industry that does carry significant risks. Efforts to amend the General Mining Act to adjust the fee for 150 years of inflation have failed in Congress.

Eric Buermeyer/Shutterstock.com

Once a mine has been exhausted, the land must under reclamation to establish a new community.

Metals are elements that have the following properties: lustrous (shiny when polished), malleable (can be shaped by hammering and heat), opaque (can't be seen through), and are good conductors of heat and electricity. Most metals are not found in a pure form, but in groupings of minerals call **ores**.

Once an **ore** has been mined, it has to be processed to extract the useful metal. **Smelting** involves heating the ore until in melts and then using chemicals to separate the metal from the other minerals. Smelting may also involve combining the metal with another metal or nonmetallic substance to form a mixture called an alloy. For example, steel is an alloy made by combining iron with carbon. Other common alloys include bronze (copper and tin) and brass (copper and zinc).

Smelting involves melting ore and using chemicals to separate the metal from the other minerals.

Smelting requires a lot of energy and water. It also generates air pollution and produces a slurry of hazardous waste chemicals called **tailings**. Because of their toxicity, tailings are often stored in **surface impoundments**, plastic and clay lined reservoirs. The water in the surface impoundment is allowed to evaporate, leaving behind a toxic residue that can be collected and sent to a hazardous waste landfill for final disposal. Surface impoundments pose a risk to surface and groundwater sources. The linings can crack and fail. Also, if they are not covered, heavy storm systems can cause them to overflow.

Mine tailings pond in Madagascar.

16.5 Sustainable Mineral Use

Minerals are a nonrenewable resource as they are found in limited quantities around the world. Promoting efficiency, reuse, and recycling can extend our use of these resources. Some minerals are likely to run out this century. For example, estimates indicate that our reserves of indium, which is used in LCD screens, may run out in 30 years.

For each mineral resource, there are two kinds of reserve estimates. **Technically recoverable reserves** are what we are able to extract based on current technology. On the other hand, **economically recoverable reserves** are what can be extracted profitably. Mines get shut down when they reach the end of the economically recoverable reserve. For example, if it now costs $50 a ton to extract a mineral, but the mineral can be sold for only $40 a ton, the mine will not be profitable and will be shut down.

Reserves are difficult to estimate as patterns of use can change. New resources could be discovered. Rising prices could cause older mines to be reopened as they are now profitable. New technologies for mining or processing the ore could be discovered. This happened with aluminum. Aluminum used to be one of the most expensive metals. It was even used in jewelry. This wasn't because it was a rare metal. Aluminum is actually very common. It was just very expensive to smelt the ore in order to obtain the metal. Then, a much cheaper method to smelt aluminum ore was discovered and today, aluminum is now one of the cheapest metals.

Economic recessions or boons could change consumer behaviors. New discoveries could replace current uses. For example, if a material is discovered that could replace indium in LCD screens, our reserves for indium could be extended as the demand for it goes down. Recycling rates could also increase, creating new sources for it.

Currently, only about 35% of the metals being used is being recycled. Increasing recycling will help increase the available supply and extend our reserves.

16.6 Discussion Questions

1. What happened at Centralia that resulted in the town being largely abandoned?
2. What forces cause the movement of the tectonic plates?
3. Describe the three types of plate boundaries and what geologic features tend to form there.
4. Why is coal not a mineral?
5. Describe the three main types of rocks and how each of them is formed.
6. Why are fossils most likely to be found in sedimentary rocks?
7. Why do 90% of earthquakes and volcanic eruptions occur along the "Ring of Fire?"
8. Why is subsurface mining considered the most dangerous form of mining?
9. Why are mining companies required to post a bond before receiving their mining permit?

10. What environmental issues are associated with surface impoundments?
11. What environmental and public health issues are associated with mountaintop removal?
12. What is the difference between the technically recoverable reserves and the economically recoverable reserves? Which one dictates when a mine with cease operations and why?

16.7 Key Terms

Acid mine drainage
Asthenosphere
Circum-Pacific Belt
Continental collision
Convergent plate boundary
Crust
Divergent plate boundary
Earthquake
Economically recoverable reserves
General Mining Act of 1872
Geology
Hot spots
Igneous rock
Inner core
Landslide
Lava
Magma
Mantle
Mass wasting
Metal
Metamorphic rock
Mineral
Mountaintop removal
Open-pit mining
Ore
Outer core
Overburden
Placer mining
Plate tectonics
Pyroclastic flow
Quarry
Ring of Fire
Reclamation
Rock
Rock cycle
Sedimentary rock
Smelting
Solution mining
Strike-slip fault lines
Strip mining
Subduction
Subsurface mining
Surface impoundments
Surface Mining Control and Reclamation Act of 1977
Tailings
Technically recoverable reserves
Tectonic plates
Transform plate boundary
Tsunami
Uplift mountains
Valley filling
Volcano
Weathering

Chapter 17

FOSSIL FUELS

Chapter Topics

Improvements in hydraulic fracturing have opened up deposits of natural gas that were previously inaccessible and not economically viable. **Hydraulic fracturing or fracking** involves drilling down until the desired level is reached. Then, the drill turns horizontally and drills through the shale layer that contains the natural gas. After the concrete and steel well casing is installed, an electric charge is used to fracture the rock formation. Finally, a mixture of water, sand, and chemicals is pumping into the fractures to force them open and pump the gas to the surface.

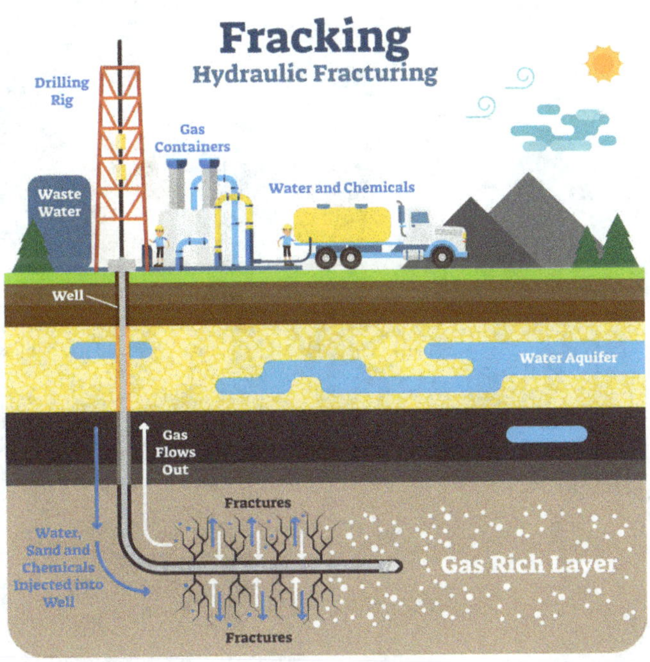

Hydraulic fracking operation.

VectorMine/Shutterstock.com

Fracking has made drilling in features like the Marcellus Shale possible. The Marcellus Shale is a geologic formation under several states in the

Northeast, including Ohio, New York, Pennsylvania, and West Virginia. The federal government even exempted fracking from laws like the National Environmental Policy Act (NEPA) and the Safe Drinking Water Act (SDWA). Gas companies were also not required to disclose what chemicals they use in fracking.

Pennsylvania in particular has promoted drilling in the **Marcellus Shale**. One town, Dimock, located near the New York state line, was approached by Cabot Oil and Gas. Cabot offered to buy the drilling rights from residents in Dimock. At first, most residents were eager to sell, but soon, some problems emerged. Residents complained about contamination in their drinking wells and chemical smells from their wells. At least one well even exploded. After some initial resistance, the Pennsylvania Department of Environmental Protection fined Cabot and required them to pay to bring an alternative supply of water to the residents of Dimock. Governor Corbett, however, soon put a halt to the shipments of Water. The Environmental Protection Agency (EPA) conducted testing and found contamination in 5 of the 64 wells they tested. In 2012, the PA General Assembly passed Act 13, which imposed and environmental impact fee on gas drilling in the state. The money collected by the impact fee is given to municipalities affected by drilling in the Marcellus Shale.

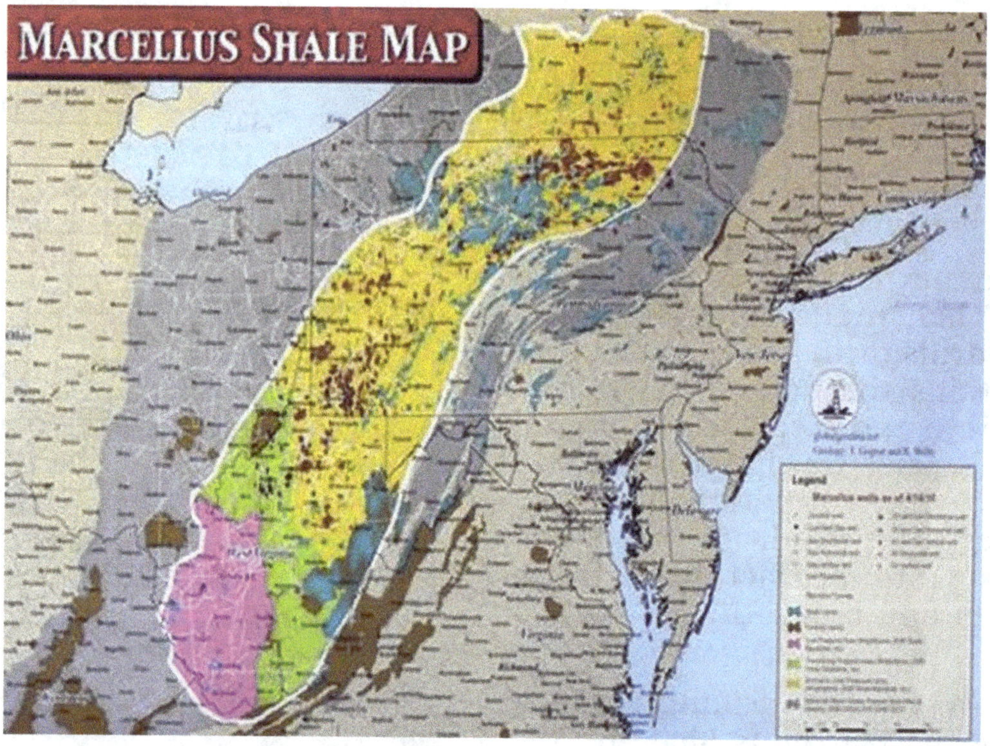

Source: https://www.usgs.gov/media/images/location-marcellus-shale

17.1 Energy Sources and EROI

We get energy from a variety of sources. Some of them are renewable. However, the majority of our energy today comes from nonrenewable sources, especially fossil fuels like coal, oil, and natural gas. As can be seen in the pie charts below,

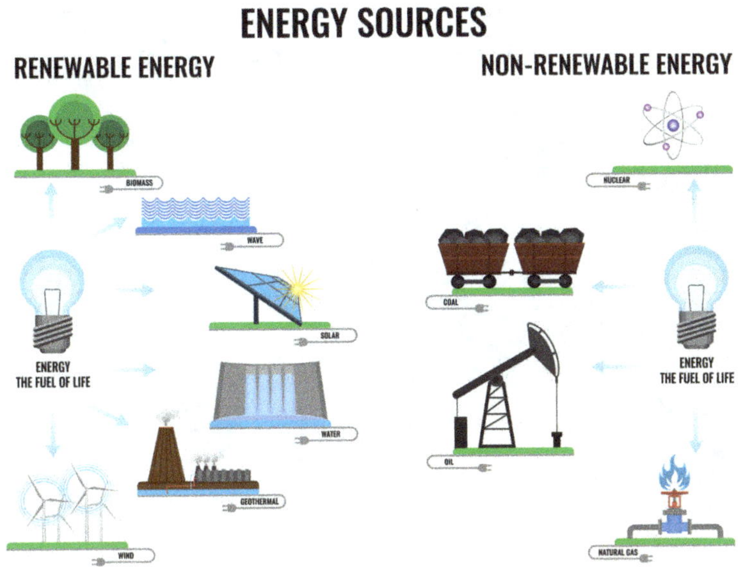

Shanvood/Shutterstock.com

nearly 80% of our total energy consumption and about 62% of our electricity in the United States comes from the three major fossil fuels. Note however, that only a small percentage of our electricity comes from petroleum. That's because petroleum is primarily used for making transportation fuels like gasoline, diesel, and jet fuel as well as for manufacturing of plastics and other synthetic materials.

Data from the Energy Information Administration in the graph below also indicates some interesting trends in our energy consumption since 2000. Coal production has begun to decline while dry natural gas has risen. Renewables have also begun to rise while nuclear power has remained flat.

As a general rule, industrialized countries use more energy than developing nations. Also, developing nations tend to use more energy on subsistence activities, while industrialized countries use more for industry and transportation.

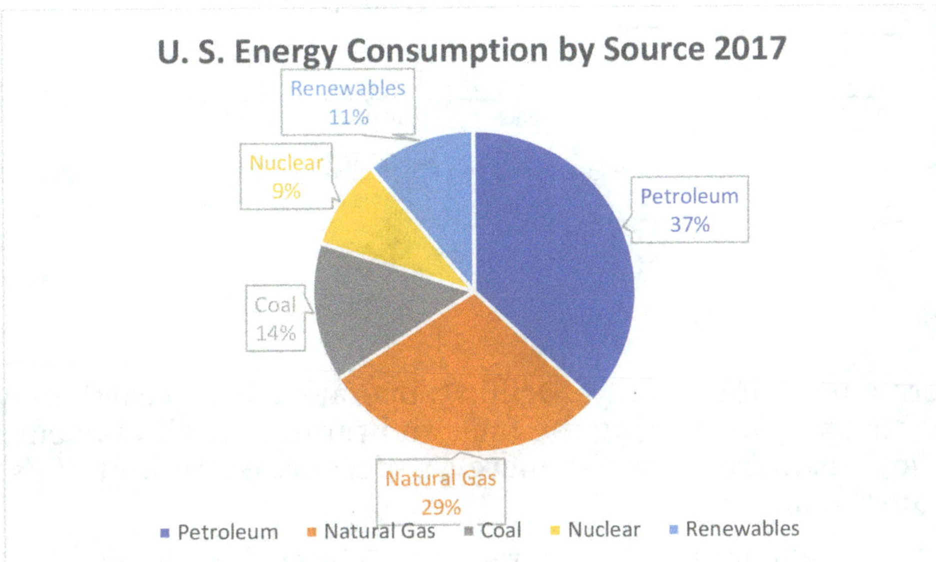

Source: US Energy Information Administration

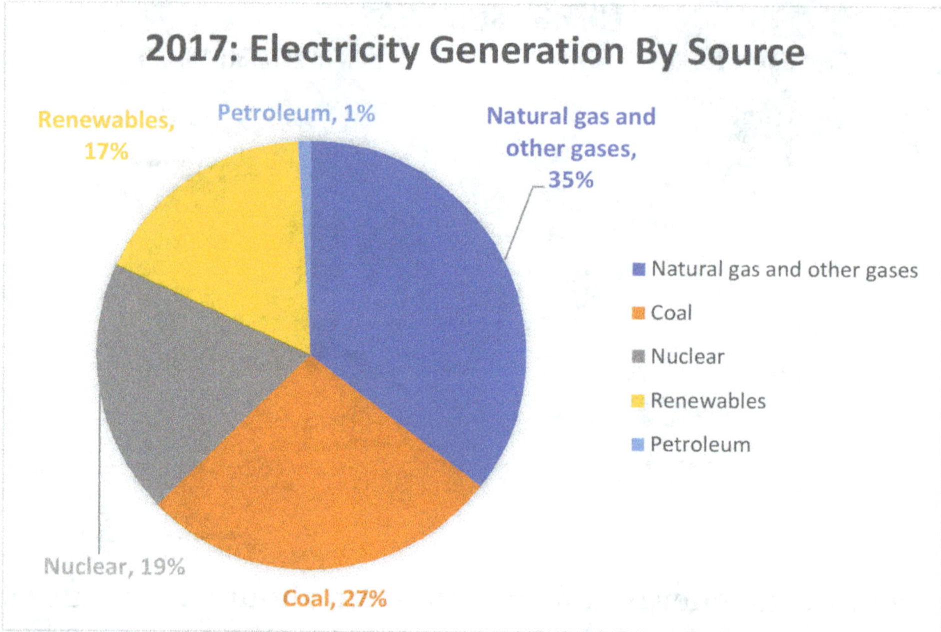

Source: US Energy Information Administration

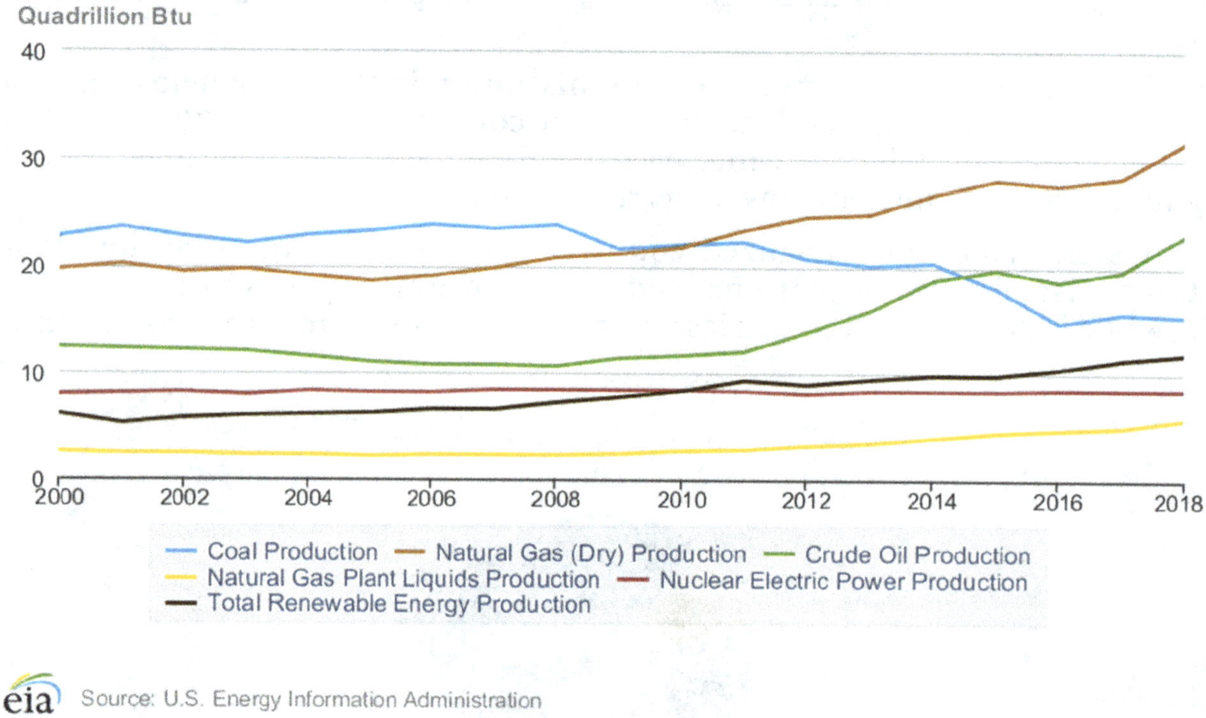

Source: US Energy Information Administration

It takes energy to produce energy. Drilling, operating heavy earth moving equipment, exploration, transporting the fuel, and refining it all take energy. So, for any energy source, there is the net energy, a measure of amount of useful energy obtained from the source.

$$\text{Net energy} = \text{Energy returned} - \text{Energy invested}$$

Obviously, a positive net energy is necessary for any source to be worthwhile. A negative net energy means more energy is being invested that is being returned. Another measure more commonly used in industry is the **Energy Returned on Investment (EROI)**, which is the ratio of the energy returned divided by the energy invested. This ratio can change over time. For example, decades ago, we still had abundant and easy to extract deposits of oil and natural gas. Today, those sources are largely depleted and now we rely on deposits that require more effort to extract the same amount of fuel. As a result, the EROI of oil and natural gas in the United States has declined from 24:1 in the 1950s to 11:1 today.

To get the same amount of energy as in the past, we have to expend more effort. This means drilling in deeper waters of the Gulf of Mexico and off the coast of California, as well as drilling in Arctic regions that have opened up as sea ice has melted. We're also exploiting other sources like oil sands and oil shale, which have a lower EROI.

All fossil fuels formed from the remains of living organisms that died millions of years ago and were buried under sediments. Over time, the remains were subjected to heat and pressure, converting them into coal, oil, and natural gas. The energy in these fuels originally came from the sun, which was captured by photosynthetic organisms, like plants and phytoplankton. This energy was stored chemically in the tissues of these organisms and remained after they died. In way then, fossil fuel energy is a form of solar energy that was turned into chemical energy millions of years ago.

As with other resources we mine, fossil fuels have both technically recoverable reserves and economically recoverable reserves (Section 16.4). Mining and drilling operations continue until the economically recoverable reserves are exhausted and the source is no longer profitable. Mines and wells, however, may be reopened if there is a surge in price or if new technologies for extraction (such as fracking) are developed that make it cheaper to obtain the remaining deposits.

17.2 Coal

Coal is a blackish, solid combustion material that was formed from woody plants that were compressed into a dense structure. The most common ways to mine coal have been subsurface mining and strip mining, as well as mountaintop removal in the Appalachian Region.

The United States was once the top producer of coal and still has the largest reserves of coal in the world. However, in recent years, it has dropped to third place behind China and India, with Australia and Indonesia rounding out the top five. The reasons for the decline in coal production are largely economic. Fracking has greatly lowered the price of natural gas and many power plants have switched from coal to natural gas to take advantage of the savings.

In fact, a study conducted by the business consulting firm Lazard in 2017 found that the Levelized Cost of Energy Analysis, which takes into account the cost of building and operating a power found that while the cost of coal has remained steady, the cost of natural gas, solar, and wind power have all dropped significantly in recent years. They put the current costs in dollars per megawatt hour for energy sources as

Source	Cost per Megawatt Hour
Coal	$102
Nuclear	$148
Natural gas	$60
Solar	$50
Wind	$45

Coal was once the primary energy source of the Industrial Revolution. It drove the steam engines that powered the trains that crossed the continent and the ships that traveled the oceans. Today, trains are either diesel or electric power and most ships are also diesel powered except for nuclear-powered naval vessels. Today, coal is mostly used to generate electricity and its share of the electricity market in recent years as shrunken from 50% down to less than 30% of the US market.

Coal burning powers start by loading the coal into a bunker where it is pulverized into a powder. Then, it is burned in the furnace and the heat generated is used to boil water. The force the steam is used to rotate a turbine which in turn turns a magnet inside the generator. As the magnet spins, it induces a current in a series of wires inside the generator.

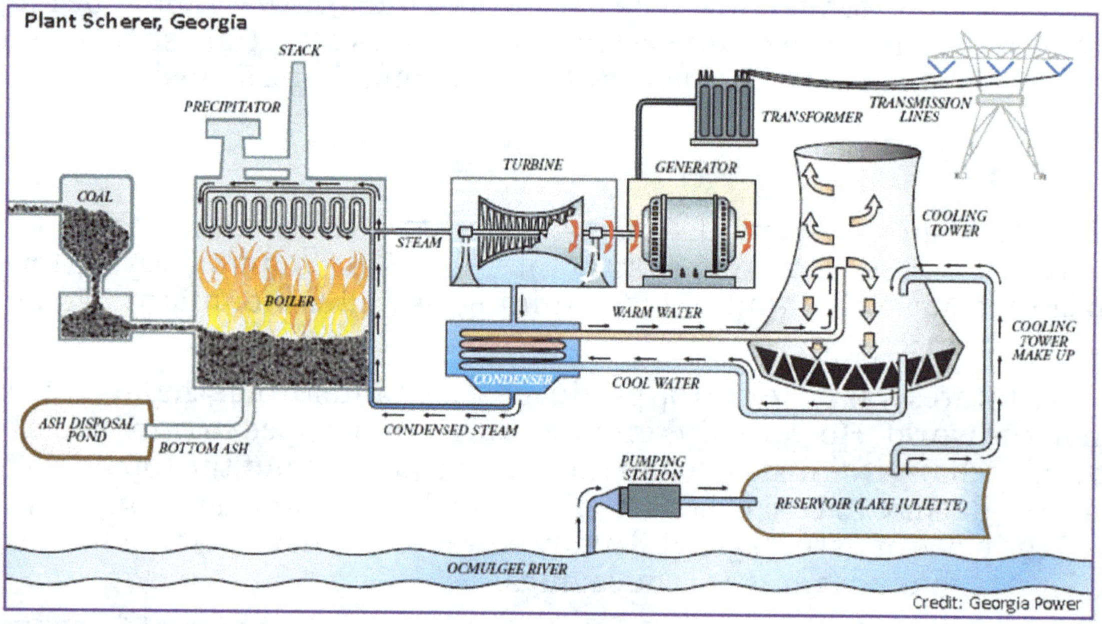

Source: Dennis Wilson

A coal burning power plant in Georgia, USA.

Meanwhile, the steam is recondensed by transferring its heat to the coolant loop so that it can be boiled again and repeat the cycle.

Coal plants produce two kinds of ash. **Bottom ash**, which as the name implies is collected at the bottom. The **fly ash** is collected by the electrostatic precipitator before it goes up the smoke stack. Both forms of ash are then sent to a landfill.

Coal is the dirtiest fuel. It generates more carbon dioxide, sulfur dioxide, and particulates than any other fuel. It also produces other pollutants including mercury. Electrostatic precipitators and other air pollution control technologies can reduce these pollutants but not eliminate them entirely. **Clean coal technologies** include other efforts, such as using calcium or magnesium to absorb the sulfur before burning the coal. **Gasification**, the conversion of coal into a gas called syngas can further reduce its pollutants, but it also reduces its EROI. Research continues to make carbon capture and storage a viable technology, but it still isn't proven effective and will further reduce the EROI of coal.

17.3 Petroleum

Crude oil or **petroleum** is a mixture of numerous liquid hydrocarbons that are formed from phytoplankton from millions of years ago. Crude oil refers to the unrefined liquid as it is extracted. Oil is trapped inside permeable rock layers under pressure. When a drill penetrates that layer of rock, the pressure is released and a gusher of oil bursts from the ground. Soon, that pressure equalizes and pumps are implemented to force the oil out using water or some other fluid. In a typical well, about one-third of the oil is released through primary production. After which, secondary production begins by pumping water or solvents into the ground to get as much of the remaining oil as possible.

Everett Historical/Shutterstock.com

Oil gusher.

Until recently, the United States had lagged in oil production behind several other nations in terms of oil production. Over the past decade, however, the United States has surged ahead and is now the largest producer of oil. This is not because we have discovered new reserves, however. We

Thaiview/Shutterstock.com

Oil pumps bringing petroleum to the surface.

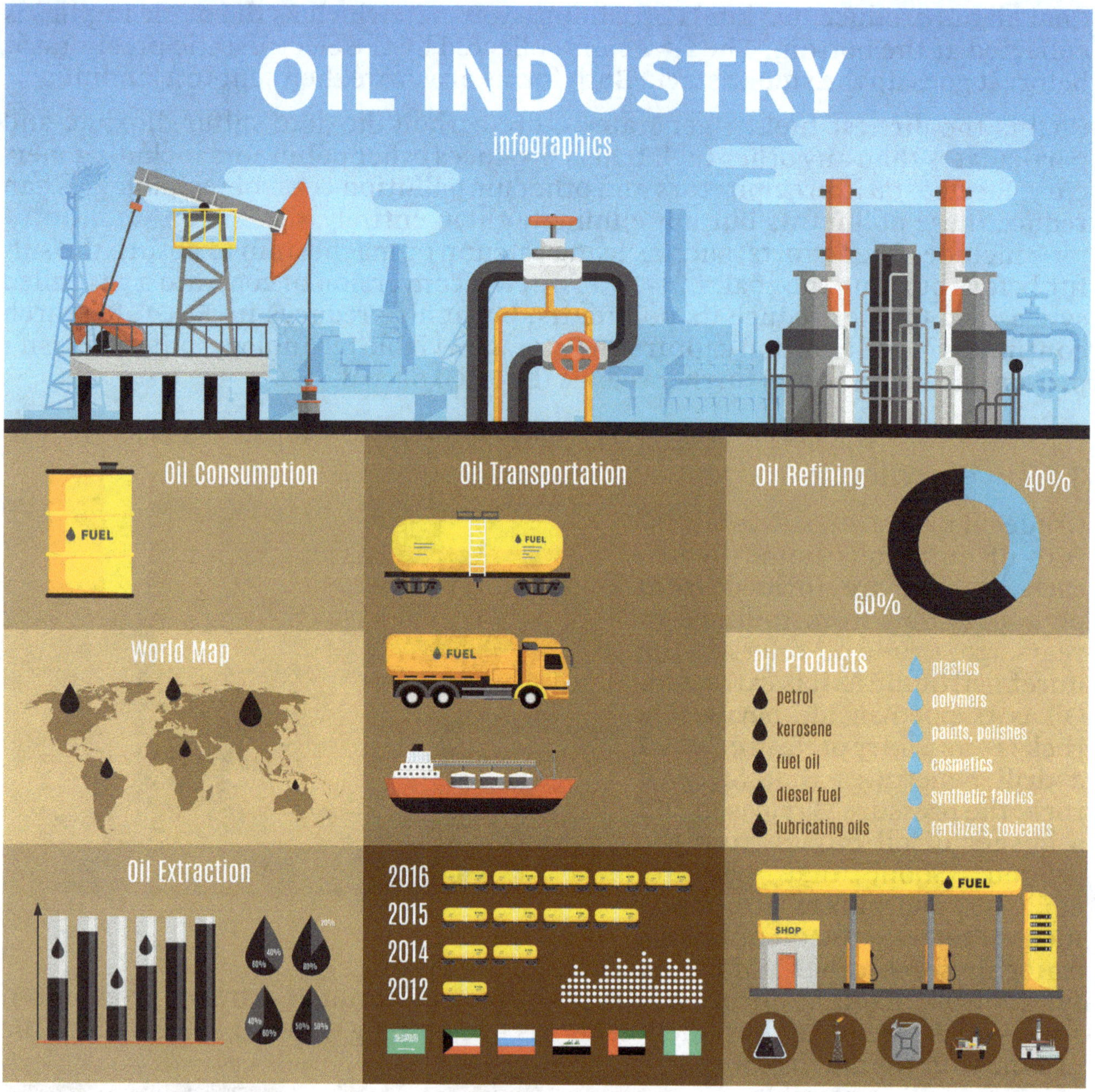

Macrovector/Shutterstock.com

are just more aggressively attacking the country's remaining reserves, including drilling out in deeper water. This of course, means that these domestic reserves will run out even sooner.

Before it can be used, crude oil must be refined. An oil refinery heats the oil and then uses a big distillation column to separate the various hydrogen carbons by weight, with the lightest ones rising to the top. About half of a typical barrel of oil is gasoline. A quarter is diesel and heating oil and about 8% is jet fuel. What is left is used for lubricating oil, other fuels, and for the manufacturing of plastics, pesticides, pharmaceuticals, and the plethora of other synthetic materials people use every day.

How much oil remains to be extracted is not clearly known. Nations and companies are reluctant to advertise their remaining reserves. However, it is estimated that there is about 1.7 trillion barrels of oil still available for extraction. At the current rate of production of 33.6 billion barrels per year, it is estimated that we have around 50 years of oil left. Changes in consumption patterns or improvements in hydraulic fracturing that would enable us to access the harder to extract tight oil could affect these projections.

It is inevitable, however, that we will hit the point of **peak oil**, where production will slow down and the price of oil will continue to rise.

In the 1960s, geologist M. King Hubbard predicted that US domestic production would peak in 1970. His prediction proved correct as the peak was named after him. After **Hubbard's peak**, US production steadily declined until the recent surge in production began in the early twenty-first century. The global peak in oil production is expected sometime in the midpoint of this century.

Oil is also a source of greenhouse gas emissions and other pollutants. Drilling for oil also disrupts a lot of land. However, the development of directional drilling, in which drills are able to curl horizontally has enabled a single drilling pad to spread out and tap a much larger area that used to require several drilling pads. This has helped to reduce the overall surface footprint of oil drilling.

About 35% of the oil produced in the United States is from offshore drilling. In 2010, the **Deepwater Horizon** drilling rig exploded, killing several workers and spilling thousands of gallons of oil into the Gulf of Mexico. The cause of the explosion was attributed to cost-cutting efforts by the rig's operator, BP plc, and the lack of a proper blowout preventer.

Elliott Cowand Jr/Shutterstock.com

Absorbent booms and chemical dispersants, which turned out to be toxic to wildlife, were used to clean up the spill.

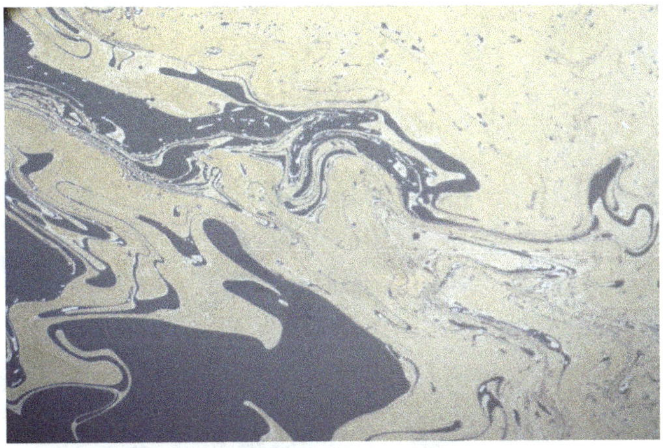

baon/Shutterstock.com

Oil floating in the water of the Gulf.

Breck P. Kent/Shutterstock.com

Cleaning a pelican contaminated with oil from the Deepwater Horizon disaster.

TuckerBlade/Shutterstock.com

Cheryl Casey/Shutterstock.com

Cleanup of a beach in Pensacola, Florida.

Oil from the spill leaked for almost two months before it could be plugged. Dealing with an oil leak a mile underwater was not something anyone had dealt with before and no one really knew how to close it. The government threw everything they could down the hole, including municipal waste in an effort to plug it. Finally, after two months, the hole was plugged, but not until after hundreds of thousands of gallons of oil had spilled into the Gulf.

The spill devastated the fishing and tourism industry in the Gulf region. Thousands of animals were also impacted, including pelicans, dolphins, and sea turtles. Cleanup of the spill took months and BP was required to pay $18.7 billion in fines and had to compensate people and industries impacted by the spill. Tar balls and oil sheens continued to be found on the beaches of the Gulf until as late as 2014.

Efforts are also underway to drill in the Arctic, which has its own unique hazards. Ice floes and heavy winds can damage rigs. A Royal Dutch Shell rig recently ran aground in a heavy storm. Also, the frigid temperatures and the lack of sunlight for months at a time will slow the breakdown of oil in the event of a spill.

While oil tanker spills have declined over the past few decades, transporting oil over land has been fraught with controversy. Two proposed pipeline projects have been met with opposition. The first was a proposed expansion of the Keystone XL pipeline by TransCanada that would pass through some sensitive lands, including

Rena Schild/Shutterstock.com

Protester against the Keystone XL pipeline in Washington, DC.

NYCStock/Shutterstock.com

A poster from a protest against the Dakota Access pipeline that would cross the Standing Rock Reservation.

areas that are part of the Ogallala Aquifer, an important source of irrigation water in the Midwest. The other was proposed Dakota Access pipeline that would run through the Standing Rock Indian Reservation. In both cases, there were concerns about the damage a leak from the pipeline would cause.

The alternative to pipelines is transporting the oil by tanker trains, which could derail and explode.

Marek Szandurski/Shutterstock.com

A derailed tanker train in Poland.

Because much of the world's oil comes from the Middle East, any conflict or instability in the region can have ramifications for the oil market. For example, in 1973, the Organization of Petroleum Exporting Countries (OPEC) voted to embargo the sale of oil to the United States and other countries that supported Israel. This set in motion a series of gas shortages that became known as the **Energy Crisis**. Other events including the First Gulf War, the US Invasion of Iraq, Arab Spring, and the downfall of Libyan leader Qaddafi have all led to increases in the price of oil.

In order to hedge against a future energy crisis, the United States has adopted a number of strategies. For one, it created the Strategic Petroleum Reserve, a month's supply of oil for the nation that is stored in a cave in Louisiana. The United States also diversified where it imports oil so that it no longer relies on the majority of its oil from the Middle East. Still, international tensions have an impact on what US consumers pay at the gas pump.

17.4 Natural Gas

Natural gas is mostly methane and, like oil, is formed from plankton that was buried under sediments. Methane can also be generated through biological processes, such as in **anaerobic** (very little available oxygen) environments such as in stagnant swamp water or in landfills. In such conditions, anaerobic bacteria will decompose organic matter and produce methane as a byproduct. People often refer to methane produced in a dank swamp water as "swamp gas."

Stephen B. Goodwin/Shutterstock.com

A natural gas-powered bus in Arizona.

Today, methane is used for heating, cooking, and generating electricity. It is also used for transportation by many bus systems. Fracking has created a boom in natural gas production, which has made it much cheaper and today, the United States is the world's leading producer and consumer of natural gas. Many power utilities have switched from coal to natural gas and this has resulted in a 14% reduction in America's greenhouse gas emissions. Note that less carbon dioxide does not mean zero and natural gas consumption still contributes to the total greenhouse gas emissions.

While natural gas emits less carbon dioxide and other pollutants than oil or coal, fracking has generated controversy wherever it has been practiced. Local residents have complained about well water contamination with methane (you can go to YouTube and see a few examples of residents lighting their tap water on fire) and other chemicals. Fracking also uses a lot of water, which is a concern in drought-plagued states like California. Also, injecting fluids underground may dislodge deep, ancient faults and trigger minor earthquakes. Both Ohio and Texas have reported an increase in tremors after fracking was introduced into those states.

A natural gas drilling rig in Cogan Station, Pennsylvania.

Proponents of fracking argue that it creates jobs and produces a domestic energy source that is less polluting than coal. Also, industry is continuing to conduct research to reduce the negative impacts of fracking.

An oil shale rock.

17.5 Unconventional Fossil Fuels

Oil shale is a sedimentary rock that contains a hydrocarbon material that can be extracted and used to produce the same fuels as petroleum. Oil shales are mined in much that same way as coal and involve a lot disruption of land.

chris kolaczan/Shutterstock.com

Oil sands mining in Alberta, Canada.

Similarly, **oil sands** are sandy materials that contain a tarlike material called bitumen that can be converted into oil. Oil sands are usually mined through strip mining and also result in severe land disturbance and oil contamination. Today, the biggest producers of oil from oil sands today is Alberta, Canada, and Venezuela. The oil transported through the Keystone Xl pipeline is from oil sands.

Irina Koledova/Shutterstock.com

Baikal Lake in Russia showing bubbles of methane gas trapped in the frozen water.

Methane hydrate or methane ice is methane gas trapped in ice crystals in Arctic sediments. This could be a significant source of methane. However, we currently do not have a safe means of extracting it.

All of these unconventional fossil fuels have negative environmental impacts, which contribute to greenhouse gas emissions, and have a lower EROI than conventional oil. In addition, experience in dealing with leaks from pipelines transporting oil from oil sands indicates that such spills are harder to clean up than spills from conventional oil.

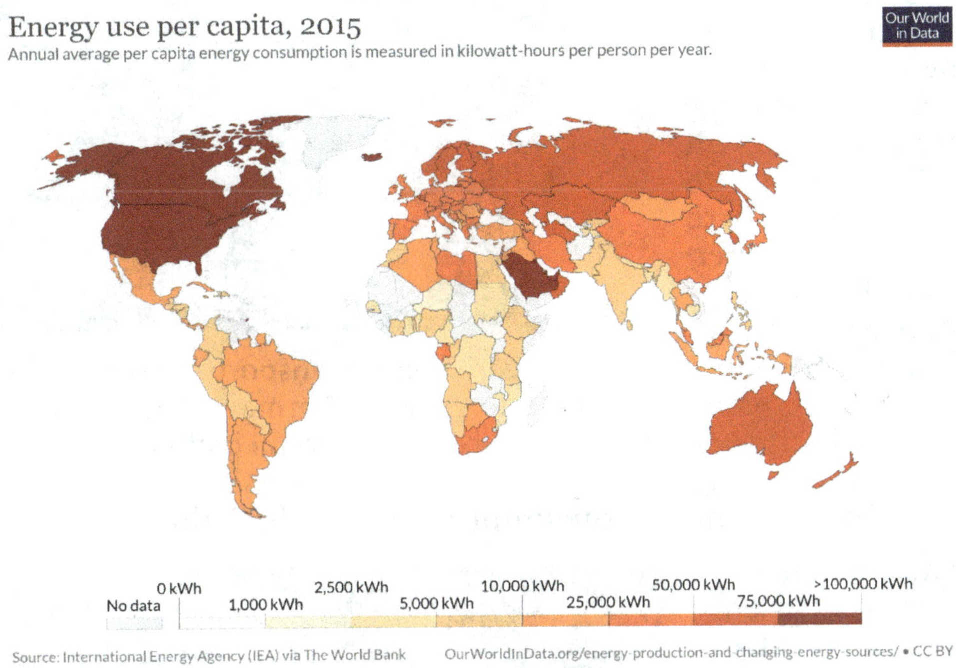

Source: Our World in Data. https://creativecommons.org/licenses/by/4.0/

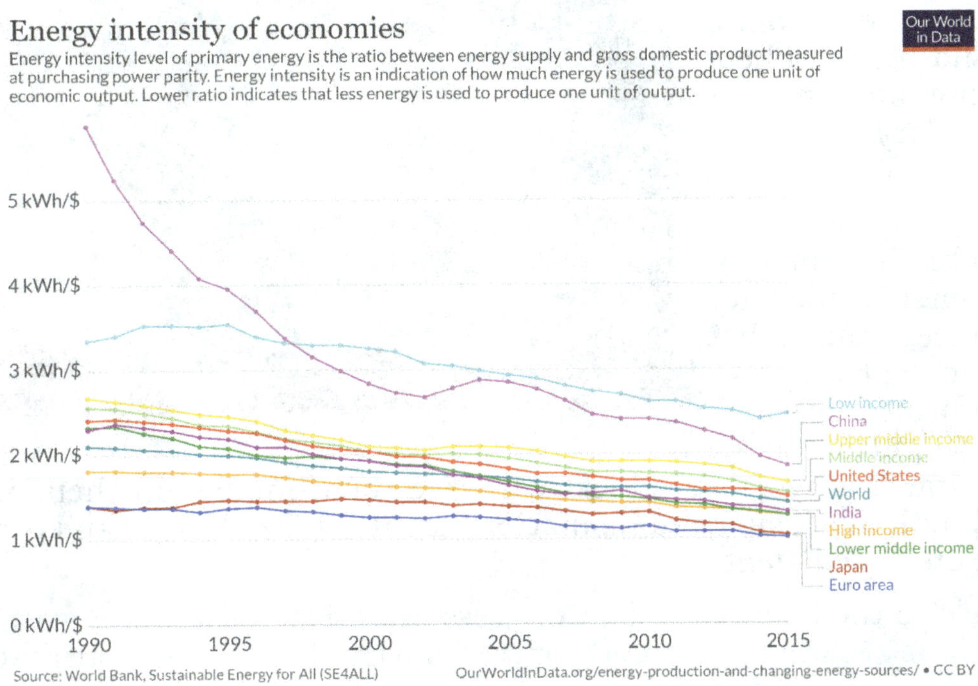

Source: Our World in Data. https://creativecommons.org/licenses/by/4.0/

17.6 Energy Efficiency

District heating pipe transporting steam from a cogeneration plant in Germany. The heat transported by these pipes can heat several buildings.

guentermanaus/Shutterstock.com

Energy efficiency describes any effort to achieve the same output with less input of energy. **Energy conservation** refers to efforts to reduce energy consumption by eliminating wasteful or unnecessary use. America still rates as one of the highest nations in per capita energy consumption, although our **energy intensity** (ratio of energy consumption to gross domestic product) has been declining steadily. A lower intensity indicates greater efficiency in energy use. So, while we are using energy more efficiently, we still have a long way to do to significantly reduce our energy consumption. See tables below.

David Tonelson/Shutterstock.com

There are a number of ways to increase our energy efficiency. **Cogeneration** involves capturing some of the excess heat from a turbine and using to heat neighboring buildings or run a secondary turbine.

The US EPA's Energy Star Program provides consumers with information about the energy efficiency of appliances and other industry so that they can make informed choices about how much energy they consume and which products will save them money on their power bill. Improved building design and insulation can also make homes and other buildings more energy efficient.

In response to the Energy Crisis, Congress established the **Corporate Average Fuel Efficiency Standards (CAFE)** for cars and trucks in 1975. Note that SUVs are classified as trucks and are not required to meet the same standards as are cars. These standards languished for decades and by the early twenty-first century, the average fuel efficiency of vehicles on the road had begun to decline. Finally, in 2007, Congress updates the CAFE standards. The current standards call for SUVs and light trucks to achieve a mileage of 35 miles per gallon and for passenger cars to reach 54.5 miles per gallon by 2026. However, the Trump administration considers those goals to be "unattainable" and has pressed Congress to find an alternative. The administration has also sought to block states such as California from adopting more stringent standards.

Rick Beauregard/Shutterstock.com

Close up of the sales sticker for an automobile showing fuel efficiency.

One of the big problems with efficiencies stems from gains being offset by increased consumption. Cars become more fuel efficient, so people drive more, thereby erasing the energy savings gained by increased efficiency. This is like thinking you can eat a whole box of sugar-free cookies thinking you won't gain weight, even though the cookies are still loaded with calories.

17.7 Discussion Questions

1. Describe the process of hydraulic fracturing
2. Why has there been an increase in drilling in the Marcellus Shale over the past 15 to 20 years?
3. What is EROI and why does it vary from different sources of energy?
4. Why has coal use declined in recent years?
5. How do fossil fuels form?
6. What is peak oil? How will we know when we've passed it?
7. What were the impacts of the Deepwater Horizon Spill?
8. How do oil shale and oil sands differ from petroleum?
9. Compare the two methods for transporting oil over land, pipelines, and tanker trains. What are the risks of each?

10. What is methane hydrate?
11. What is cogeneration?
12. Describe some energy conservation efforts.
13. What are the political impacts of depending on importing oil from the Middle East?
14. How has the United States risen to becoming the leading producer of oil and natural gas?

17.8 Key Terms

Anaerobic
Bottom ash
Clean coal technology
Coal
Cogeneration
Corporate Average Fuel Efficiency Standards (CAFE)
Deepwater Horizon
Energy conservation
Energy Crisis
Energy efficiency
Energy intensity
Energy Returned on Investment (EROI)
Fly ash
Gasification
Hubbard's peak
Hydraulic fracturing or fracking
Marcellus Shale
Methane hydrate
Natural gas
Oil sands
Oil shale
Peak oil
Petroleum

Chapter 18

NUCLEAR POWER

Chapter Topics

America has a long and complicated history with nuclear power. After the **Manhattan Project** developed the first nuclear weapons, the federal government began looking at ways to use nuclear fission to benefit humanity. The **Peaceful Atoms** initiative began with investing in developing commercial nuclear power through the 1950s and by the 1960s, the first commercial power plants were under construction.

It looked like the United States was on its way to generating most of its electricity from nuclear fission. Predictions were made that by the end of the twentieth century, the country would have 1,000 reactors all producing electricity that was "too cheap to meter." The power would be so easily produced that it wouldn't be worth the effort of tracking how much electricity each home used it every month. Instead, every household would pay a flat rate, much like how cable companies charge everyone the same monthly rate, no matter how many hours of TV they watch.

Then, in 1979, the Three Mile Island (TMI) accident happened, and while its impact was minor compared to the ones that came after at Chernobyl and Fukushima Daiichi, it soured public opinion on nuclear power. In addition, far from being "too cheap to meter," nuclear power has proven to be the most expensive way to generate electricity. Power plants that were under construction at the time were abandoned and plans for new ones were dropped. No new power plants were constructed or even permitted for decades. It's only recent years that, with new, smaller, and more modular designs being approved has there been the possibility of a nuclear renaissance. Despite lingering concerns about safety and a lack of a permanent solution for the waste issue, many people are giving nuclear power a fresh look. With rising concerns about the greenhouse gas emissions of fossil fuels, nuclear power is being promoted as an alternative solution to replace fossil fuels or even just as a bridge between fossil fuels and renewable sources of energy.

As seen in the following charts, the United States has more nuclear power plants than any other nation and generates more gigawatts of electricity from nuclear power than any other. However, it only makes up less than 20% of our total energy consumption. In contrast, France is much more dependent on nuclear power, getting over 70% of its electricity from it.

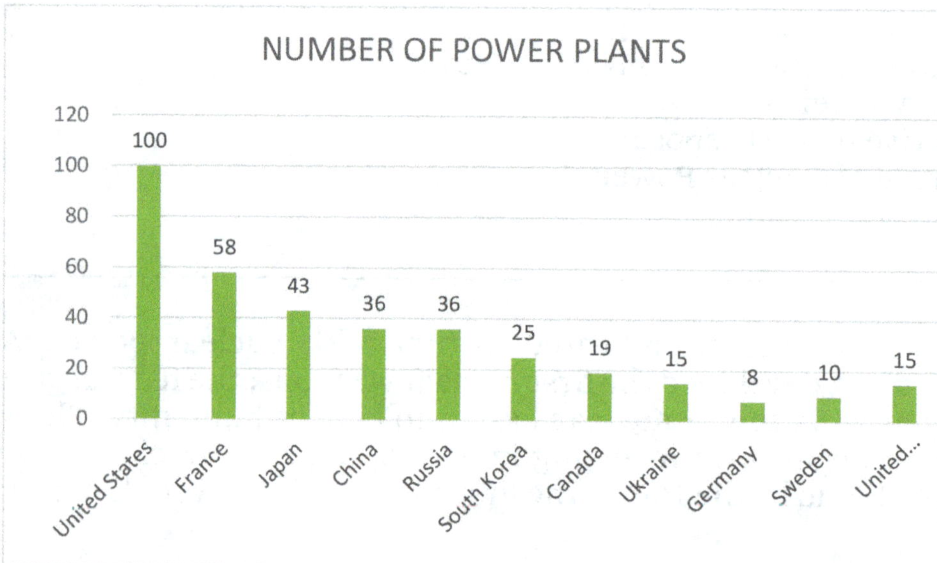

Source: Our World in Data. https://creativecommons.org/licenses/by/4.0/

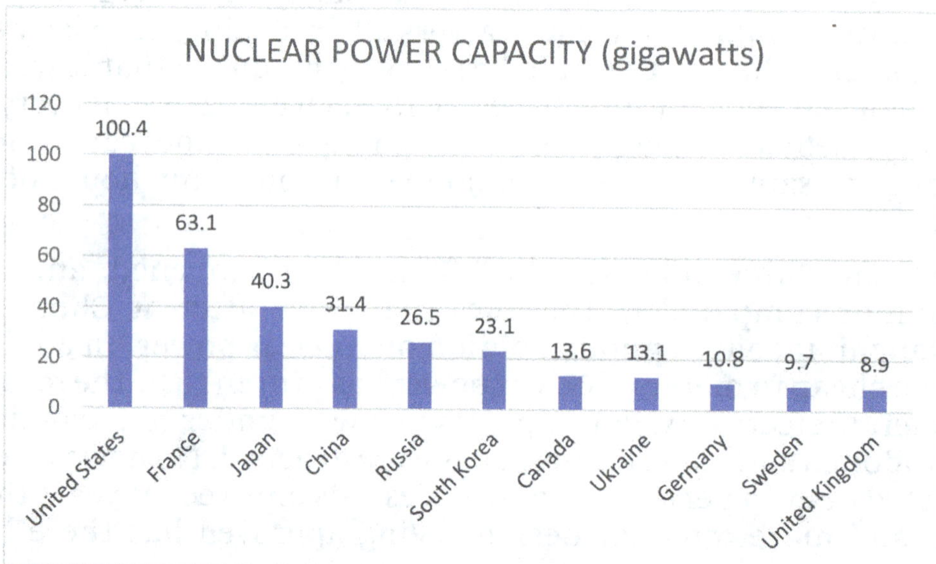

Source: Our World in Data. https://creativecommons.org/licenses/by/4.0/

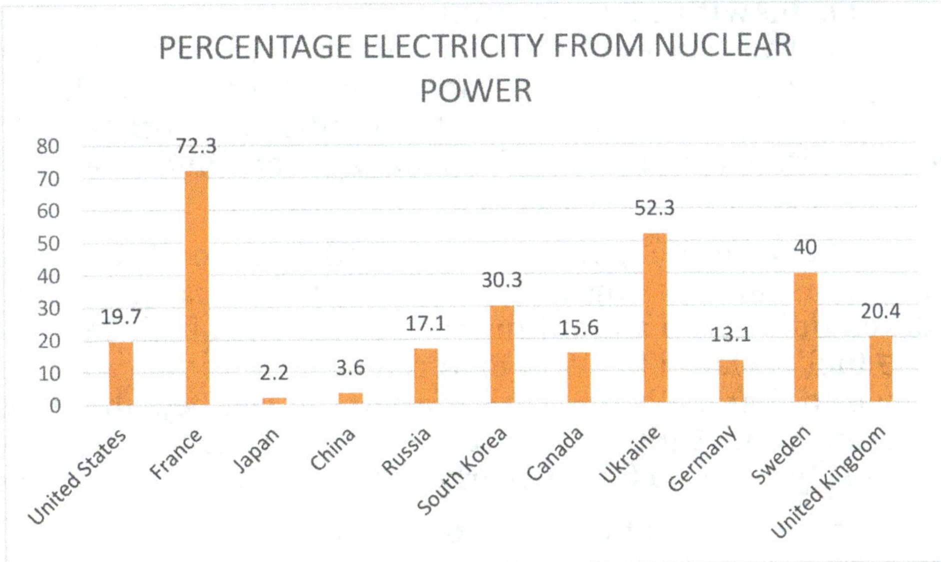

Source: Our World in Data. https://creativecommons.org/licenses/by/4.0/

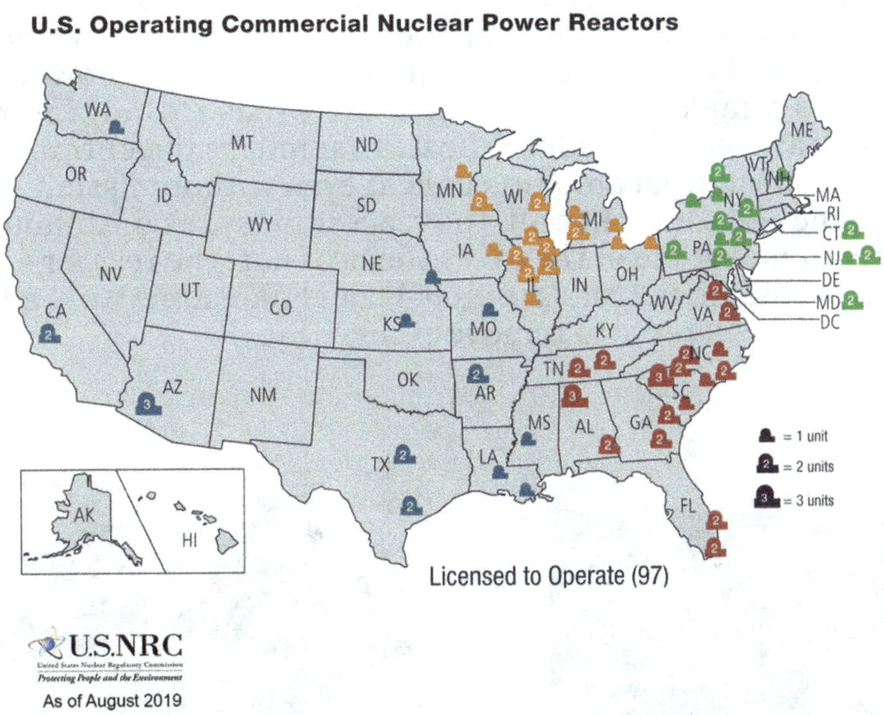

Source: US Nuclear Regulatory Commission

18.1 Energy from Atoms

Nuclear power is the process of generating electricity by splitting heavy atoms like uranium-235. The process of splitting an atom, **nuclear fission**, releases some of the energy that holds the nucleus of an atom together. This **nuclear energy** is then used to boil water to run a turbine. **Nuclear reactors** are designed to promote nuclear fission in a controlled manner in order to produce electricity safely.

A nuclear reaction begins with hitting a specific isotope of uranium, uranium-235, with a slow neutron. This causes the uranium nucleus to split into two smaller nuclei, such as krypton and barium. The reaction also emits energy in the form of gamma rays as well as some free neutrons that can collide with other uranium-235 atoms, triggering additional fission reactions that will continue on so long as there is sufficient nuclear fuel available.

Neutrons zip around at near the speed of light. In order to promote a nuclear reaction, the neutrons must first be slowed down. They move so fast that the odds of a neutron actually colliding with a uranium nucleus are extremely low. To do this, the uranium-235 fuel is placed inside a material called a **moderator** that slows the neutrons down so that collisions with uranium nuclei becomes more likely. Commercial reactors in the United States use water as a moderator. Other kinds of reactors might use graphite (a form of carbon) or even mercury as a moderator.

Nuclear fuel consists of pellets of uranium oxide that are stacked in tubes called **fuel rods**, which are assembled inside the reactor. In order to control the rate of the reaction, **control rods** made of a neutron-absorbing material are raised and lowered inside the reactor. Raising the control rods promotes more fission reactions and generates more heat. Lowering them reduces the rate of reactions, causing the reactor to produce less heat. It is important to note, however, that even when the reactor is completely shut off (the control rods are fully inserted into the reactor), the nuclear fuel still produces heat. That much concentrated radioactive material puts out substantial amounts of heat even without fission. For that reason, even when the reactor is shut off, the reactor must be continuously supplied with coolant to remove some of the excess heat. If not, the reactor will overheat. All of the major accidents at TMI, Chernobyl, and Fukushima were all caused by a failure of the **coolant systems** to remove excess heat.

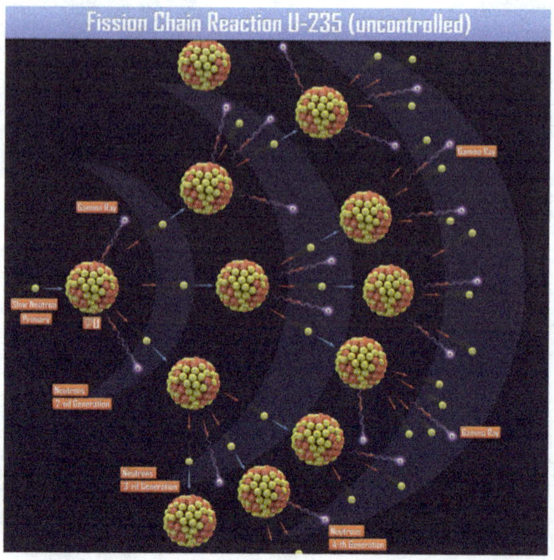

general-fmv/Shutterstock.com

Uranium-235 fission chain reaction.

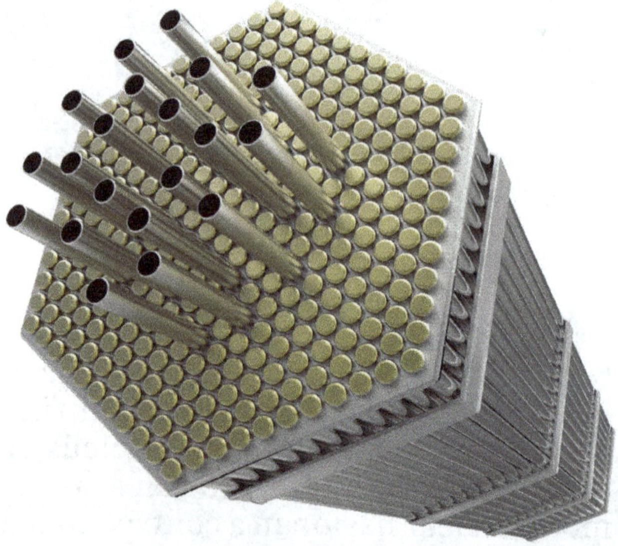

Peter Sobolev/Shutterstock.com

Nuclear Fuel Assembly.

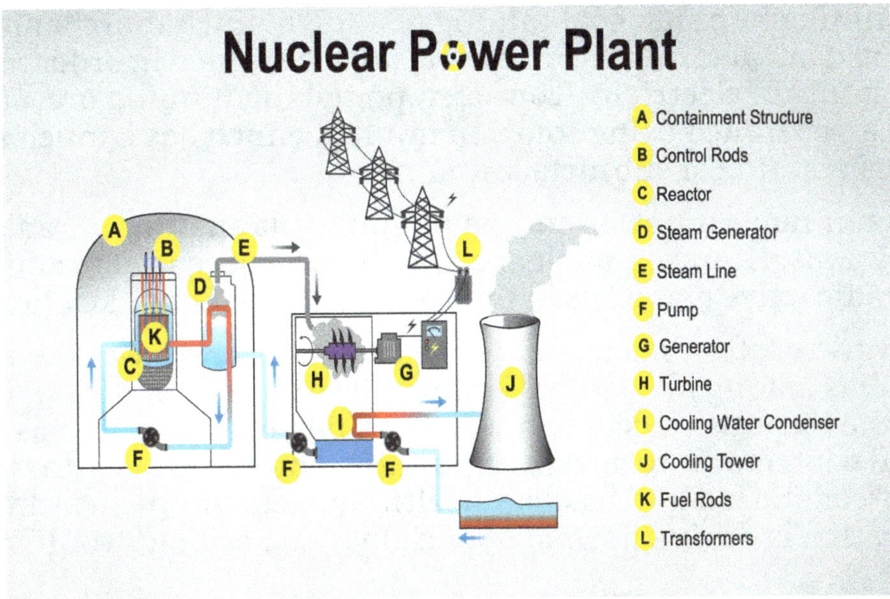

Studio BKK/Shutterstock.com

Nuclear power plant schematic.

The water pumped through the reactor is kept under pressure so that it, even though it is heated up to hundreds of degrees, it remains liquid. The heat from this primary loop is transferred to the secondary loop. The water in the secondary loop is allowed to boil and the stream produced runs the turbine to power the generator. Then, the water is recondensed using the coolant loop and boiled again. The reason for the dual primary and secondary loop system is that, as the water in the primary loop passes through the reactor, it picks up radioactive contamination. The water in the secondary loop never comes in direct contact with the radioactive material, so all of the radioactive material is kept inside the **containment building**. The concrete walls of the containment building absorb the **radiation** of the reaction and serve to contain the radioactive material in the event of a serious accident.

A viable nuclear fuel must contain at least 3% uranium-235. However, the naturally occurring uranium ore found on Earth has much less than that. It is generally 99.7% uranium-239, which does not split when hit with slow neutrons. In order to produce a useful fuel, the uranium must undergo a process using centrifuges called **enrichment** to bring the uranium-235 concentration up to 3%. This is not sufficient to be used in a nuclear weapon. To make weapons-grade uranium, it must be enriched to at least 70%, which is a much more involved process involving a cascade of centrifuges.

18.2 Benefits and Drawbacks of Nuclear Power

Perhaps the main selling point for nuclear power today is that it, unlike coal, it does not produce substantial amounts of greenhouse gases. Countries like Sweden have been able to cut their greenhouse gas emissions in half by expanding their nuclear portfolio. Coal also emits numerous other pollutants, including sulfur dioxides, nitrogen oxides, mercury, and particulates. Nuclear power does not emit any of these.

In terms of mining impacts, coal mining results in much more land disturbance and ecosystem damage. Several tons of coal must be mined in order to produce the equivalent amount of electricity from a few pounds of uranium ore. Some deposits of uranium can be mined using solution mining, which has a much smaller footprint than strip mining or subsurface mining.

Neither coal nor nuclear power emit any significant amount of radiation during normal operations. However, nuclear power has the potential to emit substantial amounts of radioactive materials in the event of an accident like Chernobyl.

In terms of worker safety, nuclear power actually has fewer worker accidents and health problems among its workers compared to coal. In addition, the pollution released by the burning of coal can cause serious health problems among local residents and contributes to acid precipitation. Nuclear power does not contribute to acid precipitation and has less health impacts on the health of local residents. Again, that is during normal operations. An accident would be a different situation.

The worst-case scenario of an accident or sabotage at a coal power plant is that the power plant is shut down for a while. For nuclear power, the worst-case scenario would be well, Chernobyl.

In terms of the waste generated, coal produces much larger quantities. However, the waste from nuclear power is highly radioactive and we do not have a permanent solution for the disposal of it.

Finally, we enough coal reserves remaining to last for many a few hundred years. Uranium is much less abundant than coal, but since we would need much less of it, the remaining reserves may last a little longer than coal.

18.3 Nuclear Accidents

George Sheldon/Shutterstock.com

Three Mile Island (TMI) Sign.

A. L. Spangler/Shutterstock.com

The TMI nuclear power plant. Of the two reactors onsite, Unit 2 has been shut down since the 1979 accident. Unit 1 has continued operation. However, the owners of the power plant have recently announced that it will soon be retired.

In 1979, a stuck valve in the Unit 2 reactor at TMI allowed all of the coolant water to boil away and escape. A technician misread a gauge and thought the reactor was full of water when it wasn't, so he turned off the pumps supplying water. The reactor began to overheat and eventually went into a partial meltdown, where the metal reactor housing melts and falls down. A meltdown will boil all of the water inside the containment building and, in this case, it caused a bubble of gas to expand that exerted pressure on the interior of the building. However, the containment building held and technicians were able to reduce the pressure by slowly venting the radioactive gas from the building over a period of several months.

Because the containment building did its job, there was no widespread radioactive contamination released from the accident. There were no deaths attributed to the TMI accident and epidemiological studies conducted by the State Department of Health did not find any significant increase in cancer rates in the region.

Pe3k/Shutterstock.com

An abandoned amusement park near the Chernobyl nuclear power plant.

TMI was a hiccup compared to what happened at Chernobyl in 1986. During a test of the backup diesel generators that supply coolant water to the reactors, the technicians, for reasons that remain unclear, disabled the safety features, drained the coolant water, and pulled the control rods completely out of the reactor. Naturally, the reactor began to overheat rapidly. In an attempt to shut down the reaction, the technicians quickly reinserted the control rods. The control rods used at Chernobyl had carbon tips on the ends, which were the first things to reenter the reactor. Remember that carbon graphite can be used as a moderator? That's exactly what the carbon tips did. Instead of shutting the reaction down, it accelerated it. The reactor went into full meltdown, which triggered a steam explosion that blew apart the containment building.

The explosion ignited a fire and the plume of radioactive material was released and traveled all the way around the globe. The countries that were most affected by the radioactive fallout were Ukraine, Belarus, Russia, and other parts of Europe.

There are 31 deaths attributed to the explosion and fire. This included firefighters who were sent into reactor building to put out the flames at gunpoint, knowing they were about to receive a lethal dose of radiation. After the accident, people in Ukraine and neighboring countries experienced a significant increase in cancer rates, especially thyroid cancer in children.

Several communities around the power plant were evacuated and are now ghost towns.

Once the fire was contained, a massive concrete sarcophagus was constructed around the remains of the reactor to contain the material. This sarcophagus eventually began to leak, so a larger containment structure was subsequently built around it.

Youproduction/Shutterstock.com

Chernobyl Unit 4 after the explosion. This amusement park in Ukraine has been abandoned since the Chernobyl disaster.

Serhii Barylo/Shutterstock.com

Abandoned bumper cars in Pripyat, Ukraine.

Fotokon/Shutterstock.com

Apartment buildings in the abandoned city of Pripyat, Ukraine.

Dmitry Birin/Shutterstock.com

New containment structure being under construction around the accident site.

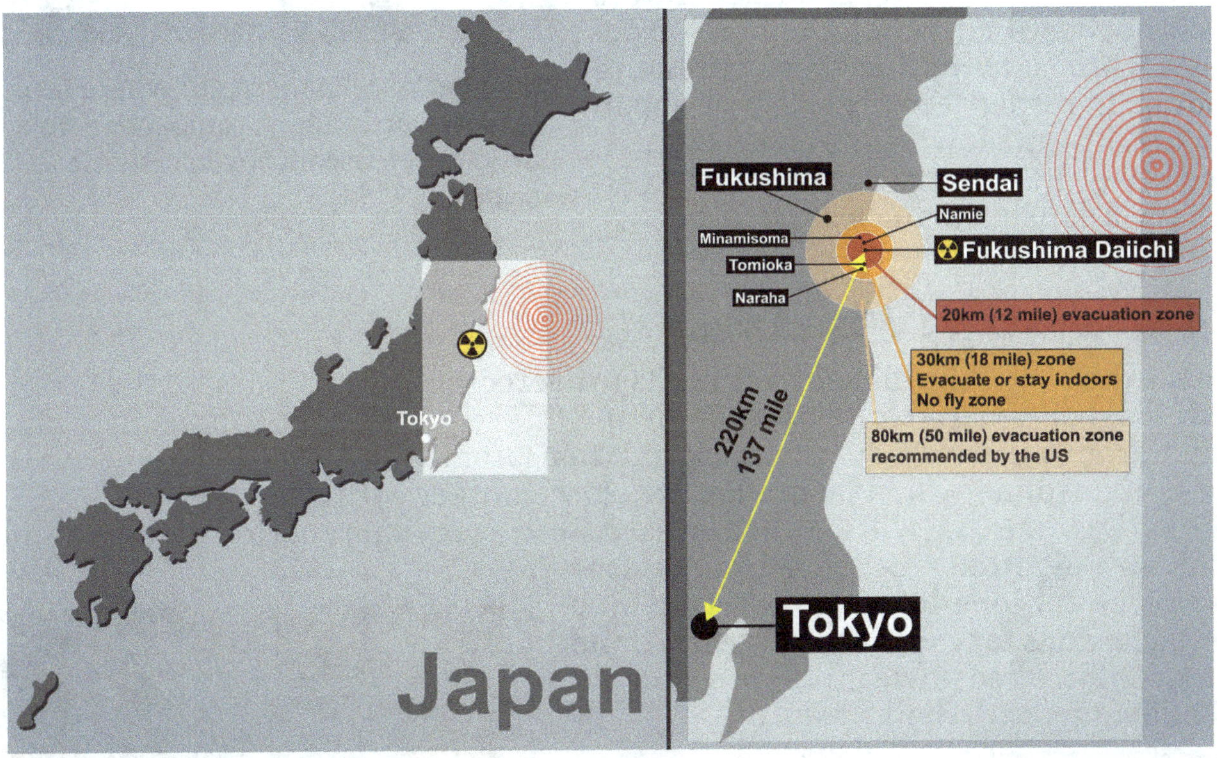

Gurgen Bakhshetyan/Shutterstock.com

Map of Japan showing epicenter of earthquake and location of Fukushima Daiichi power plant.

While TMI was caused by a combination of mechanical failure and human error, Chernobyl was the result of poor design and poor practices. The containment buildings used during the Soviet era were not built to the same standards as those used in the United States. Also, the actions taken by the technicians in shutting off safety features, pulling the control rods, and draining the coolant water were inexplicable and violated all standard practices.

The **Fukushima Daiichi** accident was initially triggered by natural disasters. In 2011, an earthquake hit Japan and then triggered a tsunami, which flooded the Fukushima region. The four reactors at the plant were shut down, but the backup diesel generators were shorted out when the basements were flooded. This left technicians without a means to supply the reactors with coolant, so they overheated. The amount of radiation emitted from this accident was about a tenth that of Chernobyl.

Studies on the health impacts of the Fukushima Daiichi accident are still ongoing. The main lesson from the incident, however, was that the backup generators should not be stored in the basements, especially on an island nation that is prone to flooding. In response to the accident, Japan temporarily shut down all 50 of its reactors. Since then, the country has been reassessing its relationship to nuclear power.

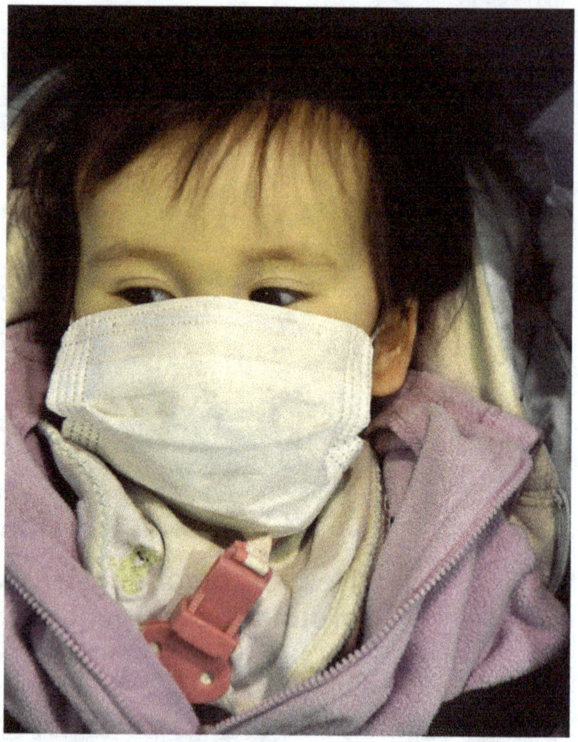

One-year-old Japanese girl where face mask to ward against radioactive dust particles released from Fukushima accident.

18.4 Radioactive Waste Disposal

Perhaps the biggest hurdle nuclear needs to overcome is the disposal of the **radioactive waste** it generates. Remember the opening sequence of The Simpsons where Homer drops a fuel rod down his back and then later throws it out his car window? In real life, Homer would be dead. That close contact with a piece of **high-level radioactive material** will deliver a lethal dose in a very short period of time. Some of the **radioisotopes** produced by nuclear power are very long-lived and will continue to be hazardous for thousands of years. If the Roman Empire had nuclear power, we would still be dealing with their waste.

Every few years, the fuel rods in a reactor are replaced. The current practice is to put the old rods into a temporary **wet storage** pool to stabilize them and allow the shorter-lived isotopes to decay away. Then, the rods are packed into concrete and steel casks for **dry storage**.

Wet storage of spent nuclear fuel.

These pools and casks remain in temporary storage sites across the country, mostly at the power plants that produce them. In the 1980s through 1990s, the federal government conducted a survey across the country to locate a site for a permanent disposal site. Eventually, the government settled on **Yucca Mountain** in Nevada. Yucca Mountain was chosen because it is in a dry climate in order to minimize water infiltration. It's also relative geologically stable. It hasn't had any earthquake activity in about 50,000 years. It's also an existing federal land. In this case, an Indian Reservation, so the tribe was generously compensated.

After several years of delay due to budget issues and legal fights, construction on the disposal facility began in the early twenty-first century.

Yucca Mountain's status as the nation's repository for high-level radioactive waste was unpopular among residents of Nevada. Nevada has no nuclear power plants. The federal government also used to conduct above-ground testing of nuclear weapons in the deserts of Nevada. Many residents felt they were again being made to take on the burden of radioactive hazards that mostly benefitted people from other states. In 2010, the federal government canceled the Yucca Mountain project, largely for political reasons. This leaves the United States without a permanent solution to the nuclear waste problem.

Spent Fuel Generation and Storage after Use

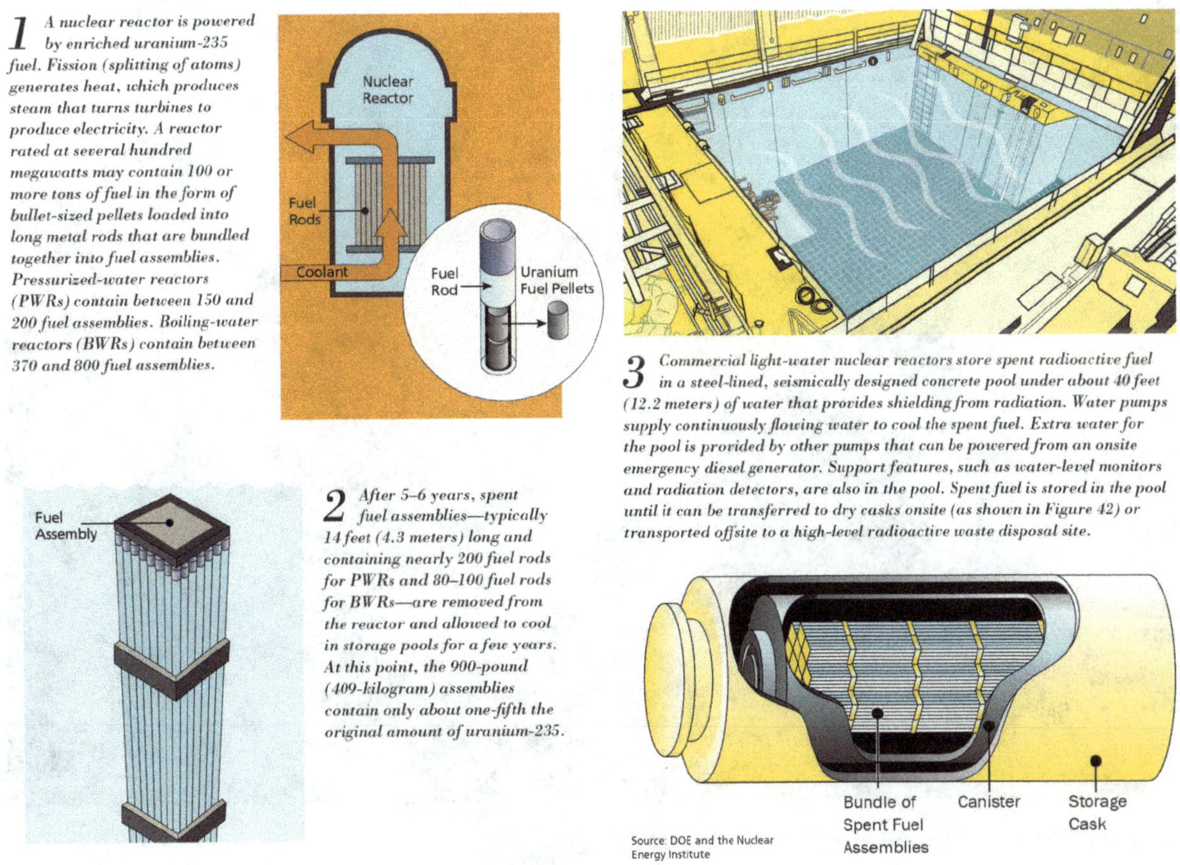

Source: ncr.gov.

Source: ncr.gov.

Entrance to Yucca Mountain.

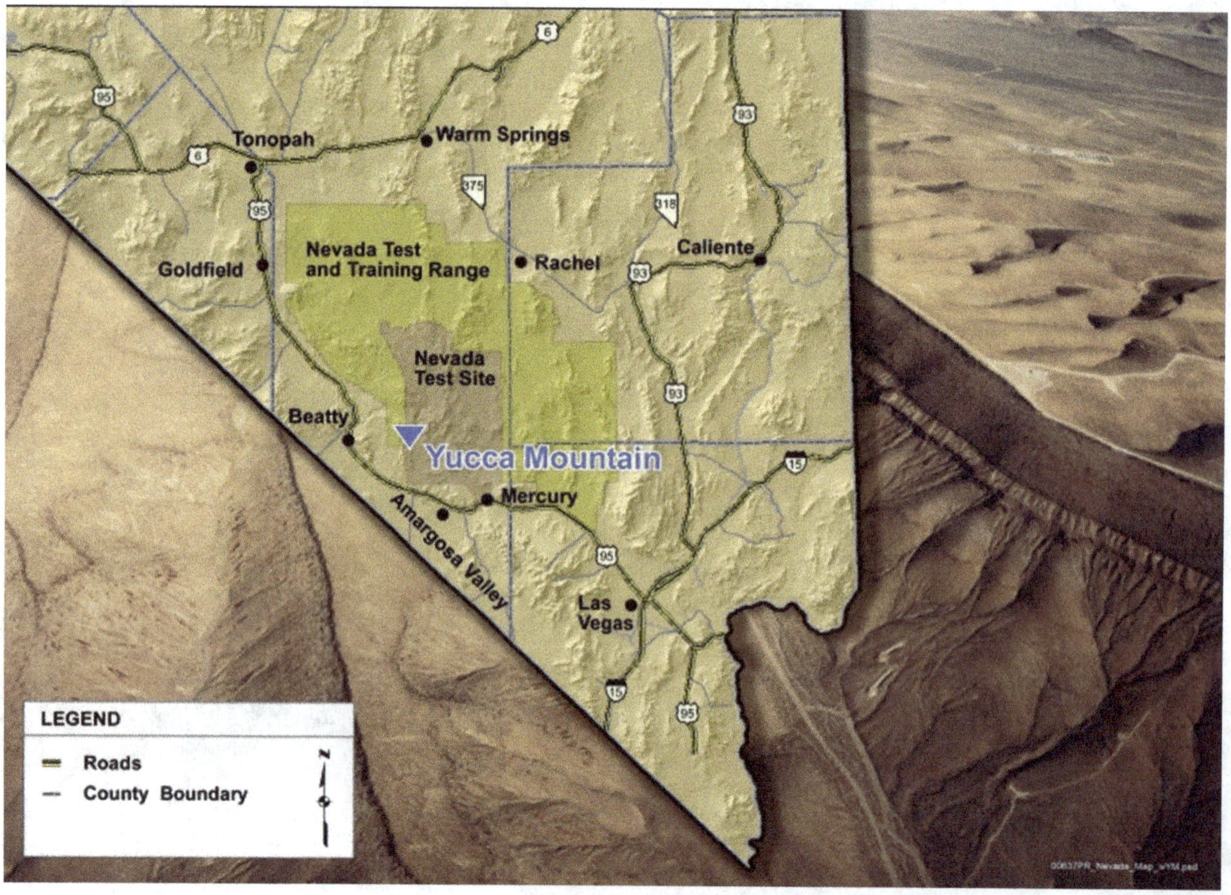

Source: doe.gov.

Location of Yucca Mountain.

18.5 The Future of Nuclear Power

Today, the nuclear power plants built in the 1960s and 1970s are reaching the end of their operating lifespan. After failing to find a buyer, the owners of TMI recently announced they were shutting down the Unit 1 reactor that had been in continuous operation since the accident at Unit 2.

Many of the components of nuclear power plants have aged faster than originally expected. It turns out that steel pipes turn brittle if they're bombarded with neutrons for several years. This has increased maintenance costs which, along with increased safety standards, has made nuclear power the most expensive means to generate electricity today. Just as these power plants have made back their construction costs, their owners and ratepayers now have to pay for the cost of dismantling them.

And a permanent solution to the waste issue still eludes us.

Despite these limitations, many feel that nuclear power is ready for a renaissance. Newer designs have been proposed that may address some of the safety concerns. The older generation of reactors were all custom designed. The newer models are manufactured in a factory to standardized designs and assembled onsite. They are also smaller, using eight reactors to generate the same amount of electricity as two larger ones did. This has the advantage of an accident at a single reactor would have less impact. Also, during refueling, the power plants could still run at 7/8 capacity instead of half capacity.

One promising design involves replacing the solid uranium pellets with a liquid salt fuel. These reactors will use thorium as the source of fission instead of uranium. Called **liquid fluoride thorium reactors** (LFTR), these models don't require water for coolant. Instead, the liquid fuel itself will circulate through the heat exchanger to boil the water in the secondary loop. These reactor designs also include a passive safety system instead of the active safety used at power plants like TMI. **Active safety** requires human intervention to work and could fail as a result of human error. For example, the technician who misread the gauge at TMI was an active safety measure that failed. In contrast, **passive safety** involves designs that will shut the reactor down without human intervention. In the case of LFTR, if the reactor overheats, a plug will melt, causing the liquid fuel to drain out of the reactor and into containment tanks that will keep the atoms separated enough to prevent a continuous reaction.

Another advantage is that waste product for a thorium reactor is much shortlived. It will only be radioactive for about 300 years instead of 10,000 years or more. This would at least make the waste disposal issue a little easier to manage.

Many of these new reactor designs are being tested out today and we will probably know in a few years if nuclear power still has a future.

18.6 Discussion Questions

1. Why did the United States never achieve its goal of producing the majority of its electricity from nuclear power that was "too cheap to meter?"
2. Why is it important to take excess heat away from a reactor, even when it is not in operation?
3. What does a moderator do?
4. How do the control rods affect the amount of energy a reactor produces?
5. What the primary advantages of nuclear power compared to coal?
6. What are the major drawbacks of nuclear power compared to coal?
7. What were the primary causes of the accidents at TMI, Chernobyl, and Fukushima Daiichi?
8. How is nuclear waste currently being handled in the United States? Is there currently a permanent solution to the waste issue?
9. Why have nuclear power plants aged faster than originally anticipated?
10. Describe the difference between active safety and passive safety designs.

18.7 Key Terms

Active safety
Chernobyl
Containment building
Control rods
Coolant system
Dry storage
Enrichment
Fuel rods
Fukushima Daiichi
Half-life
High-level radioactive material
Liquid fluoride thorium reactor
Manhattan Project
Meltdown
Moderator
Nuclear energy
Nuclear fission
Nuclear power
Nuclear reactor
Passive safety
Peaceful Atoms
Radioactive waste
Radioisotope
Radiation
Three Mile Island
Wet storage
Yucca Mountain

Chapter 19

RENEWABLE ENERGY

Chapter Topics

Germany is not known for its sunny climate. Despite this, the country has become a leader in solar power. Currently, it produces about 6% of its power from solar photovoltaic (PV) cells and 33.3% from renewable sources in general. Part of their success has come for the government implementing a feed-in-tariff system, which requires utilities to purchase electricity from anyone who generates it from renewable sources. As a result, many homeowners have installed rooftop solar panels and have sold any access power they generate.

Germany's Renewable Energy Sources Act of 2000 has created a market-based approach to encourage further development in renewables. In addition to solar, Germany is also growing wind power. The country is phasing out coal and nuclear power with the goal of reaching a capacity of 53 gigawatts from renewable sources. Already, 7 of its 15 nuclear power plants have been shut down since the Fukushima Daiichi accident. Germany is on track to meet its goal while having a growing economy. As a result of this commitment to renewables, Germany has reduced its greenhouse gas emissions by 25%.

RockerStocker/Shutterstock.com

Rooftop solar in Bavaria, Germany.

Jenson/Shutterstock.com

Engineers in China inspecting a PV field.

19.1 Reasons for Switching to Renewable Energy

The trend toward renewable sources has already begun. While they still make up a small percentage of our overall energy consumption, their share has been growing steadily. Projections for the next decade show that coal is expected to decline, nuclear will remain flat, but renewable sources will continue to rise.

The fastest growing sectors of renewable energy are solar and wind. Wind has been growing steadily for a while, but solar has also taken off in the past few years. Part of the reason for solar's surge has been the drop in price. The cost of PV cells has gone down significantly. In 1976, the cost of solar power was nearly $20/W and there was less than 1 MW total capacity worldwide. Today, the global capacity is over 100,000 MW and the cost is less than $1/W. Wind power has also come down significantly in recent years.

Today, both solar and wind are now cheaper means of producing electricity than either coal or nuclear. It is cheaper to build new solar and wind facilities than it is to keep a coal-powered plant operating. That was not true even a few years ago. Historically, renewables have been more expensive to the consumer than fossil fuels. Part of the reason for that has been that fossil fuels receive a great deal of government subsidies. In 2016, coal, oil, and natural gas together represented more than three quarters of the total energy subsidies in the United States. See the tables below. These subsidies have kept nonrenewable sources artificially cheap by hiding the actual costs of them from rate payers while requiring renewable energy to become competitive in an uneven playing field.

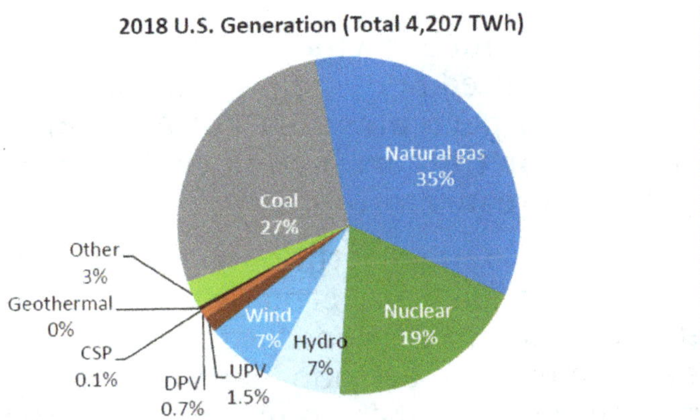

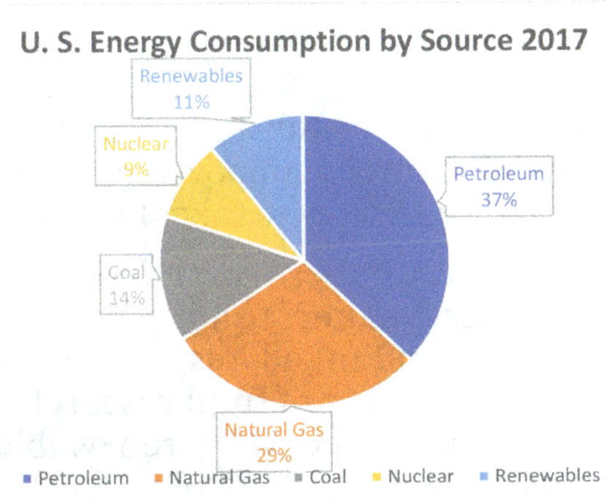

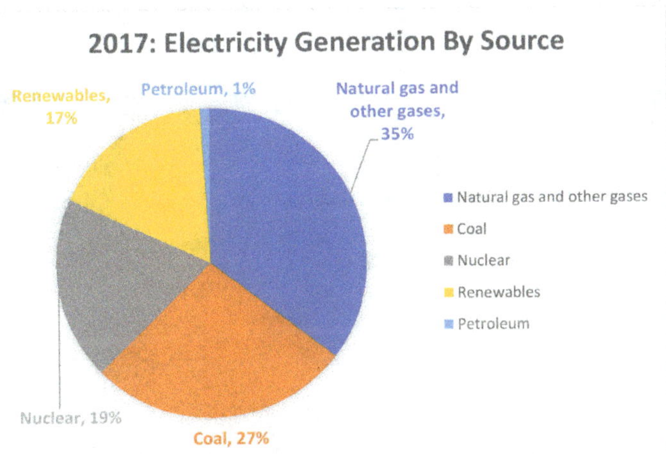

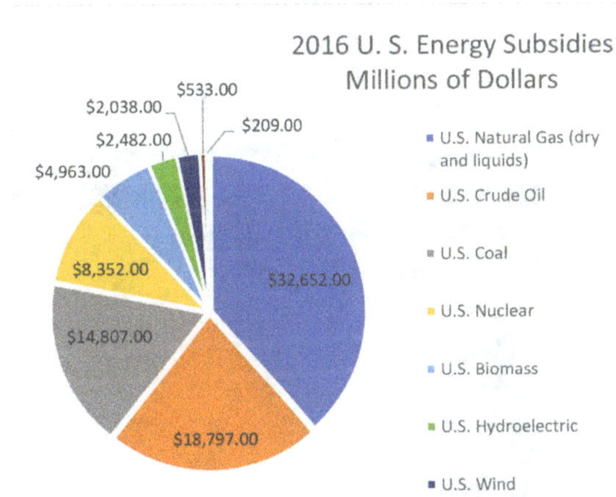

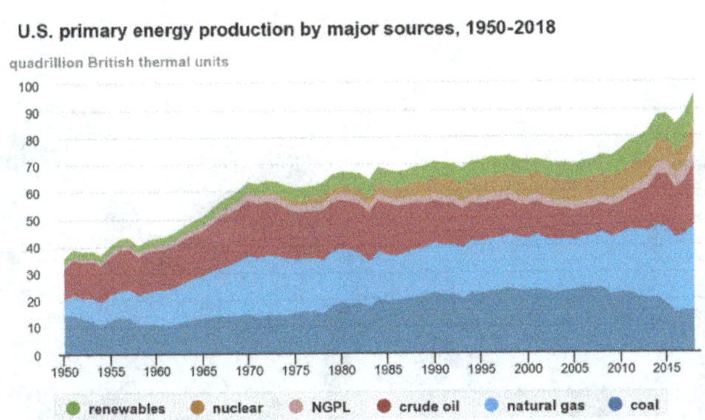

Image Source: epa.gov

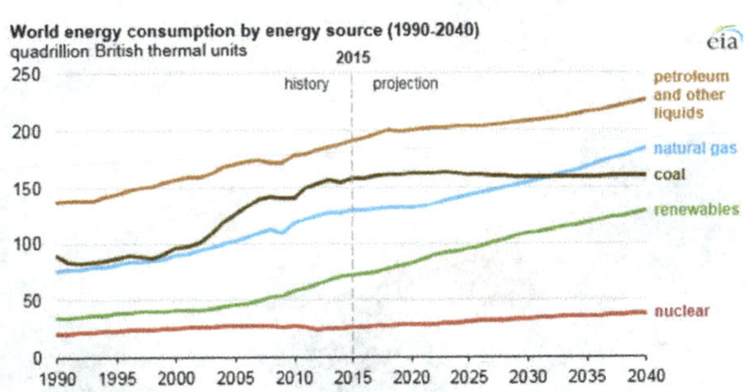

Data Source: eia.gov

The reasons nations are switching to renewable energy include a desire to reduce air pollution and greenhouse gas emissions. Nations also have good reason to diversify their energy portfolio so that they will be less vulnerable to shortages in any one source. Also, the manufacture, installation, and maintenance of renewable energy sources create **green-collar jobs**. Today, more people in the United States work in the wind and solar industry than the coal industry. Despite this, many in government today would rather try to shore up the dying coal industry rather than prepare workers for jobs in a sector that is growing.

China by far is currently the world's leader in producing wind and solar power, with the United States is in second place. The United States, however, ranks as 10th in terms of subsidies for renewable energy as a percentage of GDP. A stronger commitment to supporting renewable energy from the government could enable the United States to take the leadership role and create more jobs.

There are number of practices governments can do to support the growth of renewable energy. One is to adopt a feed-in-tariff plan modeled on Germany's. It can also set goals and mandates to reach a certain percentage of its energy from renewable sources. It can fund research and development and provide loans to start-up companies that develop renewable sources. In addition, it can offer tax incentives and rebate programs to encourage people and businesses to buy renewable energy. Finally, it could eliminate subsidies for nonrenewable sources or at least provide some equity in funding renewable sources.

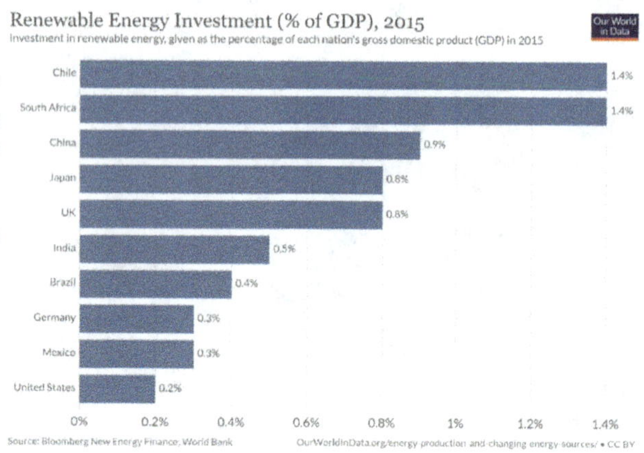

Source: Our World in Data. https://creativecommons.org/licenses/by/4.0/

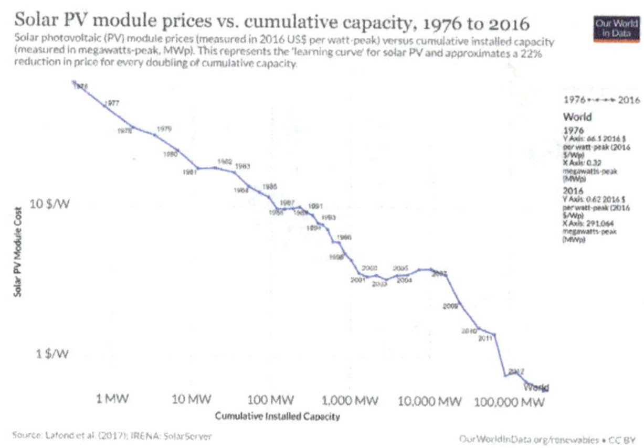

Source: Our World in Data. https://creativecommons.org/licenses/by/4.0/

Andre Nery/Shutterstock.com

only_kim/Shutterstock.com

19.2 Solar Energy

Solar energy refers to harnessing energy directly from the sun. While fossil fuels represent energy from the sun that was stored in the distant past, solar energy involves capturing and using the solar radiation reaching the Earth today. Each day, Earth receives enough solar energy to meet all of our energy needs for 25 years. We just need more efficient ways to collect and utilize it. Finding cost-effective means of achieving this has been a long-term goal for the past several decades.

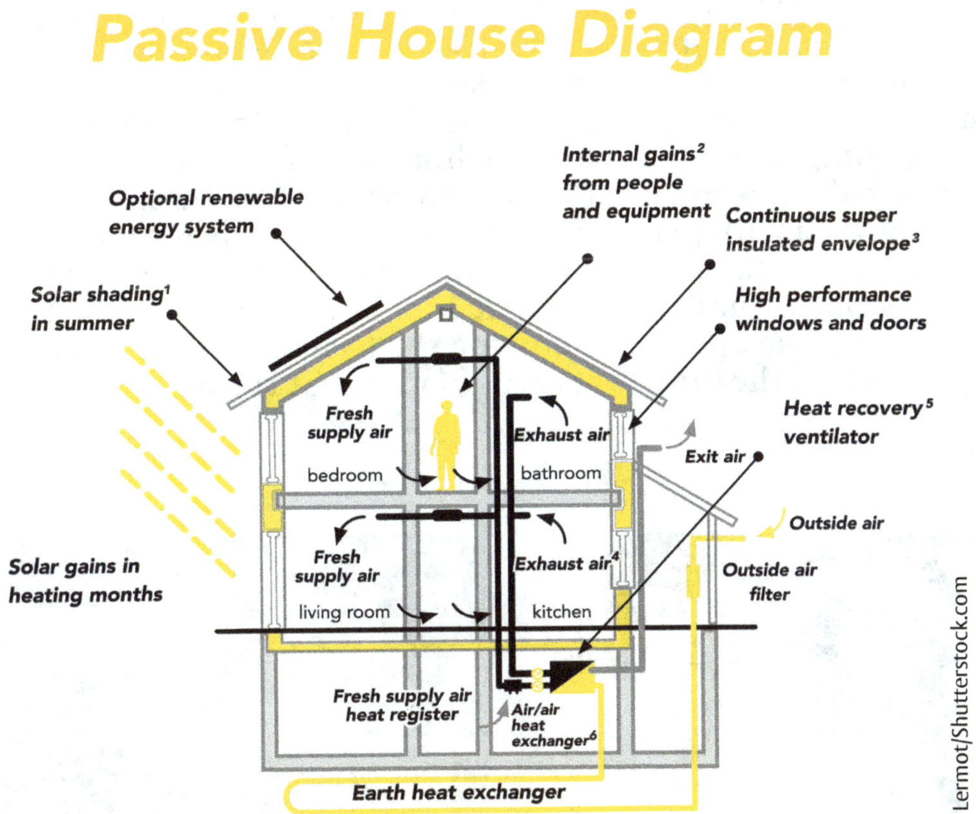

Passive solar house design.

Passive solar energy collection involves using building designs and materials to make heating and cooling them more efficiently. For example, overhangs could provide shade in the summer months while south-facing windows can help warm the building in the winter. Thermal mass materials can be including the construction to absorb heat during the day and slowly release it at night, keeping the home warm throughout the night. Trees and other plants can be used to provide temperature buffers for the home.

Solar water heater.

On the other hand, **active solar energy collection** involves technology to capture, transport, and store the sun's energy. **Flat-plate solar collectors** are dark-colored materials that can absorb the sun's energy and focus it onto pipes full of water. The water heats up as it flows through the collection and then travels into a hot water storage tank for use in a home or commercial building. **Solar cookers** are portable mirrors designed to focus the sun's rays to cook a pot of food. Concentrated solar power involves collecting solar radiation over a wide area and focusing it onto a pipe or a single point. For example, it could involve parabolic mirrors that focus the sun's rays onto a series of pipes. A water or oil would then flow through the pipes, absorbing the heat, which is then used to run a turbine. **Solar power towers** use a set of **heliostats**, or sun-following mirrors that focus the sun's rays onto a single tower. Inside the tower, water is heated up to boiling in order to run a turbine. One form of solar thermal technology that is still experimental is the inverted solar funnel. These involve covering a large area with plastic covers and allow the heat to beat down on it, heating the air under the plastic cover. In effect, this creates a large open-air greenhouse. As the air under the plastic hits up, it flows toward the inverted funnel and goes up the opening following the basic principle of hot air rising. As the air passes into the funnel, it turns blades that run a turbine and produce electricity.

Gena Melendrez/Shutterstock.com

Solar cooker.

Novikov Aleksey/Shutterstock.com

Solar power tower.

Jenson/Shutterstock.com

Concentrated solar power using parabolic mirrors.

javarman/Shutterstock.com

Solar power tower in Spain.

Of course, the form of solar energy that most people think of is the use of **PV cells**, which are semiconductor devices that create an electric current when light strikes them. PV cells are widely used in toys, pocket calculators, and satellites. They can be put together into panels and the panels assembled into arrays that can produce large amounts of electricity. The more panels in the array, the more potential electricity can be generated.

Thin-film solar cells are an ultra-thin, flexibility of PV cells that can be used to generate electricity. While they are currently less efficient than rigid PV cells, their flexibility enables them to be used in more locations, such as being incorporated into roofing tiles or embedded into asphalt, converting highways into solar power plants.

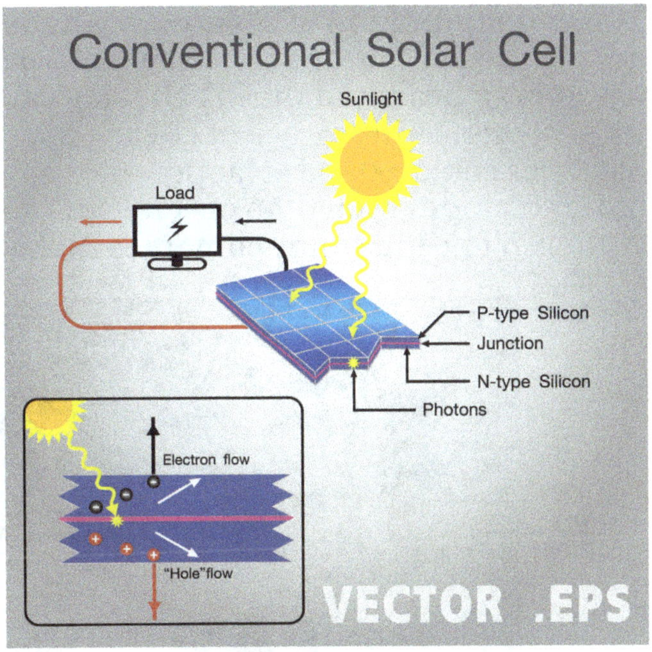

Studio BKK/Shutterstock.com

Solar PV cells use light to generate an electric current.

Solar power has a number of advantages. For one, it is unlimited. The sun is expected to continue shining for another 4 or 5 billion years, at which time it will have exhausted its usable hydrogen fuel and swell up into a red giant and engulf the Earth. Until then, it will continue to put out energy and there isn't anything we could do that would exhaust it. It doesn't involve any fuels that emit greenhouse gases or other pollutants. It has no moving parts so there is little maintenance required. It creates jobs and it is scalable. Large arrays could produce several megawatts of electricity for a large city or just a few panels could be used to provide electricity to a remote village that is too far away from a country's grid.

Jenson/Shutterstock.com

A solar PV array

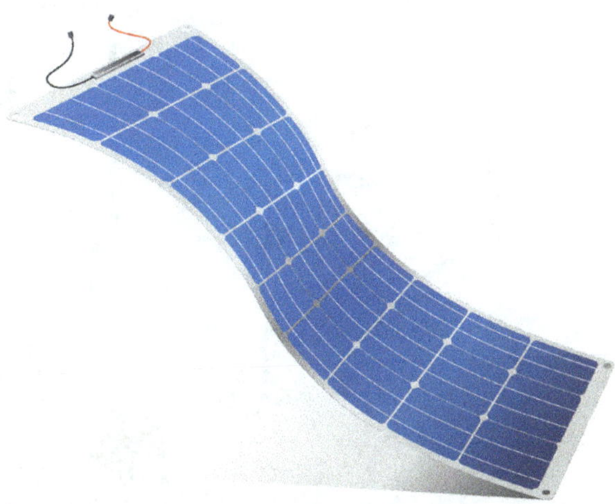

Alejo Miranda/Shutterstock.com

Thin-film solar cells can be incorporated into roofing tiles.

The main limitation of solar power, is of course, the fact that the sun only shines during the day, but that is when most people are awake and consuming electricity. Energy produced during the day can be stored in batteries for use at night. Improvements in storage of electricity, such as more efficient and longer-lasting batteries could extend solar power's capacity even more. Also, some areas receive more solar radiation than others, but Germany stands as an example that a country doesn't need a sunny climate to make solar power work.

Andrey Armyagov/Shutterstock.com

The International Space Station with its solar panels that power all of its electrical systems.

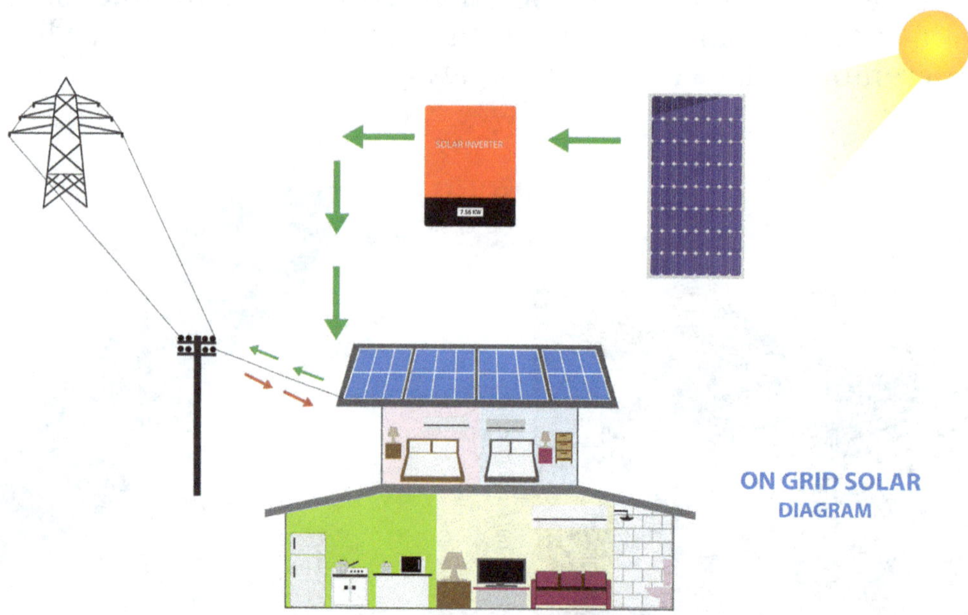

Keepsmiling4u/Shutterstock.com

Net metering enables solar power users to sell their excess electricity back to the grid.

While solar power does involve large up-front costs for installation, these costs have been coming down steadily and most solar power plants eventually pay for themselves.

Lano Lan/Shutterstock.com

Solar panels in Malaysia.

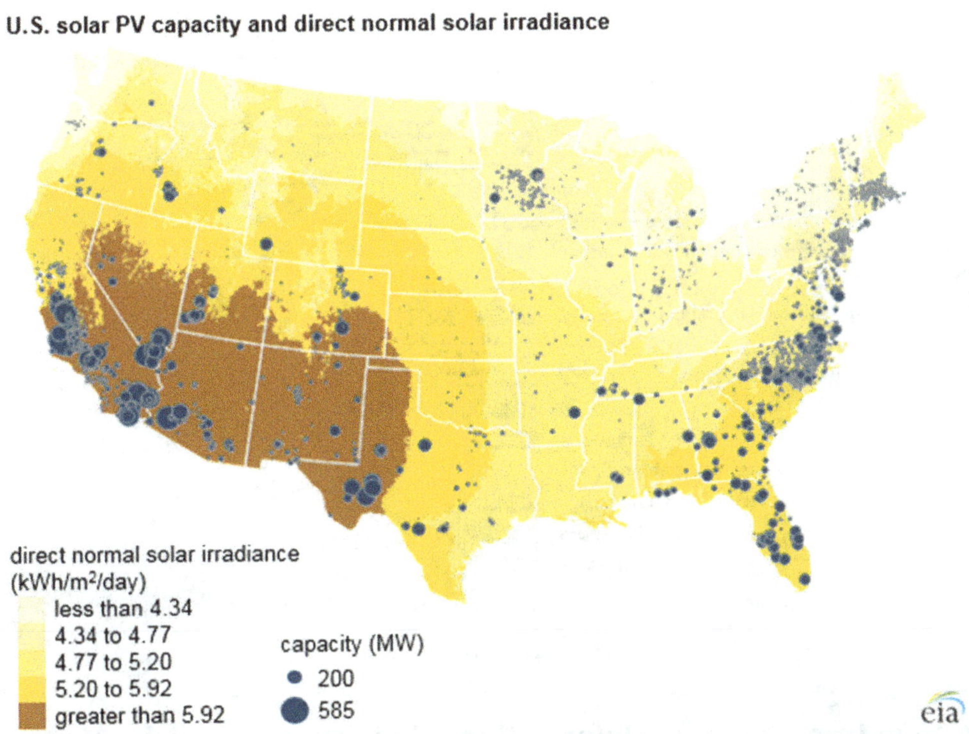

Source: US Energy Information Administration,

Preliminary Monthly Electric Generator Inventory, and National Renewable Energy Laboratory solar resource maps.

Note: kWh/m²/day is kilowatt hours per square meter per day; MW is megawatt.

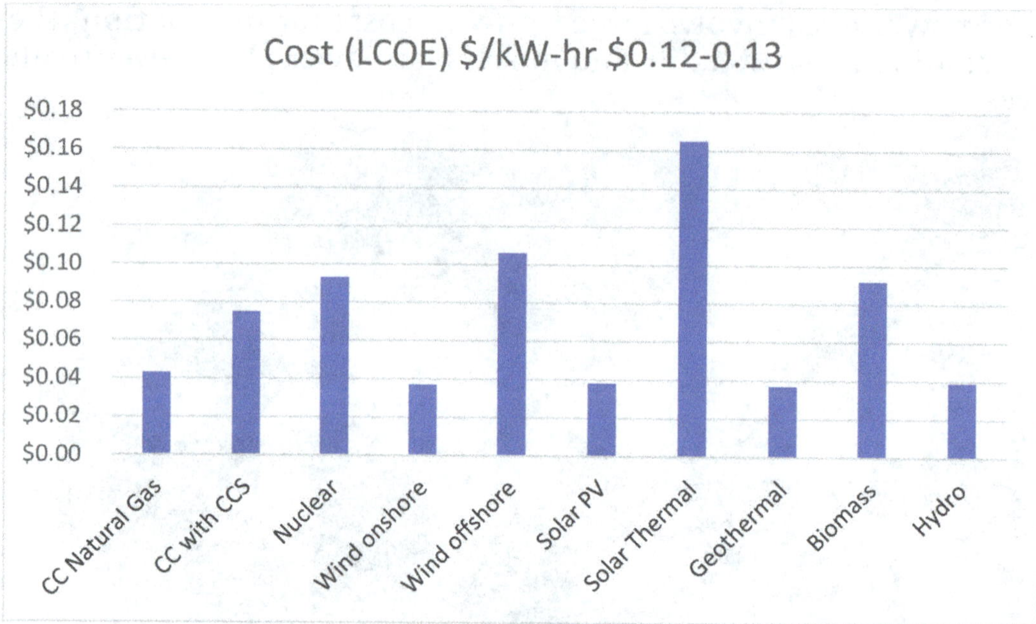

Levelized cost of electricity (costs including facility construction and operating costs) for energy shows solar PV and onshore wind have lower costs than coal (CC with CCS) and nuclear power.

Data Source: doe.gov.

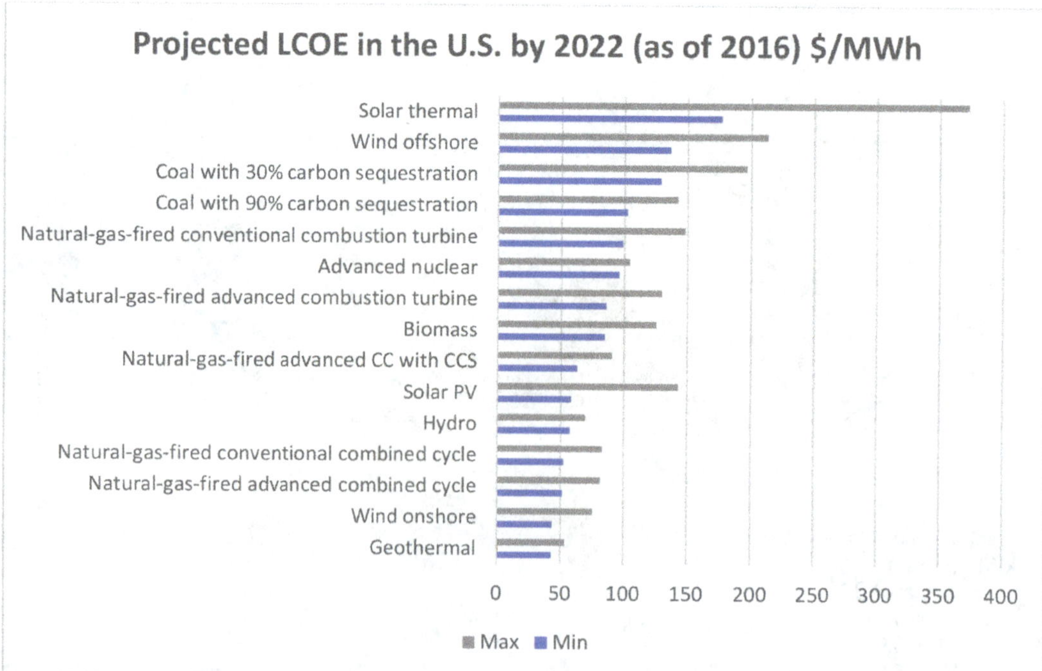

Projections from the Energy Information Administration indicate that solar PV and onshore wind power will continue to be cheaper than coal and nuclear power. *Data Source:* eia.gov.

19.3 Wind Energy

Wind power is a form of indirect solar energy that captures the motion of wind. The wind currents form from differently heated air masses. A **wind turbine** uses a set of blades to power a turbine inside of the **nacelle**. Wind turbines can be put together in groups called **wind farms**. Wind speed has a great influence on the amount of electricity generator by a wind farm. Doubling the wind speed results in an increase in electricity generation by a factor eight.

Jacques Tarnero/Shutterstock.com

Wind turbine.

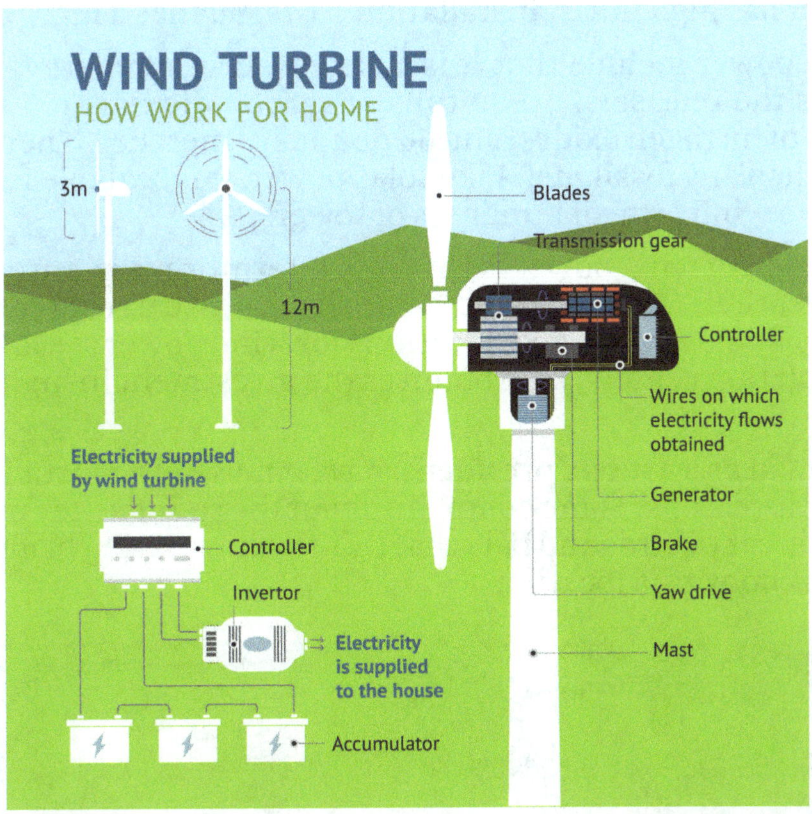

TatyanaTVK/Shutterstock.com

The top three nations in wind power production are China, the United States, and Germany. Within the United States, the top producing states are California and Texas. Not all locations have the same wind potential. Mountaintops and plains are generally the best locations on land. Off shore wind farms also have great potential as wind speeds over water are greater with less turbulence. Currently, the United

A wind farm.

States has one off shore wind farm off of Nantucket Island in Massachusetts. Like solar, wind power has high initial installation costs, but has a high profitability.

Benefits of wind power include that it is a clean source of energy. A 1-MW wind turbine prevents the release of 1,500 tons of carbon dioxide; 6.5 tons of sulfur dioxide; 3.2 tons of nitrogen oxides; and 60 pounds of mercury. They have an EROI of 20:1, greater than any fossil fuel. Like solar power, it is scalable from one single wind tower up to a wind farm of hundreds of towers.

The main limitation of wing power is that it is intermittent in nature. Some days are windier than others. Efficient storage of electricity can overcome this problem. Another is that most of the best locations for wind power are located far away from major population centers, so it requires the investment in high-voltage lines to transmit it.

Many residents (and at least one president) have opposed the installation of wind farms for aesthetic reasons. Some complain about the sounds they make. They are also a potential threat to birds and bats that fly into the rotating blades, so care has to be made when choosing a location.

Offshore wind farms have great potential.

A geothermal power plant in Iceland.

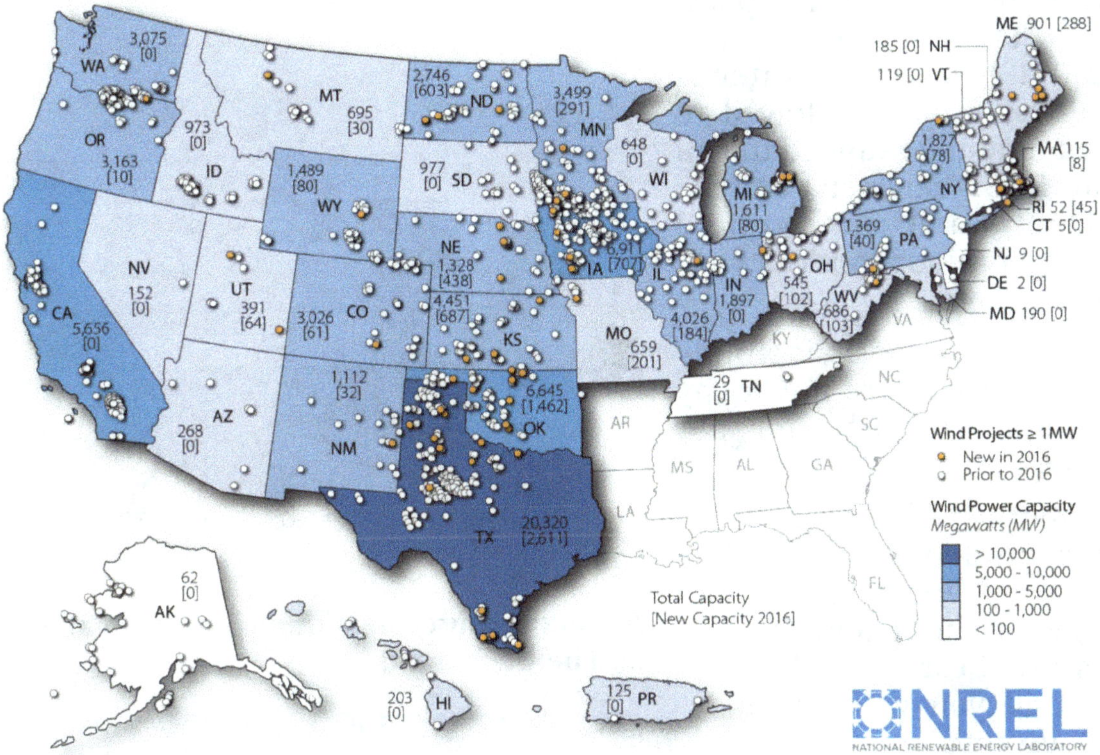

Stefano Ember/Shutterstock.com

Wind capacity in the United States.

Image Source: doe.gov.

19.4 Geothermal Energy

One of the biggest challenges in generating electricity is finding a reliable heat source to produce the steam needed to run the turbines. **Geothermal power** takes advantage of locations that are on top of hot spots, where pockets of magma are close enough to the surface to boil the water in an aquifer. This triggers geysers of water and steam that shoot out of the ground. A geothermal plant captures this steam, uses it to run a turbine. Then, the water is cooled and reinjected into the aquifer.

Geothermal power is renewable so long as the aquifer is replenished. It might not be renewable if the water is removed faster than the aquifer could be recharged. Some geothermal plants have taken to injecting wastewater into the ground to replace any that is lost in the process.

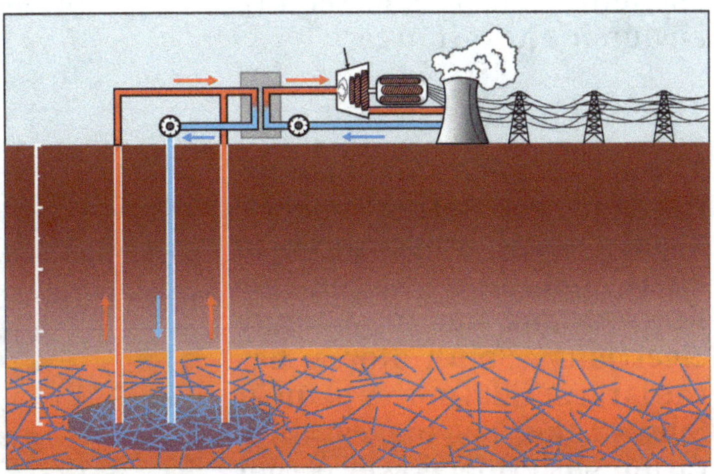

Studio BKK/Shutterstock.com

Geothermal power plant schematic.

A variation of a geothermal plant is an **enhanced geothermal system** (EGS), in which water is injected into ground, allowed to be heated, and then tapped to run a turbine. EGS is expensive and the injection of water underground might trigger minor tremors.

The biggest limitation on geothermal power, however, is simply the fact that there are not many locations on Earth with a hot spot that can be utilized.

Another method of getting energy from the ground is a **ground-source heat pump**. This takes advantage of the ground's ability to store heat. Dig deep enough, and you will find a layer of ground that is the same temperature all year round; warmer than the surface in the winter and cooler than the surface in the summer. A ground-source heat pump uses a fluid to transport heat underground to cool a building in the summer and then pump it up to the surface to warm it in the winter. These kinds of heat pumps are very efficient, as it doesn't require an outside heat source.

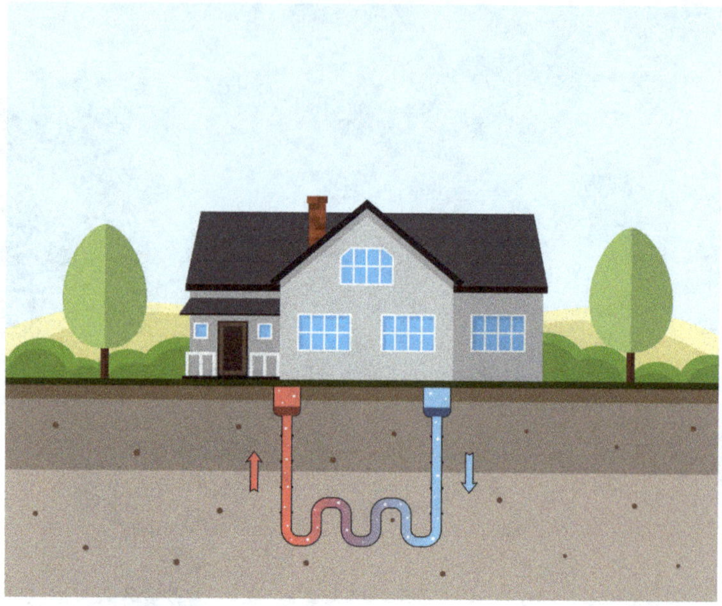

all_is_magic/Shutterstock.com

A ground-source heat pump moves heat above and below ground.

Alex Mit/Shutterstock.com

A tidal energy plant.

19.5 Ocean Energy

The ocean holds a lot of potential for generating energy. **Tidal energy** uses a series of blade that rotate as the tide flows in and out to generate electricity. South Korea has the largest tidal power plant at Sihwa Lake. The biggest limitation with tidal power is that it could interfere with boat traffic or fish migration.

Oleksandr Derevianko/Shutterstock.com

Wave energy.

There are also plans to take advantage of the up-and-down motion of waves to run a turbine. **Wave energy** has a lot of potential for the future. Ocean currents like the Gulf Stream can also be tapped as a source of electricity. Already, there are turbines off the coast of Europe tapping the motion of the currents.

Ocean Thermal Energy Conversion (OTEC) can take advantage of the difference in temperature between warm surface waters and the cool deep waters. Warm water from the surface is used to boil ammonia or some other fluid with a low boiling point to produce a vapor to power a turbine. Cold water from the depths is pumped up to recondense the fluid. There are a few experimental OTEC stations in places like Hawaii where the waters are very warm. However, this technology is still experimental and is not yet profitable enough for commercial applications.

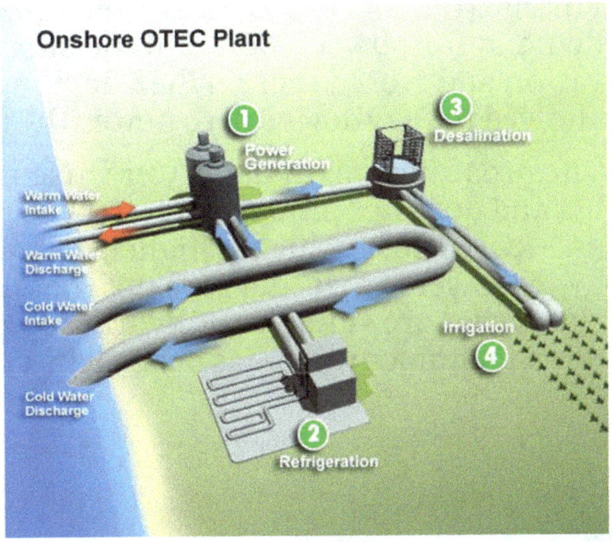

Source: oceanexplorer.noaa.gov.

An OTEC plant schematic.

Flaxphotos/Shutterstock.com

Mystic Lake in Montana has two hydroelectric plants.

19.6 Hydroelectric Energy

Hydroelectric power or hydropower is one of the oldest renewable sources of electricity. It takes advantage of the kinetic energy of the flow of a river to run a turbine. The most common technique is using a dam to build up a reservoir. Some of the water is allowed to rush through the dam, which hits a series of blades to run the turbines.

Another technique is the **run-of river** approach, which diverts some of the water from a river and allows it to flow through a powerhouse where the turbines are located. This method is less disruptive to the natural flow of the river, but produces less power.

Finally, the **pumped-storage** technique is a method used to store energy. When demand for power is low and the supply is high, water is pumped up to high elevation and stored there. When demand is higher and supply is lower, the water is allowed to flow downhill to power the turbines.

The pros and cons of dams were discussed in Chapter 13. Hydropower is renewable and does not produce greenhouse gases or pollutants during operation. It has an EROI of 80:1, which is higher than any other source. The biggest limitation for hydropower is that there are not that many places to build a new dam. Most major rivers that are suitable for hydropower already have a dam built across them. So, it is a technology that has been all but maxed out with little potential for growth.

Andrea Danti/Shutterstock.com

A schematic of a hydroelectric dam. Water flows through the dam, hitting blades that spin a turbine to produce electricity.

Roman Zaiets/Shutterstock.com

Turbine inside a hydroelectric damn.

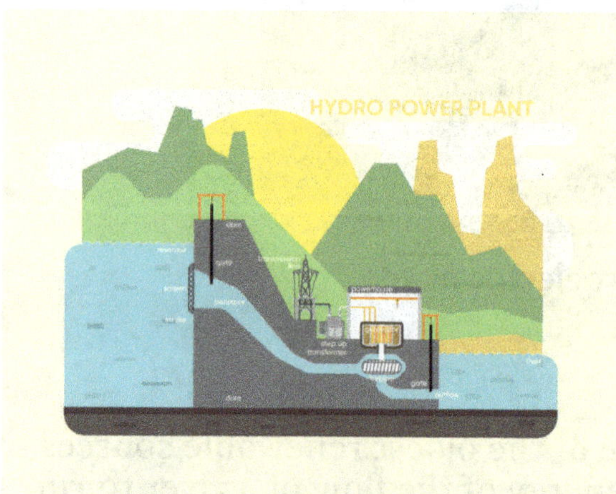

CW craftsman/Shutterstock.com

Inside a hydroelectric plant.

Ikonya/Shutterstock.com

The Tarraleah Power Station is a pumped-storage hydroelectric power station in Tasmania, Australia.

Image Source: energy.gov.

A run-of-river hydroelectric plant.

19.7 Bioenergy

Bioenergy refers to energy produced from **biomass**, which is organic material derived from living or formally living organisms. In theory, bioenergy should be carbon neutral. Any carbon dioxide emitted from burning the biomass should be offset by photosynthesis produced in creating it. In practice, there are many factors that could affect whether or not bioenergy actually is carbon neutral.

Biomass can be fuel gathered directly for heating and cooking, such as firewood, charcoal, and animal manure. **Biopower** refers to biomass produced for generating electricity. This can include crop residues, forestry residues, processing wastes from pulp mills, landfill gas, and biogas produced from animal manure. **Co-firing** combines biopower such as wood chips or pellets with coal to reduce emissions. **Gasification** converts biomass into a vapor at high temperatures and uses it to run a turbine. Finally, pyrolysis converts biomass into a liquid fuel at high temperatures that can be burned to generate electricity.

Biofuels are liquid fuels produced from biomass for powering vehicles. One of the most common biofuels is ethanol, which is usually produced from corn or bagasse (sugarcane residue). Research is ongoing to produce ethanol from cellulose. **Biodiesel** is another biofuel, which could be produced from used cooking oil, certain plants like soybeans and rapeseed, and algae.

Jam Travels/Shutterstock.com

Woman in Ecuador gathering firewood. Many developing countries still depend on firewood for heating and cooking, which can lead to deforestation, erosion, and desertification.

MariaKGraphicDesigner/Shutterstock.com

Biofuel sources.

Stefano Ember/Shutterstock.com

A pellet stove burns wood that has been compressed into dense pellets. These pellets produce heat more efficiently than conventional wood burning.

PhotographyByMK/Shutterstock.com

A biogas digester producing methane from animal manure.

francesco de marco/Shutterstock.com

Pipes for collecting landfill gas.

ThamKC/Shutterstock.com

The most common source of ethanol fuel in the United States comes fermented sugars derived from corn.

Ethanol is the same form of alcohol produced in beverages like beer and wine. It can be produced by fermenting carbohydrates like sugars. In the United States, most ethanol fuel is produced by fermenting sugar from corn. Most gasoline contains about 10% ethanol. Government regulations mandate ethanol in gasoline to reduce emissions. Corn-based ethanol is subsidized by the federal government and is politically popular, but is it carbon neutral and can it replace gasoline entirely? The answer to the first question depends on how the corn is grown. If lots of diesel-powered machinery is used, growing the ethanol could consume more fossil fuels than it displaces. Also, it is unlikely to completely replace gasoline in the United States. In order to produce enough ethanol from corn to meet our transportation needs, we would need to devote four times as much land to grown corn as we currently have. Corn-based ethanol also has a low EROI of just 1.3:1, among the lowest of any energy source.

Most cars are not designed to run on fuel blends with more than 10% ethanol. **Flexible-fuel or flex-fuel cars**, however, can run on ordinary gasoline or fuels with a higher concentration of ethanol. Most cars sold today in Brazil are flex-fuel and can run on a blend called E-85, which is 85% ethanol and 15% gasoline. Brazil produces ethanol from **bagasse**, which is crushed sugarcane residue. About half of the fuel sold in Brazil is now E-85. Unfortunately, sugarcane does not grow well in the temperate climate that makes up most of the United States. The only states that can produce sugarcane are Hawaii and Florida.

iPreech Studio/Shutterstock.com

Bagasse that is used to make ethanol.

PM8899/Shutterstock.com

Flex-fuel vehicles can run on E85 blend.

Zety Akhzar/Shutterstock.com

Even older diesel-powered vehicles can be converted to run on biodiesel and the truck will smell like french fries!

So, is there a source of ethanol that can be grown in the United States that can meet our transportation needs? Ethanol can be produced by fermenting any carbohydrate, including cellulose. A lot of research continues to find a way to make **cellulosic ethanol** a viable source. The advantage of cellulosic ethanol is that it can be produced from the parts of plants that have no food value. It can be made from the corn stalks or even from nonfood crops like switchgrass. There's more cellulose than sugar per volume in plants, so we can theoretically produce more ethanol per acre with cellulose than with sugars.

Biodiesel can be produced from used cooking oil, such as from the fryers at fast-food restaurants. It can also be produced from animal fat or from oils extracted from plants. Biodiesel has several advantages. It produces less emissions combined to petroleum-based diesel with a nearly identical fuel economy. It is also cost-effective, nontoxic, and biodegradable. It can also be produced by recycling food waste. Most diesel-powered vehicles can run or be modified to run on biodiesel.

Certain species of algae produce lipids that can be converted into biodiesel. Algae can also be used to produce ethanol. The advantage algae have over other crops is that it can be grown in large batches inside tanks or **photobioreactors**. Algae also grow very quickly, enabling near-continuous harvesting of the biofuel.

In principle, biomass can be carbon neutral. In practice, it may not be if lots of fossil fuels are burned to produce it. Also, if forest lands are cleared to grow it, it may result in a net increase in carbon dioxide emissions. As a result, most international agreements do not give credit for biofuels toward reducing greenhouse gas emissions. Critics of biofuels like corn-based ethanol also point out that we are taking food that could feed starving people and convert it into fuel.

Mr Doomits/Shutterstock.com

Many vehicles can be modified to run directly on used cooking oil.

phloxii/Shutterstock.com

Algae grown in photobioreactors can be used to biofuels.

Olexandr Panchenko/Shutterstock.com

Switchgrass and other crops high in cellulose hold the potential to produce lots of ethanol.

19.8 Hydrogen Fuel

Hydrogen is the most common element in the universe, which is highly reactive, and holds a lot of potential energy. Most plans for using hydrogen as a fuel do not involve combustion, but fuel cells. A fuel cell generates electricity by reacting hydrogen with oxygen. The hydrogen enters the cell from one side and stripped of its electrons. The electrons travel through wires, producing the electric current, while the hydrogen ions rejoin with oxygen to form hydroxyl ions and pass through a membrane. On the other side of the membrane, the hydroxyl ions meet up with their electrons to produce water vapor.

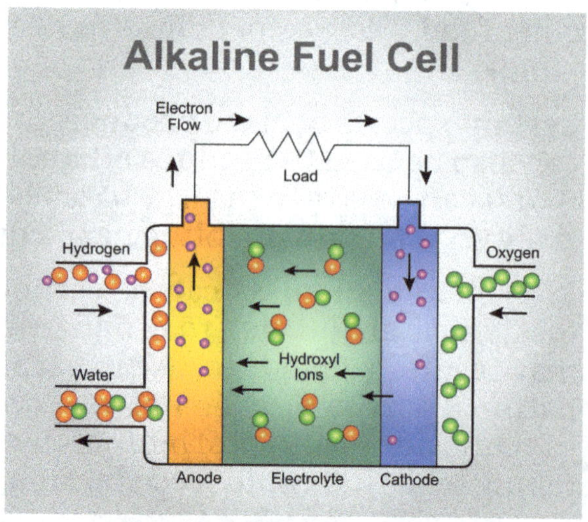

Studio BKK/Shutterstock.com

Hydrogen fuel cells produce electricity by reacting hydrogen with oxygen.

Hydrogen fuel cells are quiet and nonpolluting. Their only by-product is water vapor. Many cities have begun converting their bus fleets to run on hydrogen fuel cells.

The main limitation to hydrogen fuel is that, while the element hydrogen is abundant on Earth, hydrogen gas is rare. Hydrogen gas is so light that it eventually escapes the Earth's gravity to space. As a result, most of the hydrogen atoms on Earth are locked up in water. **Electrolysis** uses an electric current to separate water molecules into hydrogen and oxygen. Hydrogen is therefore considered more an energy carrier than an energy source. The electricity needed to produce hydrogen gas could come from an intermittent source like solar or wind. Then, the gas would be liquefied and transported to fuel stations across the country. Liquefied hydrogen gas is just as safe to transport and use as the combustible fuels used to power vehicles today.

ELECTROLYSIS

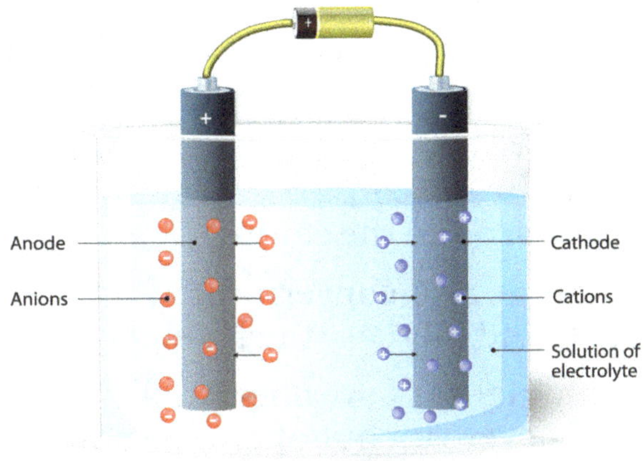

Designua/Shutterstock.com

Electrolysis used to separate hydrogen and oxygen from water.of California.

Rhonda Roth/Shutterstock.com

Hydrogen fuel cell bus at the University of California.

Rob Crandall/Shutterstock.com

Hydrogen fuel station in Washington, DC.

Takashi Images/Shutterstock.com

Fuel cell in a Toyota Mirai.

How clean is hydrogen? Fuel cells do not emit carbon dioxide and we could conceivably have unlimited quantities of hydrogen gas produced from water since the fuel cells convert it back into water. However, if fossil fuels are burned to produce the electricity used in electrolysis, it defeats the purpose of using it. Hydrogen gas can also be produced through electrolysis of methane, but the by-product is carbon dioxide which, again, defeats the purpose.

Converting to hydrogen fuel would also require a huge investment in our infrastructure to convert gas stations to supplying pressurized hydrogen and distributing around the country.

There is also concern that if space hydrogen escaped it could damage the ozone layer and increase the residence time of greenhouse gases like methane.

Still, fuel cells are energy efficient and make use of an abundant element that we will never run out of and do not emit any of the pollutants associated with fossil fuels.

19.9 Discussion Questions

1. How has Germany become a major producer of solar energy?
2. Why have renewable sources of energy been at a competitive disadvantage compared to fossil fuels?
3. Why are renewable sources of energy growing around the world?
4. What are green-collar jobs?
5. How does passive solar energy collection improve the energy efficiency of buildings?
6. What are the advantages and disadvantages for solar power?
7. What are the best locations for wind farms?
8. What are the advantages and disadvantage of wind power?
9. Why do some people object to wind farms being located in their community?
10. Why is geothermal power only available in certain locations?
11. Why is hydroelectric power unlikely to expand further in the near future?
12. Why is corn-based ethanol unlikely to replace gasoline entirely?
13. How is Brazil able to produce enough ethanol for use in flex-fuel cars?
14. Are biofuels always carbon neutral? Why or why not?
15. Why is hydrogen considered an energy carrier instead of an energy source?
16. What are the advantages of hydrogen fuel cells over fossil fuel-powered vehicles?

19.10 Key Terms

Active solar energy collection
Bagasse
Biodiesel
Bioenergy
Biofuel
Biomass
Biopower
Cellulosic ethanol
Co-firing
Electrolysis
Enhanced geothermal system
Ethanol
Flat-plate solar collectors
Flexible-fuel vehicles
Fuel cells
Gasification
Geothermal power
Green-collar jobs
Ground-source heat pump
Heliostats
Hydroelectric power
Nacelle
Ocean power
Ocean thermal energy conversion
Passive solar energy collection
Photobioreactors
Photovoltaic cells
Pumped-storage
Run-of river
Solar cooker
Solar energy
Solar power towers
Thin-film solar cells
Tidal energy
Wave energy
Wind farm
Wind power
Wind turbine

Chapter 20

WASTE MANAGEMENT AND HAZARDOUS WASTE

Chapter Topics

20.1 The Nature of Municipal Solid Waste
20.2 Sanitary Landfills and Incinerators
20.3 Recycling and Reuse
20.4 Industrial Solid Waste
20.5 Hazardous Waste
20.6 E-Waste
20.7 Discussion Questions
20.8 Key Terms

Vince360/Shutterstock.com

Recycle Mania banner at the University of Arizona.

Our modern society is often labeled a throw-away-society. We generate a lot of waste. Some of it comes from the excessive package and some of it comes from plastic shopping bags. We buy a lot of single-use and disposable products and then throw them away. Plastic water bottles, fast food wrappers, single-use coffee products, and disposable diapers now fill our landfills. Decades ago, people could make a living as a TV repairman or fixing vacuum cleaners. Today, it is often easier and cheaper to buy new products rather than try to repair them. We through them away even if parts can be salvaged.

RecycleMania is a multi-college competition in which various schools compete to see which can recycle the most waste. It began in 2001 as a contest between Ohio University and Miami University. In 2004, RecycleMania partnered with the Environmental Protection Agency (EPA)'s WasteWise program to produce a website (recyclemania.org) and further promote recycling. Today, over 1,000 schools in the United States and Canada have competed in the annual competition. The 2018 competition involved over 300 schools and 3.6 million students. It is credited with preventing 2.37 million metric tons of carbon dioxide.

Still, the amount of waste generated continues to grow with population and a long-term solution to reduce our waste stream are needed.

Waste refers to anything produced by human activity that is no longer wanted. There are several categories of waste. For example, we discussed the treatment of wastewater in Chapter 13. This chapter will discuss four other categories of waste:

Municipal Solid Waste: Any nonliquid waste that is produced by homes, small businesses, and institutions like colleges.

Industrial Solid Waste: Waste produced from industrial activities such as manufacturing, agriculture, mining, and the production and refining of petroleum.

Hazardous Waste: Solid or liquid waste that is toxic, corrosive, chemically unstable, or flammable.

E-Waste: Used electronic devices, including cell phones, tablets, laptops, and computer monitors.

20.1 The Nature of Municipal Solid Waste

Municipal solid waste, aka, trash or garbage is waste collected from homes and small businesses. It gets its name because municipal governments are responsible for the collection and disposal of the waste produced by its residents. Cities and towns may have a sanitation department to collect the waste or they may hire a private contractor to do the collection. Municipal solid waste consists of a variety of materials, including paper, plastics, metals, glass, food scraps, and yard trimmings.

Cary Kalscheuer/Shutterstock.com

Automated trash collection vehicle.

robuart/Shutterstock.com

MarcoVector/Shutterstock.com

Sorting waste for recycling.

Every community has to deal with its **waste stream**, the flow of products from production to disposal. The practice of **waste management** has three major components. The first is **source reduction**, or avoiding producing the waste in the first place. This is the most desirable practice. If we don't generate the waste, we don't have to worry about how to dispose of it. The second component is finding materials that can be reused and recycled. The final component is safe and sanitary methods of disposal.

The movement of waste from the initial production of the material to disposal is referred to as "cradle-to-grave." Tracking the life cycle of products can often be used to determine places where waste is generated and possibly discover how to reduce the waste or find ways to reuse and recycle it.

The amount of waste generated in the United States has tripled since the 1960s, reflecting the growth in the population as well as an increasing reliance of nondurable or disposal goods. However, this trend stopped around 2005, when source reduction and reuse practices began to have an impact and the rate of waste production has remained flat since then.

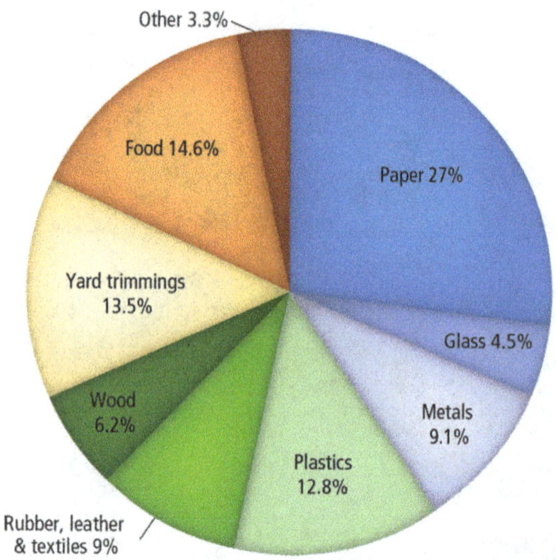

Municipal solid waste composition.

Source: epa.gov.

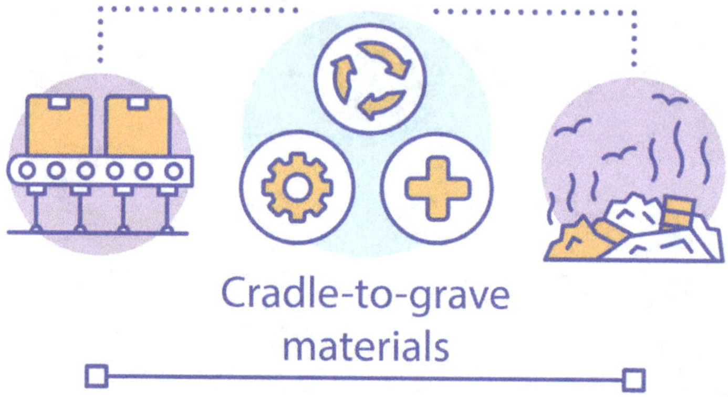

bsd/Shutterstock.com

Cradle-to-Grave tacking of waste can help us find ways to reduce waste and become more efficient.

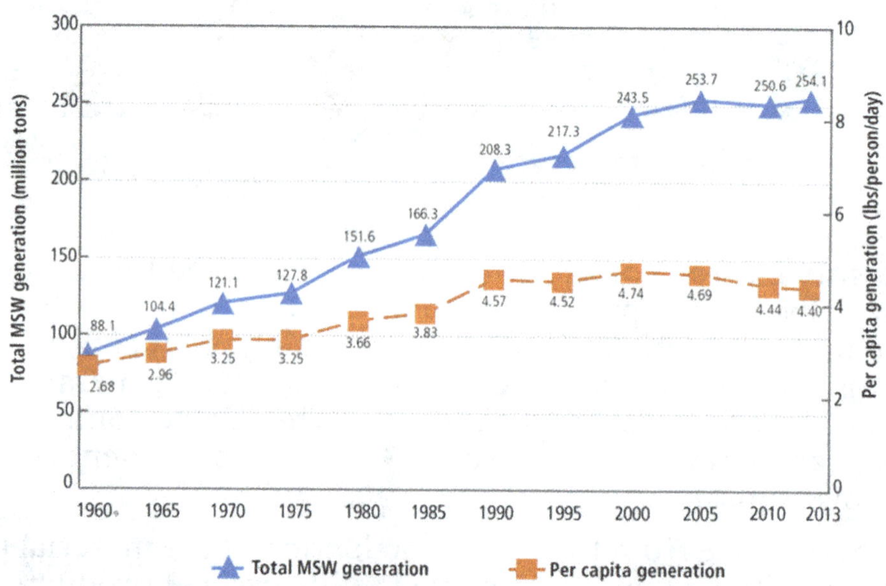

Source: epa.gov.

MSW generation rates 1960–2013.

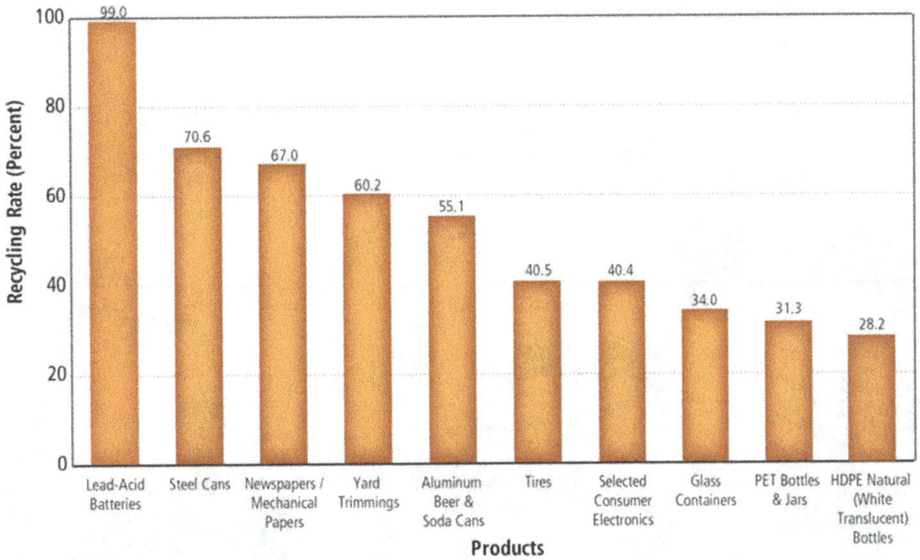

Source: epa.gov.

Recycling rates in 2013.

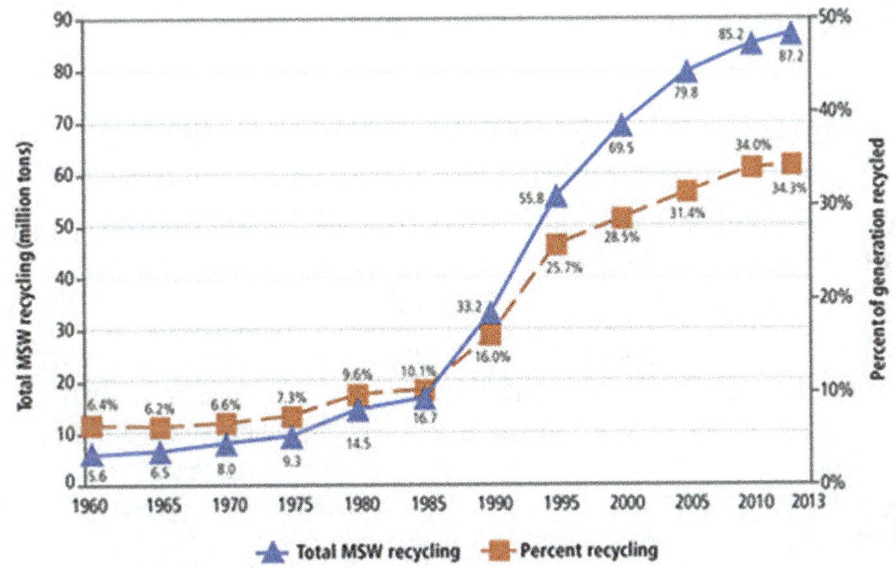

Source: epa.gov.

MSW recycling 1960–2013.

As people in developing nations have become more affluent, they have adopted many of the same wasteful practices of developed countries. As a result, consumption and waste production rates have both increased. The waste generated is often taken to open dumps that are often a breeding ground for pests and disease vectors. One byproduct of this is that poor people have become scavenging these dumps looking for items that they can either use or sell. Not only can this result in the spread of disease, but in some cases, piles of trash have collapsed,

Fehmiu Roffytavare/Shutterstock.com

People scavenging at a landfill in Vietnam. Many poor people scavenge these landfills to find things they can use or sell.

burying people alive. In response, some countries have made efforts to eliminate scavenging.

Developed countries have worked to improve the collection and disposal of waste as well as finding ways to reduce the amount of waste generated. As a result, recycling and reuse programs have increased.

Excessive packaging has become a major source of waste. This can be reduced by purchasing unwrapped produce or buying more products in bulk. Plastic shopping bags are another major component of the modern waste stream. Some local governments have become taxing or banning them in order to encourage people to use cloth bags that can be reused. Some have even started banning plastic straws in favor of biodegradable paper straws.

Some other strategies for reducing waste include:

- Double-sided photocopies
- Using electronic communications instead of paper
- Reuse cardboard boxes, foil, plastic containers, bags, fabric, packaging material, and wrapping paper
- Buy durable goods instead of disposal ones
- Buy rechargeable batteries
- Donate old clothing to charity
- Select goods with less packaging
- Compost kitchen and yard waste

Cari Rubin Photography/Shutterstock.com

Much of our waste today comes from shipping boxes and packaging.

DorotaM/Shutterstock.com

Excessive packaging is a major source of unnecessary waste.

maramorosz/Shutterstock.com

Cloth shopping bags can reduce waste from plastic bags.

Most municipalities have begun restricting yard wastes from the regular waste stream. They may use separate pickup times or encourage **composting**. Bacteria or earthworms can be used to convert yard waste and food waste into a mulch similar to humus that can be applied as a garden fertilizer. Some municipalities will collect the community's leaves, compost them, and allow residents to collect the mulch for home use.

Some communities have adopted a "**pay-as-you-throw**" strategy. For example, instead of charging residents a flat monthly or quarterly fee for trash pickup, they charge by the amount of trash generated. For example, they may require residents to purchase special bags for their waste collection and only trash in those bags will be collected. The exception being recyclable materials, which are collected in a separate bin. The most nonrecyclable materials they generate, the more bags they have to buy. The result is a motivation to either reduce the waste they generate or try to recycle more.

Some states have adopted **bottle bills**. Residents can take used plastic or glass bottles back to the store and collect 5 or 10 cents back for each bottle. The bottles can then be reused or recycled. This reduces the amount going into landfills.

rumruay/Shutterstock.com

Bottle bills promote reuse and recycling by encouraging stores to buy bottles back from consumers.

Lightspring/Shutterstock.com

Composting can convert food and yard waste into a useful garden mulch.

GEORGE STAMATIS/Shutterstock.com

Liner for a new sanitary landfill in Greece.

20.2 Sanitary Landfills and Incinerators

Bildagentur Zoonar GmbH/Shutterstock.com

Aerial view of a sanitary landfill.

In the old days, cities would often dispose of their waste in open dumps. These would be attractive breeding grounds for pests like rodents, seagulls, and pigeons. They were also fire hazards and the source of unpleasant odors.

In more recent decades, the old city dump has given way to the **sanitary landfill**, which disposes of waste in a way the isolates it from the environment and promotes decomposition. The **Resource Conservation and Recovery Act** (RCRA) of 1976 requires all landfills to obtain a permit and adhere to regulations adopted by the EPA. Landfills today are built at least 6 m above the water table and have a liner system made up of plastic, clay, and gravel to prevent water from leaching material out of the trash. A **leachate collection system** captures any water that does infiltrate the landfill and recovers it for treatment as wastewater. Each day's shipment of trash is covered with dirt to reduce odors and pest attraction, resulting in the waste being stacked in a series of cells. As the waste decomposes, the landfill settles a little, creating more space. A byproduct of decomposition is methane, a methane collection system is installed to capture the landfill gas, which can be used as an energy source.

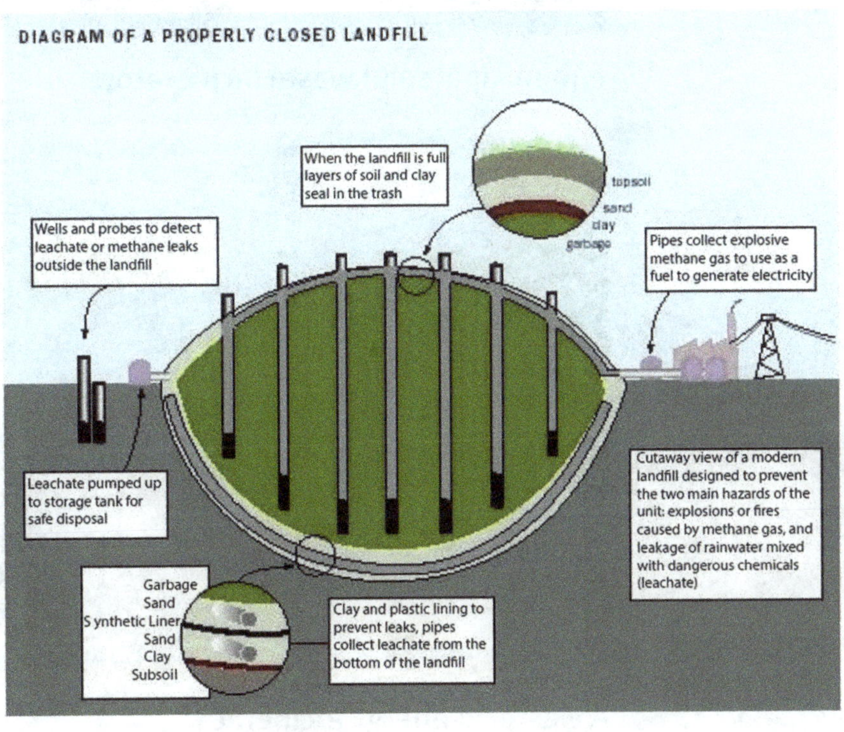

Source: epa.gov.

Modern landfills become large earthen mounds. When the landfill reaches its final permitted height, it is capped with a top liner and covered in dirt. Once the landfill has been closed, the land can be converted into a public park. For example, the largest landfill in the world, the Fresh Kills landfill in New York, was permanently closed in 2001 and is now the Freshkills Municipal Park. Some earlier landfills were converted into residential neighborhoods, but they had problems with land subsidence problems as the waste decomposed.

Incineration is the controlled burning of waste in a controlled manner. Often, the heat generated by incineration can be used to generate electricity, so these facilities are often called **waste-to-energy facilities**. Incineration can reduce the volume of trash by 95% and the weight by 85%. The ash collected at a waste-to-energy facility must still go to a landfill. So, incinerators do not eliminate the need for landfills, but they can greatly reduce the amount of landfill space needed.

Incineration begins with the tipping floor, where the trucks unload their waste. Then, a crane lowers the waste onto a conveyer belt. Magnets can be used to capture ferrous metals and then the waste is shredded before going into the furnace (If you saw the movie *Toy Story 3*, Disney did a pretty good job of illustrating the process). The heat from incineration is used to boil water and run a turbine, just like

bensliman hassan/Shutterstock.com

A municipal solid waste incinerator.

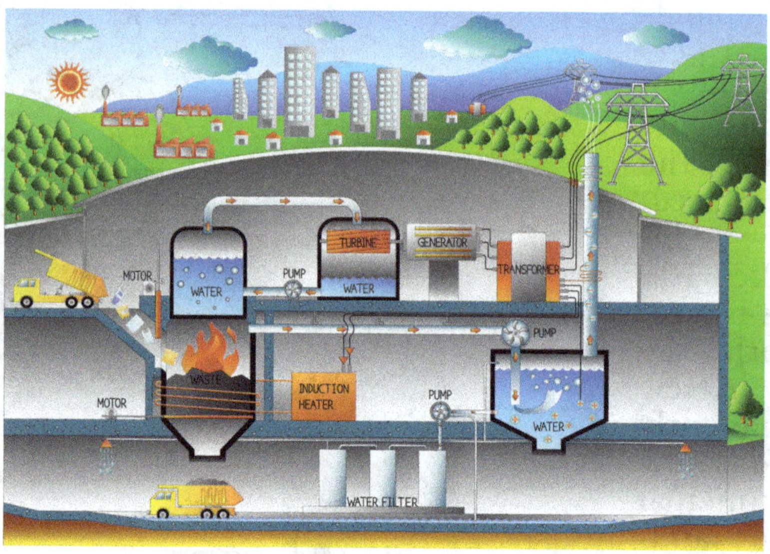

bubblea/Shutterstock.com

A waste-to-energy incinerator.

junpinzon/Shutterstock.com

A worker sorts plastics at an MRF prior to recycling.

in other power plants. Also like in other plants, waste incinerators generate both bottom and fly ash (Chapter 17), both of which must be collected and disposed of.

Some communities have begun mining old landfills for materials that could be reused or recycled. This has given us an opportunity to see how well the material has decomposed. Newspapers that were buried in the 1950s were found to be still legible and many other materials have shown little signs of decomposition, especially synthetic materials like plastic. Still, these old landfills can be a source of mineral resources and the organic material that can be incinerated.

20.3 Recycling and Reuse

Recycling consists of three major steps:

1. Collection of recycled materials for processing.
2. Converting the materials into new consumer products.
3. Consumers purchasing the products.

Recycling occurs at a **materials recovery facility** (MRF). Decades ago, residents were required to bring their recyclable materials to the MRF themselves. As this was very inconvenient for most people, recycling rates remained low. However, starting in the 1980s, communities began curbside pick-up of recyclables. Residents would be given a separate container to put their recyclables in and a truck would pick them up separate from the regular trash. As result, recycling rates began to rise, going from 6.4% in 1960 to 25.7% in 2014. Some communities have even begun "trash audits" where the contents of trash cans are examined to identify more potential recyclables.

Shredded paper prior to recycling.

alterfalter/Shutterstock.com

Recycling rates vary from one material to another. Also, different communities may decide to focus on different materials. In addition, the market demand for certain recycled products may be stronger than others. At one point, MRFs were paying so much for newspaper that some communities had rashes of thieves breaking into vending boxes and stealing the newspapers for sale at the MRFs. This problem ended when the market for recycled paper dropped. More recently, thieves have begun breaking into construction sites to steal the copper tubing for recycling.

Not all recycling programs are profitable, however. Those that aren't still receive government subsidies. However, when factoring in the increased costs of sending the materials to landfills, even with subsidies, recycling will usually save the community money. The following table gives recycling rates for various materials.

MATERIAL	PERCENTAGE THAT IS RECYCLED OR COMPOSTED
Lead-acid batteries	99
Newspapers	68
Paper and paperboard	65
Yard trimmings	61
Aluminum cans	55
Glass containers	33
Plastics	10

Car-batteries have a high recycling rate because auto parts stores will usually take the old batteries and recycle them for the consumer. These batteries contain lead-acid and should not be sent to a landfill.

Plastics have special challenges for recycling as plastic is not one single material. There are several different kinds of plastic such as polyvinyl chloride (PVC) and high density polyethylene (HDPE) each type of plastic must be separated from the others before recycling. Beverage bottles used to be made from three different kinds of plastic, but the industry eventually redesigned them, so now they are only made of two: the bottle and the cap each being a different kind of plastic.

Aluminum is another material that is very desirable for recycling. Aluminum mines and smelters have a lot of environmental impacts. Plus, the United States does not have a lot of domestic deposits of aluminum, so it must import it. Recycling aluminum cans provides a reliable domestic source of the metal.

Tires are another material that require special consideration. Whole tires are so buoyant that the rain will lift them out of the ground. They cannot be buried. Instead, tires are often shredded and made into products like playground cover, or they are incinerated in a waste-to-energy facility.

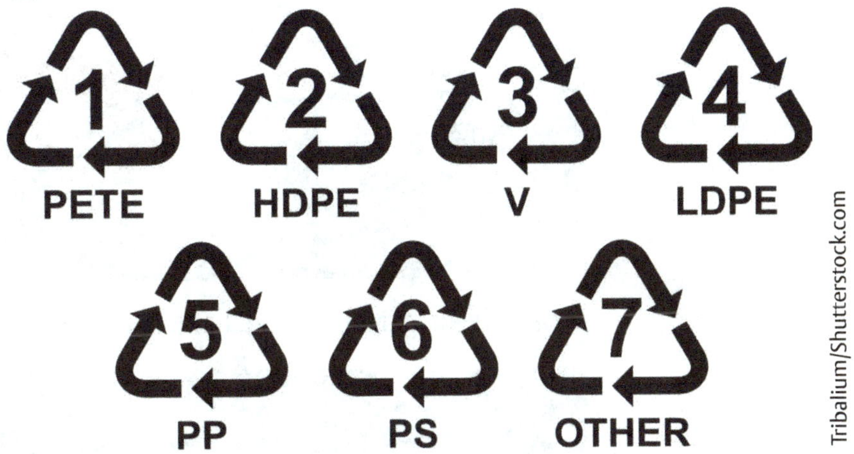

Different types of plastics must be separated before recycling.

Recycling process.

13_Phunkod/Shutterstock.com

99% of lead-acid batteries are recycled.

HollyHarry/Shutterstock.com

Clothing made from recycled materials.

sirtravelalot/Shutterstock.com

Glass is readily recyclable.

20.4 Industrial Solid Waste

Industrial solid waste is produced by manufacturing, agriculture, mining, or petroleum extraction and refining. The methods for handling and disposing of it are similar to those for municipal solid waste. However, while there are federal rules for the disposal of municipal solid waste, there are none for industrial waste. As a result, it is regulated by state governments, often with less stringent standards.

Industry often adopts whatever practices save them the most money and if that means just throwing materials away that are no longer needed, then that is what they will generally do. **Industrial ecology**, however, looks are the physical and chemical processes in the industry to look for ways to maximize both economic and physical efficiency. The goal is to reduce the physical generation of waste or, barring that, finding value-added uses for the waste product that could provide a new source of revenue for the company.

BAO-Images Bildagentur/Shutterstock.com

Industrial solid waste can also consist of a lot of packaging materials.

A **life-cycle-analysis** looks at the history of a product, starting with the raw materials, through the manufacturingprocess, sale, and final disposal. Ways are found to take the waste product in one process and incorporate it into other processes rather than just disposing of it. Environmentally harmful materials in the process may be substituted with less harmful ones. Emphasis is put on increasing the product's durability, reusability, and recyclability. One example is the Swiss XERI process for brewing beer, which takes the waste products and using them to make bread, grow mushrooms, and feed pigs and fish.

20.5 Hazardous Waste

The EPA defines hazardous waste has any waste, whether solid or liquid, that has at least one of the following properties:

1. **Ignitable**: Readily catches on fire.
2. **Reactive**: Chemically unstable and likely to react with other substances to explode or release toxic fumes.
3. **Corrosive**: Will corrode metal containers or human flesh.
4. **Toxic**: Poisonous if inhaled, ingested, or touched.

In addition to the EPA regulations, there are also regulations adopted by the Department of Transportation with regards to how these materials can be transported. Vehicles transporting hazardous materials must have the proper placards identifying what the hazard is and these vehicles are barred from certain routes such as going through tunnels.

Many synthetic materials are considered hazardous, including pesticides, solvents, and wood preservatives. They often resist decomposition from bacteria or fungi, which makes them persistent pollutants. Many of them are toxic, carcinogens, mutagens, teratogens, and endocrine disruptors. They can also be readily absorbed through the skin.

Heavy metals like lead and mercury also hazardous and can bioaccumulate and biomagnify. Heavy metals are found in many products, including paints, dyes, rechargeable batteries, and electronics.

RCRA sets standards by which states must regulate the handling of hazardous waste, including tracing them from cradle to grave.

nhumster/Shutterstock.com

Barrels used for transporting hazardous waste.

sonsart/Shutterstock.com

The Department of Transportation requires placards on vehicles transporting hazardous waste.

Silent Corners/Shutterstock.com

Midnight dumping has resulted in hazardous waste being left in places with no paper trail to identify it.

This gives a paper trail that can be followed in the event of a mishap or improper handling or disposal that occurs so that responsible parties can be found.

EPA regulations require that companies that generate, transport, or dispose of hazardous materials must report the type of material and where it came from. They must also report how it was handled and where it was disposed of. If it is transported, they must track both its origin and destination. This creates high disposal costs and to avoid those, some companies have resorted to some illegal practices, such as midnight dumping.

Midnight dumping refers to the practice of secretly leaving the waste in a remote location, like a field or abandoned warehouse, with leaving any paper trail identifying its origin or even what it is. Similarly, **orphan sites** are sites where hazardous waste has been dumped and then the owner avoids the cost of proper disposal by declaring bankruptcy and dissolving the company. One of the most famous orphan sites became known as the "Valley of the Drums" where thousands of barrels of hazardous waste were left to rust away in a field in Kentucky after the contractor who hauled the waste their declared bankruptcy. This left the taxpayers stuck with the cost of cleaning it up.

Hazardous waste can be disposed of in a number of ways. As mentioned in Chapter 16, **surface impoundments** can be used to temporarily hold slurries of waste to allow the water to evaporate and then collect the hazardous residue. They have problems such as liners failing and overflowing, resulting in contamination of surface and groundwater sources.

Deep-well injection involves drilling below the aquifer and into a porous layer of rock that is below an impermeable layer. Liquid waste is then injected into the porous layer with the expectation that the impermeable layer will keep the material from migrating into the aquifer. Deep wells, however, can corrode and leak into the aquifers and injecting liquid underground could trigger minor quakes.

While some hazardous waste can be safely incinerated, much of it is taken to a **secure landfill**. Secure landfills have more stringent standards than municipal sanitary landfills. For example, they must have double the liners and a double leachate collection system. They must also be strictly monitored.

In the 1970s, there was growing concern about many old hazardous waste disposal sites or "toxic waste dumps" as they were known. In response, Congress passed the **Comprehensive Environmental Response, Compensation, and Liability Act** (CERLCA) in 1980. This law is often more commonly known as the Superfund Law or, sarcastically as the "Welfare for Lawyers Act."

CERCLA was passed in response to two significant events. The first was Love Canal, a community in upstate New York where chemicals from an old disposal site escaped and starting seeping into people's homes. Residents complained of birth defects, miscarriages, and other health problems as a result. The federal government eventually bought up all of the residents' homes and then worked to contain the spill. By 1984, the government declared the area safe for people to live in again. You could have bought a home there really cheap back then.

The other event was Times Beach in Missouri, where people were evacuated when dioxins contaminated the soil after oil was spread on the roadways. The government then removed much of the topsoil in the area and incinerated it.

CERCLA was then passed to provide a systematic process to identify sites that have contamination and clean them up. Once a site has been identified, the most expedited methods are used to contain the material and then it is put onto the National Priorities List (NPL). Sites on the NPL are given a priority based on how close they are to people's homes, whether the material is contained or likely to spread, and whether it threatens drinking water supplies.

The government then hires contractors to do the cleanup work while it goes looking for the responsible parties. Responsible parties could include the owner of the land, the company that produced the waste, or the company that was hired to transport and dispose of the waste. The government then begins the process of trying to recoup the cleanup cost, which often results in years of expensive litigation (hence, why CERCLA is often called the "Welfare for Lawyers Act"). If no responsible parties can be identified, the cleanup costs were paid for out of the Superfund. The Superfund was funded through a tax on the manufacture of chemicals. However, that tax expired and Congress refused to renew it. The Superfund is now bankrupt and cleanup costs come out of the general taxpayer funds. This has greatly slowed the process of cleaning up these sites. As of 2017, there were still over 1,300 sites still on the NPL and only 336 have been cleaned up and removed.

CERCLA was eventually amended to encourage development on **brownfields**, old industrial sites that have some contamination. The amendments encouraged redevelopment on brownfields by reducing the liability of anyone wishing to purchase them. The purpose was to encourage redevelopment on those sites instead of building on undeveloped land.

Anna Vaczi/Shutterstock.com

Midnight dumping involves anonymously leaving hazardous waste in a remote field or old warehouse.

jfergusonphotos/Shutterstock.com

CERCLA was passed to clean up hazardous waste dumps and old industrial sites that have been contaminated with hazardous chemicals.

Thomas Alan Schneider/Shutterstock.com

A superfund site being cleaned up.

While industry produces most hazardous waste, many products in the home can also be hazardous, Examples of **household hazardous waste** include:

- Paints
- Batteries
- Oils
- Solvents
- Cleaning agents
- Lubricants
- Pesticides

Many communities restrict household hazardous waste from the regular municipal waste stream and only allow pickups at certain times of the year or require residents to bring them to special sites for disposal.

Joseph Sohm/Shutterstock.com

Many products used in the home could be hazardous waste and require special handling for disposal.

Kamonkanok/Shutterstock.com

Fluorescent bulbs are considered hazardous waste because they contain mercury.

20.6 E-Waste

Every year, Americans throw away thousands of electronic devices, even though many of them still work. The tech industry encourages this through planned obsolescence, in which they make their earlier products obsolete and encourage customers to buy the latest model.

Electronic waste or E-waste contains heavy metals like lead and flame retardants like polybrominated diphenyl ethers (PBDE). These substances are in such quantities that E-waste can be categorized as hazardous waste. Most E-waste ends up in landfills or incinerators. A lot of it is shipped to developing countries in Africa and Asia where they are dismantled and burned in conditions that lack any safety or environmental standards, putting the residents at risk for various health problems.

Recently, the amount of E-waste in the waste stream has decreased due to a combination of recycling and making the products smaller.

Many electronic devices that are thrown out can be refurbished and sold to consumers in developing countries. Or, they could be dismantled and the valuable metals recycled. Many shopping malls and other locations now have kiosks where people can drop off their old devices and collect a small amount of money. In order to encourage the recycling of E-waste, the medals awarded at the 2010 Olympics were made from metals recovered from recycled electronic devices.

KPixMining/Shutterstock.com

E-waste contains a number of heavy metals, including lead and other chemicals like flame-retardants.

zlikovec/Shutterstock.com

Americans throw out thousands of electronic devices every year, many of which still work.

wk1003mike/Shutterstock.com

E-waste can be collected for refurbishing and reuse or to recycle the materials in them.

katja kodba/Shutterstock.com

The medals awarded at the 2010 Olympic Games were made from E-waste.

20.7 Discussion Questions

1. How does municipal solid waste differ from industrial waste and hazardous waste?
2. Describe a number of ways you could reduce your personal waste production.
3. How does a pay-as-you-throw system differ from paying a flat rate for municipal solid waste pickup?
4. How are sanitary landfills designed to reduce the leaching of waste into groundwater?
5. What efforts were adopted starting in the 1980s to encourage more recycling?
6. Why are some municipalities taxing or banning plastic grocery bags?
7. What is meant by "cradle-to-grave" tracking?
8. What problems are associated with deep-well injection?
9. List and define the four categories of hazardous waste.
10. What is midnight dumping?
11. What problem was CERCLA passed to address?
12. Why does E-waste require special consideration?

20.8 Key Terms

Bottle bills
Brownfield
Composting
Comprehensive Environmental Response, Compensation, and Liability Act
Corrosive
Deep-well injection
E-waste
Hazardous waste
Household hazardous waste
Ignitable
Incineration
Industrial ecology
Industrial solid waste
Leachate collection system
Life-cycle-analysis
Materials recovery facility
Midnight dumping
Municipal solid waste
Orphan site
Pay-as-you-throw
Reactive
Recycling
Resource Conservation and Recovery Act
Sanitary landfill
Secure landfill
Source reduction
Surface impoundment
Toxic
Waste
Waste management
Waste stream
Waste-to-energy facilities

Chapter 21

SUBURBAN SPRAWL

Chapter Topics

21.1 Urbanization
21.2 Suburban Sprawl
21.3 Urban Planning Strategies
21.4 The Urban Community
21.5 Urban Ecology and Sustainable Cities
21.6 Discussion Questions
21.7 Key Terms

After World War II, the United States began a population shift. The migration of people from the rural communities to the cities that had begun with the Industrial Revolution halted. Instead, people began moving from the cities to the surrounding communities. Long Island that had been mostly farmland soon filled in with lower density suburban housing. Many New Yorkers relocated to New Jersey, taking advantages of the bridges and tunnels to easily commute to work. New York's suburbs swelled with people. The edges of what was considered the New York Metro Area extended, pushing further out. Eventually, with Northern New Jersey and Long Island running out of undeveloped land, some New Yorkers took to building their McMansions in the Poconos region of Pennsylvania. Today, while New York City has a population of about 8.5 million people, the entire New York Metro Area has over 18 million residents.

Other cities experienced similar shifts. Philadelphia lost 500,000 residents between 1950 and 2010, dropping to a low of 1.5 million residents. It fell from being the second largest city in America to sixth place. Other cities took even bigger hits. Detroit's population in 1950 was about 1.8 million. Today, it has lost nearly two-thirds of that population, down to about 670,000 residents. Baltimore lost nearly 350,000 residents since 1950. Pittsburgh lost over half of its population in that time period, dropping form about 680,000 residents in 1950 to about 300,000 today.

The reasons people give for moving from the city to the suburbs are similar in most cases: Less crime, less congestion, less pollution, better schools, cheaper real estate, and more peace and quiet. Many Baltimore residents have moved over 30 miles to the North to live York County, Pennsylvania, in search of affordable houses and lower taxes. People have been willing to endure longer and longer commutes to access these things. The average American commute today is over an hour.

There are signs that these trends are reversing. New York's population continues to grow through immigration. Philadelphia has seen some modest gains since 2010. Of the 10 largest cities in America, only Chicago reported a loss in population in 2018. On the other hand, Detroit has continued to bleed residents. Pittsburgh lost residents in every year this decade except 2012 and 2014. On the other hand, some cities have managed to avoid declines altogether. Phoenix and Las Vegas have been growing steadily as two of the fastest growing cities in America. Portland has been growing steadily since 1870, having experienced losses in population only in two decades since, in the 1950s and the 1970s.

Portland's city planning has been held up as a model for other cities. In 1973, Oregon passed a law requiring its urban areas to establish **urban growth boundaries** (UGBs). UGBs establish designated areas in the region that separate rural from urban areas. Under the law, residential and commercial development is encouraged in the urban areas but more restricted in the rural areas. The effect of the law has been to direct redevelopment into Portland while preserving undeveloped

Ink Drop/Shutterstock.com

Portland's urban planning is often held up as a model for other cities.

Mihai_Andritoiu/Shutterstock.com

land, including farmland. Critics of the law have called it excessive regulation and that land owners should have more freedom to develop their property as they see fit. In 2004, voters approved a resolution that required the state to compensate landowners if the value of their property has declined due to the UGB regulations.

21.1 Urbanization

The year 2007 saw a major demographic tipping point for the global population. For the first time, a majority of the world's population lived in urban areas instead of rural ones. **Urbanization**, the movement of people from rural communities, began with the industrialization. Mechanization of farming reduced the need for farm workers, while the growth of manufacturing spurred people to move toward the cities looking for work.

In developed countries, urbanization peaked in the 1950s and the population began shifting toward the **suburbs**, lower density communities that surround the urban center. The GI Bill, which made low interest home loans accessible to many working-class veterans, helped spur this shift in the United States. In developing countries, urbanization has been accelerating. People in those countries are moving to the cities looking for work or, in some cases, to flee wars that ravage the countryside.

What makes a successful city? Many are situated near water or transportation networks, making them important points of trade. Most of the major cities on the East Coast, New York, Boston, Philadelphia, and Baltimore, began as port cities and still serve as important points of entry for trade. Even Pittsburgh, with its access to the Three Rivers that are tributaries of the Mississippi acts as port. On the other hand, the Dallas-Fort Worth area exists on a major highway intersection and has experienced growth with the spread of the internal combustion engine.

Today, the South and Western regions of the United States are experiencing the greatest growth. People are relocating to these regions in search of warmer weather, more job opportunities, and comfortable places to retire.

Roschetzky Photography/Shutterstock.com

Dallas-Fort Worth is located at a major highway intersection.

James M. Davidson/Shutterstock.com

Philadelphia harbor. Many cities are located at bays that serve as ports for the transportation of goods.

21.2 Suburban Sprawl

As cities remained economic centers throughout the nineteenth and twentieth centuries, people have remained living in them, despite growing problems of crime and overcrowding. However, things began to shift after World War II. In addition to new home ownership becoming more accessible, the spread of the automobile made commuting easier. People no longer had to live within walking or a subway ride's distance from where they worked or shopped.

The 1950s also saw the growth of the interstate highway system. While serving in Europe during World War II, Dwight Eisenhower became very impressed with the German army's ability to use the autobahn system to move troops and materiel around quickly. At the time, most American cities were connected by narrow rural roads, many of which were not even paved and became impassable in the winter months. As the Cold War began in earnest, many people feared the Russians would invade the United States. In preparation of civil defense, President Eisenhower urged the funding for the construction of highways modeled on the autobahn. A side of effect of these highways is that it became easier for people to commute from the suburbs to the city for work and back to home by car. This spurred the growth of suburban "**bedroom communities**," so named because they were small towns with little industry of their own and mainly served as places for people to go home to at night.

Portland saw some exiting to the suburbs in the 1950s as its population began to flatten, but with the passage of land usage laws in the early 1970s, it concentrated more development in the city and less in the suburbs.

Everett Historical/Shutterstock.com

New York's Little Italy, circa 1900.

VICTOR TORRES/Shutterstock.com

As suburban communities grew, development continued to spread outward from the urban center. This became known as **sprawl**, the spread of low-density housing out from the urban center. Sprawl has happened in curious locations. Las Vegas began as a sleepy little desert town, until Bugsy Siegel and other mobsters conceived of turning it into a mecca for gambling in the 1940s. Despite its location in a resource-poor desert, the population of Vegas surged and its metropolitan area nearly tripled in size from 1986 and 2013.

robert cicchetti/Shutterstock.com

Sprawling suburban and commercial development in Las Vegas, Nevada, is pushing up against the mountains in the western desert as the population increases.

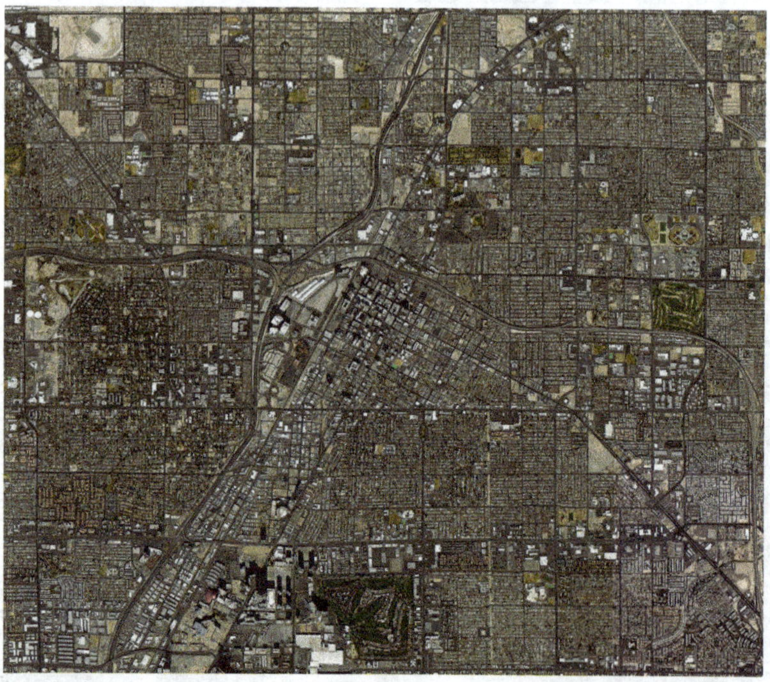

Aquatic Images/Shutterstock.com

Satellite image of Las Vegas.

Economists and local officials have historically promoted sprawl, in the belief that all growth is good. However, sprawl can have many effects. For one, it uses up more land per person. For example, the average resident of the Chicago suburbs takes up 11 times as much land as the average resident of the city. Sprawl also limits our transportation options, making us prisoners of our cars, needing them to get to work, school, and for errands. More car traffic results in more air pollution and greenhouse gases. More paved roadways cause more runoff and water pollution from motor oil and road salt. The disappearance of the corner grocery store means we spend more time driving that walking. Sprawl promotes physical inactivity and studies have found suburban residents have higher rates of obesity that urban residents. It also results in less land for farmland, forests, and recreation.

As residents leave the urban center, they take their tax dollars with them. Tax money is spent on new infrastructure, new schools, and new roads. Meanwhile, the city government becomes starved for revenue and is no longer able to maintain its existing infrastructure and may even have to cut down on basic services, like schools, fire, and police.

Kent Weakley/Shutterstock.com

A sprawling suburban community.

p-jitti/Shutterstock.com

Urban planning involves efforts to make a city an attractive place to live.

21.3 Urban Planning Strategies

The practice of **city or urban planning** is to provide advice to municipal leaders with regard to development options, resources use, and strategies in making cities safe and attractive places to live. Most cities developed organically, starting as small settlements and expanding as the population grows. One of the first cities in the United States to be planned was Washington, DC. Architect Pierre Charles L'Enfant envisioned a city with a large, stately feel and plenty of space for monuments. He also includes many crisscrossing roads that run diagonally with grid streets. This made the city's layout deliberately confusing, which he thought would stymy invaders. Given that when the British invaded in the War of 1812, they burned the White House, it was probably unsuccessful in that endeavor.

Regional planning has similar goals as urban planning, but involves coordination among multiple municipal and county governments. One tool used in planning is **zoning**, the practice of designating certain areas in a community for specific use. For example, one area might be zoned as residential where only residential development would be allowed. An area zoned as industrial would concentrate manufacturing and other business, while a commercial area would concentrate retail businesses. Critics of zoning argue that it puts unfair restrictions on property owners, while proponents argue it helps people understand what kind of development they can expect in their community and provide places for peace and quiet away from heavy industry.

Oregon has been in the forefront of using UGBs as part of their regional planning. UGBs have helped preserve farms, ranches, and forests while saving money by prioritizing maintenance of existing water and sewer lines rather than building new ones needed for sprawl. While UGBs can revitalize downtowns, they can also raise property prices, resulting in **gentrification**. As urban areas are revitalized, they become more attractive for people with higher income. While this can be beneficial as the wealthy can contribute to the tax base, it can have the effect of pricing the low-income residents out of the community.

New York City has experienced a lot of gentrification in recent decades. Even neighborhoods once considered "distressed" like Harlem, Hell's Kitchen, and Williamsburg have been revitalized. Real estate on Manhattan Island has become prohibitively expensive. Things are little better in Brooklyn. As the borough has become a "hip" place to live, home prices have skyrocketed where a tiny two-bedroom townhouse can be now worth nearly a million dollars. Portland and San Francisco have also experienced gentrification issues.

Recently, cities like Portland have begun promoting **smart growth**, which has five basic tenets:

- Mix land uses
- Take advantage of compact building design
- Create a range of housing opportunities and choices

- Create walkable neighborhoods
- Foster distinctive, attractive communities with a strong sense of place
- Preserve open space, farmland, natural beauty, and critical environmental areas
- Strengthen existing communities and direct development toward them
- Provide a variety of transportation choices
- Make development decisions predictable, fair, and cost-effective
- Encourage community and stakeholder collaboration in development decisions

A related approach to smart growth is **new urbanism**, which aims to recreate walkable neighborhoods where residents can work, shop, and go to school without the use of the automobile. Instead, residents will either walk or travel by light rails. Providing a variety of transportation options is a key approach to new urbanism. Bike lanes, bike racks, and bike sharing programs provide a means of getting around short distances without a car and promote more physical activity. Overall, smart growth and new urbanism practices are intended to make cities attractive places to live with a sense of community and high quality of life.

A variety of transportation options can reduce traffic jams, which are estimated to cost the US $7.5 billion a year in lost productivity and increases air pollution. Any effort to take cars off the road can reduce traffic jams. Buses and commuter rails, or **mass transit**, also use less energy per passenger mile and cost less money when vehicle maintenance, roadway maintenance, and parking costs are taken into account. Many major cities like New York, which has an estimated 12 million commuters, could not function without their mass transit systems. There are simply not enough bridges and tunnels to accommodate all the vehicles needed. Also, there is not enough parking facilities possible either.

Despite these advantages, most mass transit systems are not profitable and require taxpayer subsidies to operate. The United States lags behind other nations, especially more densely populated nations in Europe and Japan, in commuter rail. America's sole regional passenger rail line, Amtrak, was established in the 1960s because the commercial railroad companies wanted to dump passenger service in favor of the more profitable cargo transportation. Amtrak's trains do not even come close to meeting the service and speeds of the **high-speed rail** found in Europe, Japan, and China. Some entrepreneurs, such as Elon Musk, have proposed high-speed rail connecting cities like Los Angeles and San Francisco, but they have found little support from the public or government officials. Americans seem to prefer the freedom of driving where and whenever they please as opposed to following bus or train schedules to get around. The United States is also less densely populated that Japan or Europe, so many rural communities see little benefit in devoting their tax dollars toward commuter rail lines they are unlikely to use.

Dr. Alan Lipkin/Shutterstock.com

Washington was designed to have a large, stately feeling with spaces for monuments.

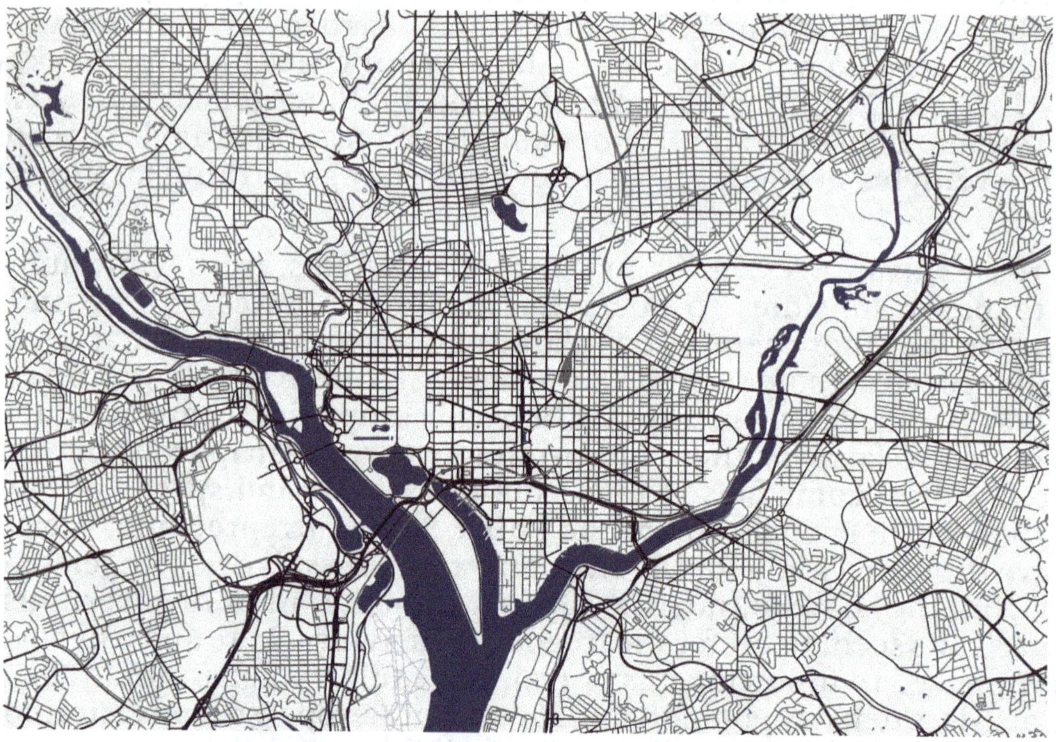

Tish11/Shutterstock.com

Map of Washington, DC, showing its crisscrossing streets.

tektur/Shutterstock.com

Old and new façade. Gentrification often involves restoring and remodeling older buildings.

HipKat/Shutterstock.com

Gentrification can drive up housing costs and force poorer residents out of the community.

Luciano Mortula - LGM/Shutterstock.com

In the 1990s, New York City "cleaned up Times Square" to make it more attractive to tourists.

Stas Walenga/Shutterstock.com

Bike sharing in New York.

bibiphoto/Shutterstock.com

Americans waste hours of productivity in traffic jams

cyo bo/Shutterstock.com

China's high-speed train.

21.4 The Urban Community

City parks provide important amenities for urban residents. They provide green spaces where residents can get away from the noise and congestion of city life. One innovative park is New York's High Line Park, where an abandoned elevated rail line was planted with grass and trees to create an extended, above-street level park.

Urbanization brings a number of positive and negative impacts. Urban residents are exposed to more air pollutants, noise, heavy metals, and thermal pollution. Cities import resources from the countryside and export waste, including municipal solid waste and air pollution. Concrete and blacktop also absorb more heat that other materials, causing the **urban heat island effect**, in which cities tend to have higher temperatures than the surrounding countryside. Cities have to import food and even water. Los Angeles gets it drinking water through aqueducts that travel hundreds of miles, while New York uses aqueducts to bring water down from the Catskill Mountains because no one wants to drink out of the Hudson River. As a result, cities have ecological footprints greater than just their land area.

On the other hand, dense populations make delivery of services such as drinking water, electricity, sewage collection more efficient. They also leave more land outside the cities undeveloped, preserving farmland and forests. Cities also bring people together from diverse backgrounds, enabling them to exchange ideas and innovation and develop solutions to problems through cooperative action. Wealthy urban residents tend to buy low-impact products like organic produce.

T photography/Shutterstock.com

Aerial view of New York's Central Park.

Stuart Monk/Shutterstock.com

New York's High Line Park.

21.5 Urban Ecology and Sustainable Cities

Urban Ecology seeks to halt the one-way flow of resources into cities and wastes out in order to make cities more sustainable. The idea is to promote more circular systems where resources are recycled and green technologies are developed. New York's PlaNYC, released in 2007, aims to make New York a true twenty-first-century city. PlaNYC sets goals such as

- Improved energy efficiency of 174 buildings
- Planted 950,000 trees
- Opened or renovated 234 school playgrounds
- Established 129 community gardens
- Acquired 36,000 acres to protect water supplies in upstate New York

Green building design is another strategy to make cities more sustainable. Green buildings are constructed using sustainably produced materials and are designed to use energy and water more efficiently. They are also trying to control pollution and recycle waste as much as possible. The US Green Building Council promotes these design ideas through its Leadership in Energy and Environmental Design (LEED) program, which awards new buildings silver, gold, or platinum status based on how well they meet these goals.

Other ideas promoted to make cities more sustainable include urban and rooftop gardening, rooftop solar, and even urban wind power. For example, Philadelphia's Lincoln Financial field uses solar panels to produce about 4 MW of electricity while also shading the parking lots from too much sun. The stadium has earned gold LEED status. The stadium also used to maintain a few wind turbines, but these were taken down in July 2019 as the managers reevaluated their usefulness. Recently, the city has passed a resolution setting a goal of 100% of the city's energy needs coming from renewable sources by 2030. Not to be outdone, Pittsburgh has begun installing vertical wind turbines in the city.

Franck Boston/Shutterstock.com

Green building.

thanis/Shutterstock.com

Air pollution in Bangkok.

ValentinaKru/Shutterstock.com

Urban Heat Island Effect.

Jay Yuan/Shutterstock.com

Pittsburgh, Pennsylvania, United States – May 20, 2018: A Carnegie Mellon Graduate wearing graduation gown Looking the campus at Flagstaff Field. Carnegie Mellon University is ranked top 25 universities in the United States.

21.6 Discussion Questions

1. What factors led to people in the United States migrating from the cities to the suburbs?
2. What are urban growth boundaries and how do they affect sprawl?
3. List some of the environmental, health, and social impacts of sprawl?
4. Why did President Eisenhower pushing for developing the interstate highway system in the 1950s?
5. What is gentrification and how does it affect the makeup of a community?
6. How is zoning used in urban planning?
7. What does the United States lag behind many European and Asian countries in the developing high-speed rail for commuters?
8. What causes the urban heat island effect?
9. Why have many local leaders encouraged sprawl?
10. Why do cities have larger ecological footprints than just the land area they occupy?

21.7 Key Terms

Bedroom communities
City or urban planning
Gentrification
Green building
High-speed rail
Mass transit
New urbanism
Regional planning
Smart growth
Sprawl
Suburbs
Urban ecology
Urban growth boundaries
Urban heat island effect
Urbanization
Zoning

CPSIA information can be obtained
at www.ICGtesting.com
Printed in the USA
LVHW062113190122
708468LV00003B/3

9 781792 419522